SECOND EDITION

Survey Research Methods

Earl Babbie

Wadsworth Publishing Company
Belmont, California
A Division of Wadsworth, Inc.

To
Sam Stouffer
Paul Lazarsfeld
Charles Glock
and the Old Turks

Sociology Editor: Serina Beauparlant
Editorial Assistant: Marla Nowick
Production Editor: Donna Linden
Managing Designer: Andrew Ogus
Print Buyer: Barbara Britton
Designer: Cynthia Bassett-Bogue
Copy Editor: Mary Roybal
Technical Illustrator: Guy Magallanes
Compositor: Interactive Composition Corporation, Pleasant Hill, California
Cover: Andrew Ogus and Stephen Rapley

Printed in the United States of America   1998

10

Library of Congress Cataloging-in-Publication Data

Babbie, Earl R.
  Survey research methods / Earl Babbie. —2nd ed.
    p.   cm.
  Includes bibliographical references.
  ISBN 0–534–12672–3
  1. Social sciences—Research. 2. Social surveys. I. Title.
H62.B23 1990
300'.723—dc20   89–29866
                                                                          CIP

Contents

3
Survey Research as a Method of Social Science   36

PART TWO

Survey Research Design   49

4
Types of Study Design   51

5
The Logic of Survey Sampling   65

6

Examples of Sample Designs   *102*

7

Conceptualization and Instrument Design   *118*

8

Index and Scale Construction   *147*

12
Pretests and Pilot Studies 220

PART FOUR

Survey Research Analysis 237

13
The Logic of Measurement and Association 239

14
Constructing and Understanding Tables 247

PART FIVE

Survey Research in

Social Context 337

Preface

This book is the second edition of a book first published sixteen years ago. It was my first textbook and has had a special place in my heart—though it has been long neglected.

I began teaching survey research at the University of Hawaii in 1968. At that time, I learned something other survey research instructors already knew: there wasn't really a good text available for the course. I chose to use a British textbook by C. A. Moser, but my students were often put off by the British examples and by the strange way the British speak English. (Not until much later did it occur to me that non-American students just might have a similar problem when they use American textbooks.)

As I began meeting other survey research instructors (it's amazing how many have to stop in Hawaii on their way to almost anywhere), I discovered an interesting pattern. We always asked each other two questions and almost always gave the same answers.

The first question was, "Are you going to do a survey in your class this year?" The answer was usually some version of "No, I did one recently, and I don't want to go through that again right away."

The second question was, "What textbook are you using?" Here the books that were being used varied, but everyone always added, ". . . but I don't like it."

My own frustration in this latter regard led me one day to doodle a table of contents for the "ideal" survey research textbook. That table of contents was essentially the one you have before you.

The title I gave this imaginary textbook reflected a more specific frustration I had with the larger body of materials published about survey research. Some of them were quite abstract, leaving the reader at a loss as to how they might *do* a survey. Others were just the opposite, offering a very concrete recipe of steps for doing one specific survey. The latter approach was fine if you wanted to do exactly the same survey as the writer, but you were lost if your desires and/or circumstances differed from those presented.

I entitled my fantasy textbook, *A Survey Research Cookbook and Other Fables*. That title was to appear on chapter drafts and correspondence throughout the creation of the book. When the manuscript finally went into production, the title was assigned to the graveyard of unused prose, whose massive accumulation of verbal waste has a half-life of forever, if my own experience is any guide.

So I doodled my table of contents with its funny title and set the whole thing aside—for one day. The next day, I received a letter from Jack Arnold, sociology editor at Wadsworth Publishing Company, indicating their interest in publishing a textbook in survey research. Moreover, he said, one of their consulting editors, Rod Stark, had suggested that I might be the person to write it.

Only a couple of years before, Rod and I had been best friends in graduate school at Berkeley, and I considered the possibility that the letter was a hoax. Not that Rod and I had ever, ever done anything like that, but I always thought *he* might be capable of such a dastardly act. Such concerns notwithstanding, I responded to the letter from Wadsworth by return mail, enclosing a table of contents for such a book. Shortly thereafter, I signed a contract for the book and was busily writing the chapters that went with the table of contents.

Writing the book was a mixed bag. Sometimes the words just seemed to flow onto the page, and my job seemed to be mostly that of feeding blank paper into the typewriter. At times like this, I felt I was reading the book more than writing it. In my best day, in the midst of my normal activities as director of the Survey Research Office, I wrote two chapters, which needed little or no editing.

Writing other chapters was more like torture. I would spend hours painfully putting one word after another, like a drunk attempting to put one foot after another along a straight line. Like the drunk, I fell down a lot. Sometimes I would stumble over every word, spend hours creating a single paragraph, and then throw the whole thing away only to start over. It was not a pretty sight.

> A little boy is late for school, see, and his teacher says, "Johnny, why are you late?"
> Johnny says, "It was so icy this morning that for every step I took forward, I slid two steps back."
> "Oh yeah," his teacher says. "If that's so, how did you ever get here?"
> "I gave up and tried to go home."
> [Drum roll, rim shot.]

The chapter on statistics was the worst. It seemed like a black hole designed to suck all the joy out of life. Although the prepublication reviewers generally fell all over themselves in praising the manuscript (I took to writing "Thanks, Mom" in the margin of the reviews), they were unanimous in regarding the statistics chapter as some disgustingly malformed freak that deserved to be put out of its misery. So I revised. And revised some more.

Ultimately, the statistics chapter was pruned down to only three calculations: *lambda, gamma,* and *chi square.* In the first printing of the book, two of the three were correct. That was not the end of the statistical shakiness, however.

I had decided it would be useful to include a table of random numbers in an appendix—to assist readers in sample selection. Rather than buying an existing table of random numbers, I decided to exercise my newly developed programming skills to create my own. After a few false starts, I was able to provide Wadsworth with a table of random numbers suitable for photo-offsetting, so there would be no risk of typesetters messing things up.

As it turned out, there was one problem, reflected in letters from several instructors who used the book: "My students have asked me a question I can't answer. Why doesn't the table of random numbers have any 9's in it?" In an early attempt to create the table, I had noticed there were no 0's, later discovering that the system I was using looked for random numbers between 1 and 9. I found out how to set the lower limit to 0, but I did that in such a way as to automatically bring the upper limit down with it. Instead of picking random numbers between 1 and 9, I had it pick them between 0 and 8. Of course, nobody said anything about the wealth of 0's, 1's, 2's, etc. They just wanted to complain about the missing 9's.

My recommendation to Wadsworth was that we print up an Errata sheet filled with 9's, with the instruction to "Insert at Random." Instead, we bought a random number table from someone else.

The book was published in mid–1973 and was very well received. Soon, Wadsworth was receiving suggestions from instructors that I write a more general research methods text—one that would include methods other than survey research. *The Practice of Social Research,* published in 1975, was my response. It became the focus of my research methods writing in the years that followed.

Although many of the instructors who had initially used *Survey Research Methods* switched over to *The Practice of Social Research,* as we expected, the sales of the survey book persisted over the years. Periodically, Wadsworth and I would consider my revising the book, but some other project always took precedence. Finally, in 1988, we made a commitment to doing it the next year, no matter what came up.

The first obstacle was a technological one. Whereas I had been writing on microcomputers for ten years, with all the materials on disk for easy revision, the survey book had been written prior to the advent of word-processing systems: on a typewriter with no memory.

Initially, Wadsworth agreed to have a typist put the existing book on disk, so I could do my revision in that environment. Later we decided to experiment with a more high-tech solution: we would have the book *scanned* electronically. In this process, a scanning machine would take a picture of each page, but whereas a photocopy machine would put such a picture on paper, the scanner stored it on a computer disk. A translation program would then review the picture and convert the letters and words into text that could be edited like any word-processing file.

Text-scanning such as I've described it is clearly one of the more important technologies of the future. It is not necessarily a technology of the present, we discovered. Wadsworth hired a company to do the work, and the resulting files were, in the proud claim of the company, "85% accurate." Put differently, one character in seven was wrong! (I've subsequently discovered that other companies and other systems are much more accurate. Still, text scanning is not yet "an exact science.")

Through the heroic efforts of my secretary, Adrienne Alexander, I eventually had a set of disks containing the 1973 version of the survey book. Revision could at last begin. In case you've worked with the earlier edition, let me make a few comments on the major changes you will find in this one.

The 1973 book, a product of its time, I suppose, was packed with sexist language: most specifically, the use of masculine, third-person pronouns when gender was indefinite: e.g., "A researcher uses stratified sampling when *he* wants to. . . ." I had been aware of this problem in the 1973 edition, particularly when I wrote about interviewers and key-punchers: two predominantly female groups in the survey process. In a bold stroke, I used female pronouns in discussions of these two groups, called the research directors "he," and wrote a lengthy footnote to explain the procedure and to criticize the general exclusion of women among project directors.

As the issue of sexist language heated up in subsequent years, I sometimes found myself blaming "them" for creating the third-person masculine convention in the first place. Eventually, I began asking who "they" were in this regard, realizing there was no formal group empowered to legislate the English language (unlike the French Academy, charged with maintaining the linguistic purity of *français*). In the case of English, however, such matters are loosely controlled by English teachers, editors, dictionary publishers, and . . . well . . . people like me. As a consequence, I made a commitment to eliminate sexist writng in my own work.

In the years since 1973, I have actually found it quite easy to avoid sexist language. I share some of these solutions and urge you to adopt them. First, there are some "solutions" I don't like and have avoided:

- The use of "she" in place of "he" is linguistic cross-dressing: the problem still exists but in drag.
- The random or alternating use of "he" and "she" is confusing to me: seeming to suggest there is a reason for one reference being female and another male.
- The repeated use of "he or she" is clumsy, though it can work when used sparingly.

- Linguistic mutants like "s/he" have always struck me as ugly, not to mention unpronounceable. (You *can't* mention it.) Moreover, they somehow seem like a sarcastic concession to what the writer may regard as an unreasonable concern: like sarcastic references to "personagement" in place of "management," "personual" in place of "manual," etc.

Here's what *has* worked for me. Writing in the first and/or second person instead of the third person produces more of a personal communication while avoiding the sexist dilemma. For example, "You and I would choose systematic sampling if we wanted to. . . ." Where I need to add a third person, I use androgynous names like Jan and Pat. Plural pronouns offer another solution: e.g., "Researchers choose systematic sampling when they want to. . . ."

Sexist language can also be avoided through the use of a colloquial device, generally accepted in oral English: using the third person *plural* when gender is indefinite. For example, "When a student comes to my office, I have them sit in the chair." Over the years, however, I have had a lot of trouble smuggling this latter construction past copy editors.

After discovering these and other devices for avoiding sexist language, I finally hit upon the most powerful solution of all: I stopped *thinking* in sexist terms. When I thought of a researcher, I stopped having a male come to mind; similarly, I stopped thinking of only female interviewers. As a consequence, I have much less frequently painted myself into a sexist corner that requires escape through some linguistic device.

In addition to cleaning up the sexist language of the 1973 edition, I needed to deal with some pretty moldy research examples. For example, the present book no longer trumpets the success of survey researchers in forecasting the Nixon election of 1968. Instead, I've provided data from the 1988 exit polls to illustrate the power of survey sampling.

The fact that I did not buy my first personal computer until six years after the publication of the first edition gives some indication of the updating required in the area of research technology. The present edition does not assume researchers will be using punch cards and counter-sorters or wood-burning computers. Instead, I've focused heavily on the use of microcomputers, computer-assisted interviewing, and the like. If anything, my description of survey technology is probably a bit ahead of the norm at present, but I'm confident general practice will quickly catch up and the current descriptions will soon be out of date.

Quite aside from research hardware, a lot has changed in the sixteen years since this book first appeared. Survey researchers have learned new tricks, developed new techniques, etc. I've tried to reflect much of the new literature, though it would be impossible to do justice to it all in one book.

In my own teaching and writing over the years, I think I've gotten better at explaining various aspects of the research process, so this edition has given me an opportunity to share what I've learned. In particular, the unfolding computer technology has helped me portray many concepts graphically, which is the way I understand them in my own mind. Whereas I always used to translate pictures and diagrams into words, I'm now better able to present them *au naturel*.

Finally, the new book gives more explicit attention to market research. It is my understanding that the 1973 edition has been widely used as a supplemental reference in marketing courses and agencies, since marketing research textbooks, in my view, tend to be more theoretical than practical. Certainly, survey research is a widely used technique among marketing people. While this book is not intended as the main textbook for a course in marketing research, I have attempted to make it more relevant and useful as a supplement.

As with the original edition of this book, I would like to acknowledge a number of people who were instrumental in creating it. First, I have rededicated it to Samuel A. Stouffer, Paul F. Lazarsfeld, and Charles Y. Glock, whom I regard as my survey research lineage. The odd comment about "old Turks" is an insider reference to Paul Lazarsfeld's dedication of *The Language of Social Research* (1955) to "Charles Y. Glock and his 'Young Turks' at Columbia University's Bureau of Applied Social Research."

The following colleagues offered comments on drafts of the revised edition, and I am indebted to them for their expertise and kindness: Alan Acock, Louisiana State University, and Tom Guterbock, University of Virgina.

Serina Beauparlant, the new sociology editor at Wadsworth, had the task of filling the shoes Steve Rutter had worn sixteen years before. She did a great job, as you can tell from the finished product. She was assisted in that by production editor Donna Linden and managing designer Andrew Ogus.

Finally, I would like to thank my wife, Sheila, who was initially to have been co-author of the first edition. While she eventually chose to forego that role, she has remained my partner and collaborator forever. A constant source of inspiration, she has always urged me to look over the hill, around the corner, and under the rock. I invite you to look with me in the pages that follow.

Preface to
First Edition

Survey research is probably the best known and most widely used research method in the social sciences today. Its use in the academic world grows daily. It is now taught and used in departments of sociology, political science, psychology, business administration, public health, and geography to name a few. Increasingly, social science graduate students are encouraged to conduct surveys to satisfy thesis and dissertation requirements for original research. Undergraduate classes frequently conduct surveys, and faculty members publish reports of hundreds of surveys every year.

Outside the academic world, almost everyone has heard of public opinion polls, election predictions, consumer market studies, and censuses. To some extent, everyone in the United States at least has been affected by surveys. Politicians launch or scuttle campaigns and dreams on the basis of voter surveys. Manufacturers mass-produce or discontinue products on the basis of market research findings. Federal aid programs often hinge on the results of population surveys in distressed cities and states. In addition to professional survey research activities, nonprofessional surveys are conducted frequently. Social clubs poll their members on club policies and picnic dates. Libraries and cafeterias poll their users about desired services and hours. Radio stations invite listeners to call in votes on community issues.

Such widespread use and acceptance of survey research would seem to indicate that the technique is easily learned and used. Everyone who has read the report of a public opinion poll in his local newspaper is likely to feel he could conduct a poll of his own. After all, anyone can ask questions and count the answers.

Why This Book Emphasizes Logic and Skills

This book is addressed to three problems related to the misconception that survey research is simple. First, the faddish popularity of survey methods has inevitably resulted in a large number of bad surveys. Sometimes survey techniques are used poorly; often surveys are conducted when some other research method would have been more appropriate. Hopefully, by discussing the logic and skills of survey research in some detail, the text will help raise the quality of surveys conducted.

Second, the widespread overuse and misuse of survey methods has led to the wholesale rejection of survey research by many people, including a growing number of young social scientists. In this regard, it is hoped that the book will indicate that survey research can be an extremely useful method of scientific inquiry in certain situations. Used correctly and in appropriate situations, surveys can provide needed information that could not feasibly be obtained by any other method.

Third, the assertion that a given survey was poorly conducted presupposes an established body of scientific standards against which to evaluate survey activities. Unfortunately, this is not really the case. To be sure, in some aspects of survey research, such as sampling and the statistical manipulation of data, some very rigorous standards do exist. But in other aspects, such as the wording of questions and the codification of responses, the standards are not as clearly defined. Such standards that do exist are generally not formalized, and, moreover, they tend to be transmitted as an oral tradition from researcher to student-apprentice. However, beginning researchers who do not have access to this apprenticeship system may not understand even the informal standards. This book attempts to point out such informal standards and to place them within the logical context of scientific inquiry.

The general orientation for this text is reflected in the title, *Survey Research Methods*. As a survey methods instructor and as a survey consultant, I sorely felt the need for a single volume that would provide students and other prospective researchers with a practical guide to survey research. Book and essays discussing the theoretical logic of scientific inquiry are plentiful enough, but typically they do not specify how scientific norms are applied in practice. At the other extreme, a handful of "survey cookbooks" are available, providing step-by-step guidelines for making surveys, but their utility diminishes rapidly as the researcher's field situation differs from those discussed directly in the texts. Whenever a researcher faces a field problem not covered by the "cookbook," he will probably be at a loss. This book attempts to provide a solution to such a problem.

This text focuses on the *logic* and the *skills* of survey research. Relatively little attention is given to statistics, since several excellent statistic texts are available. In covering the various aspects of survey design and analysis, I have used the following format. First, the theoretical logic of, say, sampling is discussed. The reader should understand the basic logic of selecting a sample of respondents whose answers can be taken to represent the population from which they were selected. Then the most typical methods of sampling are discussed step by step. At this point, the reader should understand not only *how* to select a conventional sample but also *why* conventional sampling methods follow from the theoretical logic of sampling. Then a variety of practical problems frequently faced by researchers in the field are introduced, and the typical methods for dealing with those problems are discussed within the context of the logic of survey sampling. Here, the reader should understand why the solutions suggested offer the best correspondence with sampling logic and ideal sampling methods. While it will not be possible to consider all the practical problems that readers may eventually face, it is hoped that they will understand the logic of sampling in practice sufficiently well to permit them to arrive at good solutions of their own. Even when the research situation does not permit a *good* solution, the reader should be able to evaluate the significance of his actual decision in regard to the conclusions to be drawn from his study.

No survey fully satisfies the theoretical ideals of scientific inquiry. Every survey represents a collection of compromises between the ideal and the possible. The primary goal of this text is to help readers arrive at the best possible compromises. *Perfect* surveys may not be possible, but *good* surveys can and should be done.

This book may seem rather elementary. It focuses almost exclusively on *basic* logic and skills. For example, relatively little attention is given to scaling techniques, whereas much attention is devoted to constructing simple indexes. Similarly, the discussions of survey analysis focus primarily on percentage tables. There are two reasons for this orientation. The existing methodological literature already covers advanced techniques rather fully, while the more basic techniques have not been treated as extensively. More important, a firm grounding in the basic logic and skills of survey research is a prerequisite for fully understanding the more advanced techniques. Thus, if the reader can fully understand the logic of analysis through percentage tables, I feel he will be better equipped to undertake correlations, regressions, factor analysis, and so forth. At present, unfortunately, many students leap directly into more complicated forms of analysis without really understanding the logic of research in general.

How This Book Can Be Used

This book is aimed at a rather broad audience: anyone interested in conducting a survey or in using and/or evaluating the results of a survey. More specifically, it is aimed at three distinct audiences: methodology students, beginning researchers, and research consumers.

First, *Survey Research Methods* is intended for undergraduates taking their first course in research methods. While a number of instructors prefer to emphasize survey research in the general course, other methodologies may be covered in addition. Hence, this book is published in softcover to allow for a flexible use of materials. The organization of the material grew out of my own classroom experiences in teaching survey methods, and an early draft of the text was tested in that setting. In addition to the materials in the body of the text, most chapters are accompanied by a list of additional readings that might be used for more detailed examinations of specific topics.

Second, the book is intended as a practical and realistic guide for the beginning researcher who has had little or no previous experience in survey research. The book dwells on the likely problems, decisions, and compromises that are the day-to-day substance of field research. I have attempted to provide the broadest possible range of examples from my own research and consultation in a wide variety of research projects and research conditions and from the experiences of colleagues and other researchers. I hope that the number of examples will increase the likelihood that a researcher will find guidance in the text. At the same time, I have sought to present such examples within the basic logic of scientific inquiry. Rather than asking the reader to memorize specific techniques—even a broad variety of techniques—I ask that he understand why those particular techniques are recommended and used. Thus, when the researcher must compromise or approximate, he should be in a position to know what he is compromising and what the effect of compromise is likely to be.

Finally, I hope that consumers of survey research will find in this book a methodological grounding to aid them in enlightened, critical evaluation. I hope I have presented a sufficiently complete picture of the practical realities of field research to improve the perspective of the more casual survey critics. By discussing honestly the merits and possible uses of survey research, I would hope to convert some critics into enlightened supporters. But no less importantly, I discuss the disadvantages, shortcomings, and misuses of survey research in the hope of helping some of its more overzealous supporters to become a bit more critical.

Acknowledgments

I was extremely fortunate to have had the opportunity to learn survey research in essentially an apprenticeship setting under the tutelage of Charles Glock. At a time when faculty members are accused of exploiting students, Charlie consistently treated me as a junior colleague. I was encouraged to take on projects of my own and became more independent, although I was always able to turn to Charlie for more advanced guidance. I feel a debt to Charlie Glock that cannot be discharged by the standard acknowledgment in a preface. While I have subsequently learned things about survey research that Charlie did not teach me, I am still somewhat embarrassed to call this book my own.

At the same time, I am indebted to a number of other teachers and colleagues. Foremost among these is Rod Stark, who served as my big brother during the early years at Berkeley. Later at Berkeley, I had the good fortune to meet Gertrude Selznick, who forced me to look ever deeper into the logical grounding of survey analysis, and our comparing notes on empirical research gave me a much better understanding of what scientific inquiry was all about. Other teachers and colleagues who have improved my own understanding of survey research as science include George von Bekesy, Lin Freeman, Dave Gold, Ren Likert, Bill Nicholls, Jay Palmore, Steve Steinberg, and Charlie Yarbrough. And I am especially grateful for the detailed and thoughtful comments of three reviewers of the manuscript: Andy Anderson, Joseph Spaeth, and Billy J. Franklin.

By the same token, I should acknowledge my debt to the many researchers with whom I have worked in a consultative capacity. Some have taught me new techniques; some have reviewed and commented on early drafts of this text. Among them are Jim Dannemiller, Dave Ford, Dennis Hall, Dave Johnson, Jan LeDoux, Heung-soo Park, Dinny Quinn, Francoise Rutherford, Yongsock Shin, Dave Takeuchi, Chuck Wall, and Choon Yang.

Despite all this, the textbook itself would never have materialized without the help of several people involved in the manuscript itself. Jack Arnold, Steve Rutter, and Rod Stark organized and provided important editorial inputs from Wadsworth. Barbara Higa undertook an extensive bibliographic project. Pat Horton did yeoman service in helping me find time to write and in typing the rather lengthy manuscript.

Because of her special expertise in data collection and general experience in survey research, my wife, Sheila, was asked to coauthor the book, and she played an important role in the initial organization of it. Shortly after writing began, it became impossible for Sheila to devote sufficient time to the book, but she continued to serve as an active consultant, confidante, and critic.

PART ONE

The Scientific Context

of Survey Research

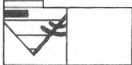

This book addresses the subject of scientific research. It has the dual purpose of assisting you in the execution of your own research and in understanding research conducted by others. Although the book focuses on a particular research method—survey research—it should be read within the general context of science. Whereas survey research employs specific techniques peculiar to that particular method, each of those techniques may be seen to fit into the general norms of scientific inquiry.

Although science has been with us for centuries and countless books have been written on the topic of scientific research, it remains a somewhat enigmatic subject for nonscientists and beginning scientists. The purpose of Part One of this book is to provide an overview of the logic of science so that the specific techniques of survey research may make more sense later in the book. Three chapters address this topic.

Chapter 1 considers the basic logic of science. It begins with a discussion of the traditional image of science—an image I believe is generally misleading and inappropriate in the context of actual scientific research. My purpose in this regard is not to debunk but to make science relevant and realistic.

Science in practice is neither mystical nor pure. Like all human activities, scientific research is a compromise between the ideal and the possible. It is often guided, in practice, by emotion, error, and nonrationality. Nevertheless, scientific research is importantly different from other human activities, and I will attempt to elucidate the differences in Chapter 1.

Chapter 2 addresses the even knottier topic of *social science*, about which a great deal has been written in recent years. Some writers insist that social science is every bit as "scientific" as natural science, while others debunk such a notion. Still others see social research as progressing toward, but not yet achieving, the status of science.

I suggest that this debate is meaningless. Just as it is impossible to provide a definitive statement of what "science" really is, it is similarly impossible to state whether social science is or is not scientific. Instead, we will consider the differences and similarities between the social and natural sciences and explore the implications of those differences and similarities for the logic and techniques of social scientific research.

Chapter 3 is devoted to a specific examination of survey research. The chapter begins with a consideration of the history of surveys and then turns to the ways in which survey research fits into the general norms of science and social science.

The ultimate purpose of these chapters, then, is to prepare you to understand the logical bases that underlie the specific skills and techniques of survey research. More generally, I hope you will also become better grounded in the basic logic of science per se. This is especially important because, as noted earlier, all scientific research is a compromise between the ideal and the possible. The bulk of this book deals with such compromises.

We begin by examining what ought to be done under ideal circumstances. Then we consider what compromises are most appropriate when the ideal cannot be achieved. Unless you fully understand the basic logic of science, you will be unable to understand why one compromise is accepted in lieu of another. If you do understand this logic, when you set about conducting your own research you will be able to achieve the best possible compromises under actual field conditions.

1

The Logic of Science

Science is a household word. Everyone has used it at one time or another, yet images of science differ greatly. For some, it is the same as mathematics, for others it is white coats and laboratories, others confuse it with technology, and still others equate the term with difficult high school or college courses.

Science is, of course, none of these things per se. Specifying exactly what science is, however, is difficult. This chapter begins by summarizing the image of science that is frequently presented to students in introductory science courses. Then we examine the negative image of science that some people hold. Finally, we describe the logical components of science as it is *practiced*.

The Traditional Perspective

Introductory courses in scientific disciplines often present an image of science that makes it appear straightforward, precise, and even routine. While I will criticize this perspective later in the chapter, it provides a useful starting point.

Scientific Theory

Scientists, we are told, begin with an interest in some aspect of the world around them. You might be interested in knowing how blood pressure is regulated in the body, why one strain of rice is hardier than another, what determines the paths of comets, or what causes cancer. You would then rigorously examine that topic and structure it in logical, abstract terms.

You would identify all the phenomena relevant to the subject under study. On the basis of existing knowledge, you might interrelate those phenomena in a network of **causal relationships**[1]—what elements cause or influence other elements. In this fashion, you would develop a **theory**, a set of interrelated, logical **propositions** that explain the nature of the phenomenon under study.

To test the validity of your theory, you would then presumably derive **hypotheses**, which are predictions about what will happen under specified conditions. Often, these hypotheses are of an if-then form. If Event A were to occur, then Event B would follow. For example, if you study hard in this course, you will live a happy life. (That's a hypothesis, remember, not necessarily the way things are.) Since this causal relationship is warranted by the general theory, the failure of Event B to follow Event A would call into question the validity of the theory itself.

Operationalization

Theories are by nature abstract and general. Hypotheses, while more specific, are also typically somewhat abstract. (After all, I didn't specify what kind of studying or what kind of happiness in the earlier example.) Thus, hypotheses must be converted into operational terms, that is, into the concrete operations you will perform in testing them—a process called **operationalization.** You must specify which real-world phenomenon will constitute Event A and which will constitute Event B.

If your hypothesis concerned the effect of temperature on the growth rate of a certain plant, you would specify the manner in which both temperature and growth would be measured, that is, what operations would be equated with those somewhat abstract concepts. Having specified the operations, you would then describe an **experiment** that would test your hypothesis, specifying the duration of the experiment, the frequency with which temperature and growth would be measured, whether temperature would be artificially controlled (and how) or natural variations would be noted, how the measurements would be recorded and analyzed, and so forth.

On the basis of your hypothesis, you might predict growth rates that should correspond with different temperatures and might specify how closely your predictions would have to match the observed growth rates in order to constitute a confirmation or rejection of your hypothesis. Laying all this out in advance can avoid a lot of arguments later on.

Experiment

Finally, the operations specified are undertaken. The data are collected and manipulated as prescribed, and the hypothesis is

---

[1] Words and phrases **boldfaced** like this are defined and discussed in the Glossary at the end of the book. If you aren't completely comfortable with a term, it's well worth the time to stop immediately and look it up in the Glossary.

thereby tested. If your hypothesis is confirmed by the experiment, then the general theory from which the hypothesis was derived is supported. If your hypothesis is not confirmed, the general theory is called into question. Whatever the outcome, you would presumably publish your findings, the world would become a little better place in which to live, and you would start thinking about other topics to conquer.

While imagination and brilliance would seem called for in the construction of theories, little or none seems necessary in the collection and analysis of data. With operations and experimental methods specified in advance, technical assistants could presumably conduct and interpret the experiment.[2]

Summary

Since scientists operate in accord with **rational** and **objective** procedures, their conclusions are presumably of a higher quality than the subjective impressions and prejudices of the layperson. Scientists deal with facts and figures, and figures, we are told, do not lie.

The Debunking of Science

In recent years, some students have been given an image of science and scientists that is quite different from the traditional perspective presented earlier. This more jaundiced view of science has several dimensions.

First, it is pointed out that scientists are motivated by the same human emotions and hindered by the same human frailties as other human beings. Scientists, we are told, frequently select their subjects of study on the basis of personal biases, and some may devote all their energies to "proving" some pet hunch. Rather than objectively framing and executing experiments, they conduct research that is a continuing search for data to substantiate a prejudice.

Similarly, the cults and cliques of the scientific world have been held up to public view. It has been pointed out that a scientific paper submitted for publication may be judged more on the basis of the researcher's academic pedigree (degrees, school, and the like) than on the intrinsic merits of the paper itself. A journal editor trained under Professor X may reject all papers submitted by students of Professor Y.

Moreover, it is frequently pointed out that "accepted ideas" in scientific disciplines are very difficult to challenge. A research paper presenting a radically new perspective on an old and presumably settled issue may never see the light of publication.

---

[2] In my experience, this is more typical of psychology than of sociology, as reflected, for example, in PhD dissertation proposals. A psychology graduate student whose dissertation proposal has been accepted is usually near completion of the dissertation, whereas the sociology student is just beginning.

On another topic, we are told that "grantsmanship" has replaced scholarship in science, that is, that many researchers evaluate a prospective research project in terms of the likelihood of foundation funding rather than in terms of its possible contribution to understanding.

Since so much scientific research is conducted within universities, it is relevant to note criticisms of the "publish or perish" norm attributed to many academic departments. Faculty members who excel in teaching may nonetheless be fired, it is charged, if they don't undertake and publish enough research. Research conducted under such duress is unlikely to manifest the intellectual curiosity and search for truth we associate with the idealized image of science.

Research by coercion, moreover, sometimes produces outright fraud in science. Every now and then you will read reports in the mass media about a researcher who has falsified his or her findings for the sake of appearing to be a productive scholar. Such cheating is not only wrong in its own right, but it can sometimes mislead the research of others, thereby wasting resources, delaying discoveries, and otherwise disrupting the research process.

These criticisms of present-day science have been further fueled by a number of candid research biographies published over the years by noted scientists.[3] Increasingly, practicing scientists have attempted to present honest accounts of their research projects, to place their findings in proper perspective, and to provide better guidance to aspiring researchers. Since these accounts have pointed to errors, oversights, and other practical problems, many of the contemporary critics of science in general have taken them as inside admissions that science is bunk.

A little debunking of science now and then is probably healthy. It is too easy to regard science as a mystical enterprise and scientists as its infallible wizard-practitioners. If science is fundamentally a rational and objective activity, then it should withstand objective and rational evaluation. Those aspects that do not survive such a critique probably should not remain a part of the scientific enterprise.

The main danger in criticism of sloppy and even unsavory research practices is that it can provide an easy escape from the difficulty of understanding science as it ought to be. Students sometimes find it simpler to dismiss all science as ritualistic nonsense than to learn statistics or the logic of scientific research. It is easier to regard all science as bad than to become a good scientist.

I am, of course, biased on this topic. I believe that science is a distinctive human activity. While many activities that may be called "scientific" are not so, in my opinion, many scientific activities are importantly different from other human activities. Understanding those differences is extremely important, both

---

[3] See, for example, James D. Watson, *The Double Helix* (New York: The New American Library, Inc., 1968) and the collections of social research biographies in Phillip Hammond, ed., *Sociologists at Work* (New York: Basic Books, 1964), and M. Patricia Golden, ed., *The Research Experience* (Itasca, IL: F. E. Peacock, 1976).

for those who conduct research and for those who read about it and whose lives are affected by it.

The primary problem here lies with the inaccuracy of the traditional perspective of the scientific method as it is conventionally presented to beginning science students, in both the social and natural sciences. Science in practice does not correspond exactly to the traditional image of science, but, at the same time, it need not be as bad as its severest critics contend. The following section aims at describing science in practice as distinguished from the traditional perspective. Then we will turn to those features that make science distinct from other human activities.

Science in Practice

Whereas the traditional perspective suggests that the scientist moves directly from an intellectual curiosity about some phenomenon to the derivation of a theory, this is seldom, if ever, the case. The initial interest in a phenomenon often stems from some previous empirical research, perhaps some inconsistent findings generated by your own research or the work of others. In a sense, you may start with the "answer" and set out to discover the "question."

You might begin with a specific observation: that children from broken homes have higher delinquency rates than others. You would then set out to develop a more general understanding of why that is the case.

Theories are almost never the result of wholly deductive processes. They are more typically the end result of a long chain of **deduction** and **induction**. At one point, you may have a tentative explanation for an empirical relationship; you may test it through the collection of more data, use the new results to modify your explanation, collect more data, and so forth. Theory building, then, involves an interaction of observation and explanation.

It follows, therefore, that theories are seldom confirmed at a given point in time. There are relatively few "critical experiments" in science, that is, experiments upon which the whole theory stands or falls. Instead, evidence built up over time lends support to a continually modified theory. Eventually, some form of the theory may become generally accepted, but we can seldom point to a given time at which it was "proved" or accepted. Moreover, all theories continue to undergo modifications. No scientist has yet discovered or ever will discover "The Truth."

The operationalization of **concepts** is never as clear-cut as the traditional image of science suggests. While this will be discussed in greater depth in Chapter 7, it should be noted here that most scientific concepts lend themselves to a variety of different interpretations. Thus, you may specify tentative operationalizations of those concepts, and the results of your experiments will be used to evaluate those operationalizations as much as to test hypotheses. If improvements to a new consumer product do not appear to cause increased satisfaction, you may question whether you are measuring satisfaction appropriately.

Even when concepts are operationalized in a reasonably acceptable fashion, experimental results are seldom conclusive in any absolute sense, even with regard to specific hypotheses. Typically, a hypothesis is confirmed or rejected *to a certain extent,* but seldom completely. As noted previously, theories are normally accepted on the basis of a weight of evidence over several experiments. If a large number of empirical observations are better explained by Theory A than by any other available theory, Theory A will probably be accepted.[4]

Finally, the impression that empirical testing is a routine activity is wholly incorrect. The traditional image of science suggests that creativity and even brilliance may be required in the derivation of theories and in the design of experiments but that the execution of the experiment itself is pretty dull and unimaginative. In practice, the execution of an experiment, the collection of empirical data, requires countless critical decisions. Unexpected situations arise. Bizarre observations are recorded, suggesting measurement errors. Data may be lost or falsified. (These kinds of problems are particularly common in survey research, as we will see throughout the book.) Moreover, the operationalization of concepts is never completely unambiguous and must be further specified during the experiment. Each of these situations requires decisions that will influence the outcome of the experiment and, by extension, the evaluation of the hypothesis and the theory from which it may have been derived.

This is an important point in the context of the present volume. You may have worked as a research assistant on a project or may know others who have. If so, you can probably attest to the disparity between the project design in theory and the day-to-day work on the project. Especially if it was a poorly supervised project, you might have a rather jaundiced view of the scientific-sounding description of the project as reported in an academic journal. In such a situation, you might be tempted to conclude that all science is "unscientific."

I emphasize this point for two purposes. First, I would stress (and will do so throughout the volume) that the quality of a research project can depend in an important way on the seemingly mundane decisions and activities that go into the collection and processing of data. The project director who does not become intimately involved in such activities runs a serious risk of directing a meaningless project. At the same time, however, the implications of such decisions—even sloppiness—are not always readily apparent. What might seem a sloppy procedure to the layperson might in fact be insignificant to the ultimate worth of the data. Making good decisions and evaluating the implica-

---

[4] Do not assume that this is automatically the case. Much has been written about the tenacity with which scientists sometimes hold on to established **paradigms** (theoretical models or points of view) in the face of contradictory evidence and competing paradigms. The classic book on this topic is Thomas S. Kuhn, *The Structure of Scientific Revolutions* (Chicago: University of Chicago Press, 1970).

tions of bad ones depend on a well-grounded understanding of the logic of science as a distinctive activity. It is this latter topic to which we now turn.

What Is Science?

At base, all science is aimed at understanding the world around us. Three major components make up this activity: description, the discovery of regularity, and the formulation of theories and laws. First, scientists *observe and describe* objects and events appearing in the world. This may involve the measurement of velocity of a falling object, the wave length of emissions from a distant star, or the mass of a subatomic particle. Such descriptions are guided by the goals of accuracy and utility.

Second, scientists seek to *discover regularities and order* among the sometimes whirling, buzzing chaos of experience. In part, this may involve the coincidence or **correlation** of certain characteristics or events. Thus, for example, you might note that air pressure correlates with altitude or that the application of force to an object results in a modification of its velocity.

Third, scientists seek to *formalize and generalize the regularities* they discover into theories and **laws**. Examples of this include Newton's law of gravity and Einstein's general and special theories of relativity. Theories and laws, then, are general, logical statements of relationships among characteristics and events that provide explanations for a wide range of empirical occurrences.

It is worth noting that *non*scientists pursue these same three goals. All of us observe and describe the world around us. We seek to find regularities: The office worker may find that arriving late for work results in a reprimand. We seek to formulate laws and theories that will provide overall guidance in everyday life, such as religious perspectives holding that adherence to religious teachings will result in worldly and/or otherworldly rewards.

To emphasize this point, it might be useful to note the similarities in the respective activities of a careless scientist and a racial bigot. Both make observations about the world and may report those observations to others. For example, the scientist observes that members of a given preliterate tribe enjoy relatively good dental health; the bigot observes that a Chinese shopkeeper has cheated in a business transaction. The scientist tentatively concludes that the primitive tribe's diet may be responsible for the dental health of its members, while the bigot concludes that the shopkeeper's race was responsible for the unethical business practices.

Both the scientist and the bigot seek new observations to reinforce their tentative conclusions. The scientist checks the dental health of other primitive tribes with similar diets, while the bigot keeps an eye on Chinese shopkeepers. It is important to note that both the scientist and the bigot are *selective* in their subsequent observations. In focusing on diet, the scientist may overlook tribes' climatic environment, economic structure, and so forth. The bigot, in turn, will overlook shopkeepers' education, social class, and so forth.

Moreover, both the careless scientist and the bigot may tend to ignore observations that contradict the conclusions they have already reached. The scientist may ignore reports of good dental health among tribes with radically different diets and reports of tribes who do follow the diet in question but whose teeth are rotten and falling out. The bigot may ignore all honest Chinese shopkeepers and dishonest non-Chinese.

The careless scientist may justify disregarding such reports by attributing them to bad fieldwork. (The careful scientist, of course, would ultimately deal with such cases.) The bigot may refer vaguely to "exceptions that prove the rule."[5]

When they are faced with too many disconfirming observations, both the careless scientist and the bigot will seek additional information that might bring the troublesome observations into line with their conclusions. In the case of a tribe having good teeth but a different diet, the scientist might intensify the field work in an effort to discover that the tribe's diet is more similar to the one in question than initially appeared. Confronted with a dishonest non-Chinese shopkeeper, the bigot might engage in a search for a Chinese great-great-grandfather or a passion for Chinese food.

Despite empirical anomalies, the scientist in question may end up with a theory relating a certain diet to dental health, and the bigot will end up with a theory that Chinese people are dishonest. Both theories will be apparently substantiated by empirical observations and logical explanations.

The above comparison is aimed at making two points. First, no magical difference exists between scientific and nonscientific activities. We have seen two rather similar lines of inquiry. Second, the activities of "scientists" vary in "scientific" quality. It makes more sense to speak of more or less scientific activities than to dichotomize between scientific and nonscientific activities. Thus, a particular line of inquiry conducted by a professional physicist may in fact be relatively unscientific, while a plumber may conduct an inquiry of the highest scientific caliber.

In the remainder of this section, we will turn our attention to those characteristics that make a given activity more or less scientific. We will attempt to understand the *ideal* characteristics of science, realizing that no given activity is ever fully in accord with those ideals, whether conducted by a professional scientist or by a layperson.

Science Is Logical

Science is fundamentally a rational activity, and scientific explanations must make sense. Religions may rest on revelations, customs, or traditions, gambling on faith; but science must rest on logical reason.

---

[5] Originally, *prove* in this expression meant "test," as when we say that great obstacles "severely tested" someone's commitment. In other words, exceptions were seen as threats or challenges to a rule.

Logic is a difficult and complex branch of philosophy, and a full delineation of systems of logic is well beyond the scope of this textbook. Perhaps a few examples will illustrate what is meant by science being logical. For example, a given event cannot logically cause another event that occurred earlier in time. Your social attitudes about, say, race relations cannot determine the region of the country where you are born, although the converse can be true.

Thus, science takes a different approach from the **teleological** views assumed by some religions. For example, some Christians believe that Jesus was destined to be crucified and that this destiny thereby caused him to be betrayed and tried. Such a view could not be accepted within the logic of science.

In the logic of science, it is impossible for an object to have two **mutually exclusive** qualities. The flip of a coin cannot result in both a head and a tail. By contrast, we might note that many deeply prejudiced people argue that Armenians are both "clannish" (refusing to mix with non-Armenians) and "pushy" (forcing themselves in with non-Armenians). Faced with such conflicting assertions, the logic of science would suggest that either one or both of the characterizations of Armenians are untrue or that the two characteristics are being defined in such a way that they are not mutually exclusive.

Similarly, a given event cannot have mutually exclusive results. Thus, getting a college education cannot make a person both wealthier and poorer at the same time. It is possible for a college education to result in wealth for one person and poverty for another, just as some Armenians might be described as clannish and others pushy, but contradictory results or descriptions fly in the face of logic and are intolerable to science.

All this is not to say that science in practice is wholly devoid of illogical assertions. You might know that physicists currently regard light as both particles and waves, even though these descriptions of the nature of light are contradictory. This particular contradiction exists in science because light behaves as particles under some conditions and as waves under others. As a result of this situation, physicists continue to use the two contradictory conceptualizations as they may be appropriate in given conditions. Nevertheless, such a situation represents a strain for the logic of science.

Moving somewhat beyond this "common-sense" notion of logic, two distinct logical systems, introduced earlier, are important to the scientific quest: *deductive logic* and *inductive logic*. Beveridge describes them as follows:

> Logicians distinguish between inductive reasoning (from particular instances to general principles, from facts to theories) and deductive reasoning (from the general to the particular, applying a theory to a particular case). In induction one starts from observed data and develops a generalization which explains the relationships between the objects observed. On the other hand, in deductive reasoning one starts from some general law and applies it to a particular instance.[6]

---

[6] W. I. B. Beveridge, *The Art of Scientific Investigation* (New York: Vintage Books, 1950), p. 113.

The classical illustration of deductive logic is the familiar **syllogism**: "All men are mortal; Socrates is a man; therefore Socrates is mortal." A researcher might then follow up this deductive exercise with an empirical test of Socrates' mortality. This is essentially the approach discussed earlier in this chapter as the "traditional perspective of science."

Using inductive logic, you might begin by noting that Socrates is mortal and observe a number of other men as well. You might then note that all the observed men were mortal, thereby arriving at the tentative conclusion that all men are mortal.

Deductive logic is very old as a system, dating at least to Aristotle. Moreover, this system was predominant in Western philosophy until the sixteenth or seventeenth century. The birth of modern science was marked by the rise of inductive logic in a variety of scientific contexts. Increasingly, the general conclusions derived from careful observations contradicted the general postulates that represented the anchoring points of many deductive systems.

In astronomy, for example, Ptolemy accounted for observed variations from his model by developing an epicyclical model in which stars and planets rotated in circles around points in space that, in turn, rotated around the stationary earth in circles. As further variations were noted, the system was made more and more complicated, with new levels of epicycles added in order to retain the key beliefs of circular motion and a stationary earth.

Copernicus attacked the Ptolemaic system by suggesting that the sun, rather than the earth, was the center of the universe. He derived his radically new perspective from observed celestial motion rather than from an initial commitment to the earth as the center of the universe. Copernicus, however, did not challenge the assumption of circular motion. "The later astronomer, Kepler, said that Copernicus failed to see the riches that were within his grasp, and was content to interpret Ptolemy rather than nature."[7] Kepler, on the other hand, was determined to interpret nature in the form of the voluminous empirical data that he inherited from the Danish astronomer Tycho Brahe. As Butterfield goes on to tell us:

> We know how with colossal expenditure of energy he tried one hypothesis after another, and threw them away, until he reached a point where he had a vague knowledge of the shape required, decided that for purposes of calculation an ellipse might give him at any rate approximate results, and then found that an ellipse was right.[8]

This example illustrates the rise of inductive logic in science. Similar dramas occurred in other fields of inquiry during the fertile sixteenth and seventeenth centuries. A century or so later, the inductive, scientific research of Charles Darwin clashed with yet another tradition.

---

[7] Herbert Butterfield, *The Origins of Modern Science* (New York: The Macmillan Company, 1960), p. 24.
[8] Ibid., p. 64.

It should not be concluded from the foregoing historical account that deductive logic is inherently incorrect or that it is now outmoded. An exercise in deductive logic is as good as its internal consistency and the truth of its beginning assumptions. Inductive logic, on the other hand, is only as good as its internal consistency and the accuracy of its observations.

In practice, scientific research involves both inductive and deductive reasoning as the scientist shifts endlessly back and forth between theory and empirical observations.

Science Is Deterministic

Science is based on the assumption that all events have antecedent causes that are subject to identification and logical understanding. For the scientist, nothing "just happens"—it happens for a reason. If a person catches a cold, if it rains today, if a ball seems to roll uphill, the scientist assumes that each of these events is susceptible to rational explanation.

As we will see in the following chapter, this characteristic of science presents a special strain for the social sciences, which compete with commonsense notions about social behavior. While you might say you performed a certain act, voting for a political candidate, for example, simply because you decided to do so, the social scientist would probably argue that your vote was determined by a variety of prior events and conditions. The decision to vote might be attributed to your social class, the region of the country you live in, and so forth, even though you might deny the influence of such factors.

Several caveats should be entered in this regard, however. First, scientists do not know, nor do they pretend to know, the specific causes of all events. They simply assume that such causes exist and can be discovered. Second, science accepts multiple causation. A given event may have several causes; the voting decision may have resulted from a number of different factors. Also, one event may have one cause, while a similar event may have a different cause. Two people may vote for the same candidate for different reasons, but it is assumed that reasons exist in each case.

Finally, much science is based on a **probabilistic** form of determinism. Thus, Event A may result in Event B 90 percent of the time, or 70 percent of all Republicans may vote for a given political candidate while only 23 percent of the Democrats do so. In this sense, then, political party affiliation would be said to *determine* voting behavior, even though the determination was not complete. Other factors might be introduced to explain the discrepancies.

Science Is General

Science aims at general understanding rather than at the explanation of individual events. The scientist is typically less interested in understanding why a given ball falls to earth when released from a height than in understanding why *all* such balls tend to do so. Similarly, the scientist is less interested in explaining why you voted as you did than in understanding why voters in general voted as they did.

This characteristic of science is related to its probabilistic determinism. It is conceivable that we could *completely* explain the reasons behind a given event—why a given person voted for Candidate X, for instance. We might conceivably discover every single factor that went into the voting decision. If we were successful in this, then presumably we could predict the voting behavior of *identical* persons with perfect accuracy. In the long run, however, such a capability would not provide much useful information about voting behavior in general. First, it is doubtful that we would ever find another person with exactly the same characteristics. More important, our discoveries might help us very little in understanding the voting behavior of people with other characteristics. We would be happier, then, with less than 100 percent understanding if we were able to understand voting behavior in general.

This is the sense in which the scientist and the historian differ in their approaches to the same subject matter. The historian aims at understanding everything about a specific event, while the scientist is more interested in generally understanding a class of similar, though not identical, events. By the same token, the psychologist and the therapist differ in their approaches to human behavior. The psychologist would examine schizophrenic behavior among several individuals in an effort to arrive at a general understanding of schizophrenia, while the therapist would take advantage of existing general knowledge in an effort to help a specific individual.

Generalizability, therefore, is an important characteristic of scientific discoveries. The discovery that red balls fall to earth at a given acceleration is less useful than the discovery that balls of all colors do so. Similarly, it is less useful to know that balls fall with a given acceleration at sea level than to know that the acceleration of all falling balls can be determined from their altitude.

Science Is Parsimonious

As the previous sections indicate, scientists spend much effort in the attempt to discover factors that determine types of events. At the same time, they attempt to discover those factors that do *not* determine the events. Thus, in determining the acceleration of a falling object, we discount its color as being irrelevant.

More generally, scientists attempt to understand the reasons for the events using as few explanatory factors as possible. In practice, of course, the number of explanatory factors taken into account typically increases the degree of determination achieved. One political scientist may achieve a certain degree of explanation of voting behavior through the use of only two factors, say, party affiliation and social class. Another might achieve a more complete understanding by also taking into account other factors such as race, region of upbringing, sex, education, and so forth. Frequently, scientists are forced to choose between simplicity on the one hand and degree of explanation on the other. Ultimately, they try to optimize the balance between explanation and simplicity in order to get the greatest amount of explanation from the smallest number of factors. This **parsimonious** quality is nicely illustrated in the elegance of Einstein's famous equation $e = mc^2$.

Science Is Specific

We noted earlier that science is general in that it aims at discoveries and laws with general applicability. Most general concepts are subject to a variety of interpretations, however. Thus, while you might wish to explain the sources of prejudice in general, you would always realize that prejudice takes many different forms. In designing, conducting, and reporting your research, therefore, you need to be precise in your methods of measuring such a concept.

In conducting a research project on the topic of prejudice, you would need to generate a specific operationalization of the concept of prejudice, for example, agreement with several questionnaire statements that seem to indicate prejudice. In reporting your research, you would be careful to describe your operationalizations in detail so that the reader will know precisely how the concept has been measured. While some readers may disagree with the operationalization, they at least will know what it is.

Often the generalizability of a given discovery is substantiated through the use of several different operationalizations of the concepts involved. If a given set of factors causes prejudice regardless of how prejudice is measured, you may conclude that those factors result in prejudice in general.

Science Is Empirically Verifiable

Science at its most elegant results in the formulation of general laws or equations describing the world around us. Such formulations, however, are useful only if they can be verified through the collection and manipulation of empirical data. A general theory of prejudice would be useless unless it suggested ways in which data might be collected and unless it predicted the results that would be obtained from the analysis of those data.

There is another way of viewing this characteristic, however. In a sense, no scientific theory can ever be proved. Let's consider the case of gravity. Physicists tell us that a body falls to earth because of the general attraction that exists between physical bodies and that this relationship is affected by the mass of the bodies involved. Since the earth has a vast mass, a ball thrown out a window will move toward the earth.

Such an explanation of gravity is empirically verifiable. A researcher can throw a ball out a window and observe that it falls to earth. This does not *prove* the truth of the theory of gravity, however. Rather, the researcher specifies that if the ball does *not* fall to earth, then the theory of gravity is incorrect. Since the ball is, in fact, observed to behave as expected, the theory of gravity *has not been disconfirmed*.

Thus, when we say that a scientific explanation must be subject to empirical testing, we mean, more precisely, that the researcher must be able to specify conditions under which the theory would be disproved. In this regard, scientists speak of the **disconfirmability** of theories. As you consistently fail to disprove your theory, then, you may become confident that the theory is correct. But it is important to realize that you will never have *proved* it.

Pursuing the preceding example, another theorist might note that the experimental ball was of the same color as the ground to which it fell, suggesting that bodies of the same color are attracted to each other for whatever reason he or she might devise. The initial experiment would lend confirmation to both competing theories. The "color-attraction theory," however, suggests that if a ball *differing* in color from the earth were thrown out the window, it should not fall to earth. An appropriate second experiment would (we hope) result in an empirical disconfirmation of the theory.

Science Is Intersubjective

It is frequently asserted that science is "objective," but such an assertion typically results in a good deal of confusion as to what "objectivity" is. Moreover, it has been noted increasingly in recent years that no scientist is completely objective in his or her work. All scientists are "subjective" to some extent—influenced by their personal motivations. In saying that science is **intersubjective,** we mean that two scientists with different subjective orientations would arrive at the same conclusion if each conducted the same experiment. An example from political science should clarify this concept.

The tendency for intellectuals in America to align themselves more with the Democratic party than with the Republican party has led many people to assume that Democrats as a group are better educated than Republicans. It is reasonable to assume that a Democratic scientist would be happy with this view, while a Republican scientist would not. Yet it would be possible for the two scientists to agree on the design of a research project that would collect data from the American electorate relating to party affiliation and educational levels. The two scientists could then conduct independent studies of the subject, and both would discover that Republicans as a whole have a higher educational level than do Democrats. (This is due to the fact that the Democratic party also attracts a larger proportion of working-class voters than does the Republican party in America, while businesspeople are more attracted to the Republican party.) Both scientists—with opposite subjective orientations—would arrive at the same empirical conclusion.

Clearly, scientists often disagree among themselves. They may offer grossly different explanations for a given event. Such disagreements, however, typically involve questions of conceptualization and definition. Thus, one social scientist may report that religiosity is positively related to prejudice, while another disagrees. The disagreeing scientist will, in all likelihood, suggest that either or both of the variables have been incorrectly measured. You might conduct your own study, measuring the two variables differently, and report a negative relationship between them. If the earlier researchers had reported the design and execution of their studies in precise and specific details, however, and you were to replicate the study exactly, you should arrive at the same finding. This is what is meant by the intersubjectivity of science.

Science Is Open to Modification

It should be clear from the previous section that "science" does not provide a set of easy steps to the attainment of "The Truth." Two scientists, both adhering to the previously discussed characteristics of science, may arrive at quite different explanations of a given phenomenon. At a given time, moreover, there may be no way of evaluating their relative merits. If the two explanations contradict one another, presumably both cannot be correct. Either one or both will later be proved incorrect, or else it will be discovered that the two explanations are not mutually exclusive after all, through a shift of paradigm, for example.

Countless "scientific" theories of the past have subsequently been disproved and replaced by new ones. Everything we "know" today was previously "known" differently, and we sometimes look back on those earlier views as naive, silly, or stupid. It is worth considering, however, that everything we "know" today will probably be overturned sometime in the future, and those future folks—our arrogant descendants—will look back on us as naive, silly, or stupid. (If that bothers you, perhaps you can take some comfort in knowing that they'll suffer the same fate.)

In an important sense, science does not seek ultimate truth, but rather *utility*. Scientific theories should be judged not on their relative truth value but on the extent to which they are useful in furthering our understanding of the world around us.

In the final analysis, the characteristics of science discussed in this chapter provide a set of guidelines that enhance the utility of discoveries and theories. Inquiries that seek to adhere to these characteristics will, in the long run, produce more useful discoveries than will inquiries of another sort. Thus, one person may be able to predict the weather more accurately on the basis of a "trick" knee than all the scientific meteorologists in the world, but, in the long run, the scientists will contribute more to our general understanding of the nature of weather.

Summary

This chapter began with a review of the traditional image of science, primarily as a set of steps that inevitably lead to "The Truth." This view was balanced against a more recent, critical view of science that suggests that scientists are not much different from laypeople. In the bulk of the chapter, we have attempted to show that while scientific inquiry is not cut-and-dried, that is, is not infallible, it *is* importantly different from other human activities. While scientists are surely subject to all the human frailties of other people, science provides a set of guidelines that can enhance the utility of their inquiry.

While this chapter has been addressed to science in general, the following chapter will focus specifically on *social science*. We will see that social science is bound by the same rules as are other types of scientific inquiry. At the same time, however, its special subject matter presents special problems—and special opportunities.

Additional Readings

W. I. B. Beveridge, *The Art of Scientific Investigation* (New York: Vintage Books, 1950).

Herbert Butterfield, *The Origins of Modern Science* (New York: The Macmillan Company, 1960).

M. Patricia Golden, ed., *The Research Experience* (Itasca, IL: F. E. Peacock, 1976).

William Irvine, *Apes, Angels, and Victorians* (New York: Meridian Books, 1959).

Abraham Kaplan, *The Conduct of Inquiry* (San Francisco: Chandler Publishing Co., 1964).

Thomas S. Kuhn, *The Structure of Scientific Revolutions* (Chicago: University of Chicago Press, 1970).

Bob Toben, *Space-Time and Beyond* (New York: E. P. Dutton, 1975).

James D. Watson, *The Double Helix* (New York: The New American Library, Inc., 1968).

Alfred North Whitehead, *Science and the Modern World* (New York: The Macmillan Company, 1925).

2

Science and

Social Science

One of the livelier academic debates of recent years has concerned the "scientific" status of those disciplines gathered under the rubric of the social sciences—typically including sociology, political science, social psychology, economics, anthropology, market research, and sometimes fields such as geography, history, communications, and other composite and specialty fields. Basically at issue is whether human behavior can be subjected to "scientific" study. Whereas the previous chapter pointed to the confusion that surrounds the term *science* in general, it should come as no surprise that academicians disagree about the social sciences.

Opposition to the idea of social sciences has risen both within the social science fields and outside them. Within the fields, the movement toward social science has meant a redirection and, in some cases, a renaming of established academic traditions. Increasingly, departments of Government have been replaced by departments of Political Science. Speech departments have become departments of Communications. In many cases, the movement toward social science has entailed a shift in emphasis from description to systematic explanation. In political science, it has meant a greater emphasis on explaining political behavior than on describing political institutions. In anthropology, it has been represented by a lessening of the emphasis on ethnography. The growth of subfields such as econometrics has had a similar effect in economics, as has historiography in history. Some geographers have moved from the enumeration of imports and exports to mathematical models of migration. Quite understandably, professionals trained and experienced in the more traditional methods of these fields have objected to the new orientations.

Outside the social science departments, similar opposition has come from the physical sciences—from physicists, biologists, chemists, and so forth.

Guided by the traditional image of science discussed in the previous chapter, some physical scientists have objected that the "scientific method" cannot be applied to human social behavior.

All too often, the advocates of social science have fueled the debate through blind emulation of the trappings and rituals of the established sciences. This emulation has taken many forms: a fascination with laboratory equipment, often inappropriate uses of statistics and mathematics, the development of obscure terminology, and the wholesale adoption of theories and terminology from the physical sciences.

For the most part, these errors would seem to have grown out of an acceptance of the traditional image of science and a lack of understanding of the logic of science in practice. Would-be social scientists have too often attempted to reach understanding through methods that do not work even for physical scientists. The result frequently has been ridicule from physical scientists, professional colleagues, and laypeople.

It is the firm assumption of this book that human social behavior can be subjected to "scientific" study as legitimately as can atoms and cells. This assumption must be understood within the context of the earlier discussion of science in practice, however. From this perspective, no significant difference would seem to exist between the physical and social sciences. Like physical scientists, social scientists seek to discover regularity and order. Social scientists look for regularity in social behavior through careful observation and measurement, the discovery of relationships, and the framing of models and theories.

The Search for Social Regularities

Measuring Social Phenomena

The first building block of science is **measurement,** or systematic observation. There is no fundamental reason why social scientists cannot measure phenomena relevant to their inquiry. The age and sex of social actors, for example, can be measured. Place of birth and marital status can be measured in a number of different ways, varying in accuracy and economy.

Aggregate social behavior can be measured systematically as well. The political scientist can determine the election-day voting behavior of the entire electorate or of individual precincts. The amount of traffic over a given section of highway can be measured at different points in time. Market researchers can measure volume of sales.

Attitudes can also be measured, although this is a point of broad disagreement. For example, prejudice against women can be measured by determining individual acceptance or rejection of beliefs and perspectives representing such prejudice. Religiosity, political liberalism and conservatism, authoritarianism, and similar variables can be measured in like fashion.

Attitude measurement is frequently challenged as being "unscientific"; while later sections of this chapter and subsequent chapters in this book address this issue, some comment is in order here. It must be recognized that all such measurements (all measurements, in fact) are arbitrary at base. The social scientist cannot unequivocally describe one person as "alienated" and another as "unalienated." Rather, people will be described as *relatively more or less* alienated—that is, in comparison with one another. This characteristic is by no means unique to social science, however, as evidenced by the "hardness scale" used by physical scientists, the Richter scale for earthquakes, and so forth. No one can say that a particular metal is "hard" or an earthquake "severe" in absolute terms, only that it is "harder" or "more severe" than something else.

Discovering Social Regularities

People have a tendency to regard the subjects of the physical sciences as more regular than those of the social sciences. A heavy object falls to earth every time it is released, while a person may vote for one candidate in one election and against the same candidate in the next election. Similarly, ice always melts when heated, while seemingly religious people do not always attend church. Though these particular examples are generally true, there is a danger in going on to discount the existence of social regularities altogether. The existence of observable social norms contradicts this conclusion.

Some social norms are prescribed by the formal institutions of a society. For example, only persons of a certain age or older are permitted to vote in elections. Prior to 1920, American men were permitted to vote, while women were not. Such formal prescriptions, then, regulate or regularize social behavior.

Aside from formal prescriptions, we can observe other social norms. Registered Republicans are more likely to vote for Republican candidates than are registered Democrats. University professors tend to earn more money than unskilled laborers. Women tend to be more religious than men.

Reports of such regularities by social scientists are often subject to three types of criticism. First, it may be charged that the report is trivial, that everyone was aware of the regularity. Second, contradictory cases may be cited, indicating that the observation is not wholly true. And third, it may be argued that the people involved could upset the observed regularity if they really wanted to.

The charge that many of the social scientist's discoveries are trivial or already well known has led many would-be social scientists to seek esoteric or obscure findings that would prove that social science is more than pretentious common sense. This response is inappropriate from a number of standpoints. To begin, so many contradictions are evident in the broad body of "common sense" that it is essential to systematically weed out the existing misconceptions. Even when a proposition is unchallenged by laypeople, it must be tested empirically.

Many social science methodology instructors begin their classes by revealing a set of "important discoveries" that have come from social science, derived from studies conducted by Samuel A. Stouffer during World War II.[1] These "discoveries" include the following findings:

- Black soldiers were happier in Northern training camps than in Southern ones.

- Soldiers in the Army Air Corps, where promotions were rapid, were more likely to feel that the promotion system was fair than were those in the Military Police, where promotions were very slow.

- Educated soldiers were more likely to resent being drafted than those with less education.

When students begin to dismiss the "important discoveries" as obvious, the instructor then reveals that *each of them was disproved* by Stouffer's research and explains why the observed relationships make sense when examined below the level of surface appearances.[2] In short, "documenting the obvious" is a valuable function of any science, physical or social, and is not a legitimate criticism of any scientific endeavor. (Darwin coined the term "fool's experiment" in ironic reference to much of his own research.)

The criticism that given generalizations from social science are subject to disconfirmation in specific cases is not a sufficient challenge to the "scientific" character of the inquiry, either. Thus, it is not sufficient to note that a given man is more religious than a given woman. Social regularities represent probabilistic patterns, and a general relationship between two variables need not be true in 100 percent of the observable cases.

Physical science is not exempt from this criticism. In genetics, for example, the mating of a blue-eyed person with a brown-eyed person will *probably* result in a brown-eyed offspring. The birth of a blue-eyed offspring does not challenge the observed regularity, however, since the geneticist states only that a brown-eyed offspring is more likely and, further, that brown-eyed offspring should be expected in a certain percentage of cases. The social scientist makes a similar probabilistic prediction—that women overall will be more likely to appear religious than men. With an adequately tested measurement device, we may be able to predict the percentage of women who will appear more religious than men.

Finally, the charge that observed social regularities could be upset through the conscious will of the actors is not a sufficient challenge to social science, even though a parallel situation does not seem to exist in the physical

---

[1] Samuel A. Stouffer, et al., *The American Soldier* (Princeton, NJ: Princeton University Press, 1949).
[2] These findings are discussed in Chapter 15, "The Elaboration Model."

sciences. (Presumably an object cannot resist falling to earth "because it wants to.") There is no denying that a right-wing, white bigot could go to the polls and vote for a left-wing radical black if he or she wanted to upset the political scientists studying the election. All voters in an election could suddenly switch to the underdog so as to frustrate the pollsters. By the same token, workers could go to work early or stay home from work and thereby prevent the expected rush-hour commuter traffic. But such things do not happen sufficiently often to seriously threaten the observation of social regularities. Ironically, of course, if all the workers stayed home from work, that in itself would be a regularity and would be susceptible to explanation.

The fact remains that social norms do exist and that we can observe those norms. When norms change over time, we can observe and explain the changes. Ultimately, social regularities persist because they tend to make sense for the individuals involved in them. Whereas we might suggest that it is logical to expect a given type of person to behave in a certain manner, that person may very well agree with the logical basis for the expectation. Thus, while right-wing, white bigots could vote for a left-wing radical black candidate, they would consider it stupid to do so, just as left-wing, black voters would consider it stupid to vote for a candidate from the Ku Klux Klan.

Creation of Social Theories

Social scientists have not yet developed theories of social behavior comparable with the theories developed by physical scientists. Countless theories of social behavior have been developed dating back centuries, but none of these theories is seriously defended as adequate. Of course, a great many theories pertaining to the physical world have been discarded as well. The ultimate demise of the Ptolemaic theory of epicycles does not negate the scientific standing of present-day astronomy. Nor does the knowledge that contemporary theories in physics will eventually be supplanted deny the scientific status of that field.

Nevertheless, at present the social sciences do not have formal theories comparable with those existing in other fields. In part, this situation is due to the fact that systematic, "scientific" methods have not been applied to social behavior for as long as they have been applied to physical phenomena. At the same time, the reluctance to admit the susceptibility of social behavior to scientific study has limited the resources made available for the development of social sciences.

In addition, this textbook grows out of the conviction that the scientific development of the social sciences has been seriously handicapped by a misunderstanding of the logical nature of science in general, specifically, by a commitment to the traditional—exclusively deductive—image of science, as opposed to an understanding of science in practice. In view of this, we turn now to a discussion of the characteristics of social science parallel to the discussion in Chapter 1, which dealt with science in general.

The Characteristics of Social Science

Social Science Is Logical

The social sciences aim at rational understanding of social behavior. This is not to say that all social behavior is rational, however. Some social behavior is irrational, some nonrational, but social scientists themselves should be relatively rational in seeking to understand all forms of behavior.

The social scientist is bound by many of the same logical constraints as the physical scientist. A given event cannot cause another event that occurs earlier in time. A given object cannot have mutually exclusive characteristics, and a given event or characteristic cannot have mutually exclusive results. Also, both deductive and inductive logic are appropriate to social science, as discussed with regard to science in general in Chapter 1.

Social Science Is Deterministic

Like physical scientists, social scientists assume that events occur for reasons, that things do not "just happen." Every event or situation has antecedent determinants. This characteristic of social science often seems at odds with common sense, as some earlier discussions in this chapter have indicated. The social scientist may conclude that a group of people behave in a certain fashion because of a number of prior events and conditions, as in the voting behavior of the right-wing, white bigot in our earlier discussion. In this sense, the conditions of race, prejudice, and political orientation *determine* voting behavior. This is not to deny that the voters in question could vote for the left-wing radical black candidate; they are simply unlikely to do so.

The deterministic posture of the social sciences represents the most significant departure from the more traditional, humanistic examinations of social behavior. Whereas the humanistically oriented observer might consider the soul searching and agonies by which a given person will weigh the relative merits and demerits of a given action, arriving at a considered decision, the social scientist would more typically look for the general determinants of such a decision among different aggregates of persons. Where the humanist would argue that the decision reached by each individual person represented the outcome of an idiosyncratic process, the social scientist would say that it could be fit into a much simpler, general pattern.

Social Science Is General

As already indicated, social science aims at the observation and understanding of overall patterns of events and correlations.

The utility of a social theory or social correlation is enhanced by its general-izability. The larger the scope of phenomena it explains, the more useful it is. Thus, a theory of consumer behavior that applies only to young people is less useful than one that applies to consumers of all ages. A theory of political behavior that applies only to Americans is less useful than one that applies to people of all nations.

While we may often begin with the attempt to explain a rather limited range of social behavior or the behavior of a limited subset of the population, our goal is normally to expand the explanatory power of our findings to other forms of behavior and other subsets of the population.

Social Science Is Parsimonious

Like the physical scientist, the social scientist at-tempts to gain the most explanatory power out of the smallest number of variables. In many cases, the additional consideration of new variables adds explanatory and predictive power, but it also results in a more complicated model. In practice, the addition of more variables often reduces the general-izability of the explanation, since certain variables might have one effect among members of one subset of the population and a different effect among members of other subsets.

It should be noted that the parsimonious character of social science, like its deterministic posture, opens it to criticism from people holding a more humanistic view. Whereas the humanist would tend to explore the depths of idiosyncratic factors resulting in a decision or action on the part of a given social actor, the social scientist consciously attempts to limit such inquiry.

The market researcher, then, might attempt to explain preferences for living in the city, the suburbs, or the country on the basis of, say, three or four variables. It could be argued, of course, that everyone has many other individ-ual reasons for choosing his or her lifestyle and that a limited number of variables cannot adequately explain the depth of decision making for any of the people under study.

The humanist's assertion is correct, but the social scientist has an impor-tantly different goal from that of the humanist. The social scientist is con-sciously attempting to gain the greatest amount of understanding from the smallest number of variables. Neither the scientist nor the humanist in such a case is more correct than the other; each simply has different goals. We must fully understand the scientist's goal in order to recognize that this criticism is not valid.

Social Science Is Specific

Social scientists, like physical scientists, must spec-ify their methods of measurement. This may be especially important in the social sciences, which deal with concepts more vaguely defined in common discourse. While the physicist defines "acceleration" more rigorously than the layperson, the scientific definition is not greatly at variance with the common understanding of the term. Concepts such as "alienation," "liberalism," and "prejudice," however, have such varied meanings in common usage that their rigorous definitions are not readily apparent.

Social scientists can subject such concepts to rigorous scientific examination, but in order to do so they must clearly specify the nature of the measurements being made in each instance. Also, the definitions will be evaluated on the basis of utility—their contribution to generalized explanation and understanding— rather than on the basis of absolute "Truth."

Social Science Is Empirically Verifiable

To be useful, social scientific propositions and theories must be testable in the real world. Thus, to assert that education is negatively associated with prejudice is useless without suggesting ways in which the two variables might be measured and the proposition tested. As in the physical sciences, the social scientist must be able to describe empirical conditions under which a given proposition would be judged incorrect, that is, ways in which it might be disproved.

Religious beliefs, such as the existence of God, for example, are not susceptible to empirical verification. Similarly, the assertion that members of a religious or racial group are disloyal "in their hearts" even when they appear to act in a loyal manner is not subject to empirical verification. The same would be true of propositions predicting human social behavior in the event that the sun did not rise on a given morning.

Social Science Is Intersubjective

To the extent that a given social scientific examination has been described in adequately specific detail, any other social scientist, of whatever personal persuasion, should be able to replicate the examination with similar results. Contrary social scientific findings are more often the result of disagreements over the most appropriate research design—including the definition and measurement of concepts—than over the results obtained from a given design.

This is as true of research into highly emotional topics such as religion, politics, and prejudice as of the study of the acceleration of falling objects. In practice, of course, researchers seldom conduct studies that they believe to be designed incorrectly. The conservative social scientist, for example, is unlikely to define conservatism in a way that might make conservatives appear evil.

Social Science Is Open to Modification

No social theory is likely to survive indefinitely. Either a growing weight of disconfirming evidence will bring it down, or a newer, more parsimonious replacement will be found. In any event, no social science finding can be expected to withstand the long-term test of time.

In practice, of course, the social scientist deals with phenomena that also come under the purview of ideology—religious, political, and philosophical—and ideologies are less open to modification than science is. When social scientists explain religiosity in terms of background variables, they challenge basic religious beliefs about moral behavior, religious reward and punishment systems, and so forth. When political scientists conclude that the working class in America is more authoritarian than the middle class, they challenge left-wing political ideology. The deterministic posture of the social sciences in general flies in the face of a philosophical free-will image of human beings that has a long history in Western civilization.

The danger is that individual social scientists may be so personally committed to particular ideological stances that their commitment prevents them from maintaining the openness of their science. Thus, the committed left-wing political scientist may be unwilling to consider, undertake, or accept research activities that might lead to the conclusion that the working class is more authoritarian than the middle class.

As noted in earlier discussions, this situation is not unique to the social sciences. Physical science inquiry has challenged and continues to challenge accepted ideological belief systems, and individual physical scientists have been handicapped by ideological commitments that have reduced the openness of their scientific activities.

Methods of Social Scientific Research

Although this textbook specifically addresses a single method of social research, it is useful to place that method in the context of the other methods available to the social scientist. I do this in part to suggest that the examination of a given social phenomenon is often best accomplished through the use of several different methods—an especially important point to make at a time when survey research enjoys such broad popularity. Though survey research has special advantages, we will see throughout this book that it has shortcomings as well and is not the appropriate method for studying certain topics. Social researchers who limit themselves to a single method, survey or other, severely limit their ultimate ability to understand the world around them.

At the same time, it is important to realize that all methods of social research are guided by the general characteristics of science as outlined in this and the previous chapter. Examining the relative strengths and weaknesses of each method in that regard, therefore, is useful.

The Controlled Experiment

In many ways, the controlled experiment represents the clearest example of scientific research, at least in the popular image of science. While experimental design has many variations, we will limit our attention to the before/after design with a single control group.

Assume for the moment that you are interested in methods for reducing racial prejudice. Assume further that you tentatively believe that antiblack prejudice might be reduced by a greater awareness of the important role played by blacks in American history. To test this hypothesis, you might locate or even produce a motion picture documenting black American history. Such a film would represent the **stimulus** for the experiment.

Next, you would select two groups of subjects. In practice, you would probably ask people to volunteer for the experiment, and you might pay them for participating. If you were a university researcher, the chances are that your subjects would be university students. The single most important consideration in the selection of subjects would be the creation of two matched groups, that is, two groups of subjects as similar to each other as possible. You might accomplish this through conscious matching of characteristics (sex, age, race, and so forth), or you might assign subjects to the two groups on a **random selection** basis.

One of the groups would be designated the **experimental group**, while the other would be designated the **control group**. Both groups would be given some test to measure their initial levels of antiblack prejudice. For example, they might be asked to complete a questionnaire calling for agreement/disagreement with a number of statements reflecting antiblack prejudice. Hopefully, both groups would achieve about the same overall score on this **pretest**.

Then the experimental group would be shown the film documenting black American history. The control group would *not* be shown the film. Subsequent to the showing of the film, both the experimental and control groups would be retested for antiblack prejudice. The researcher's hypothesis would be confirmed if the experimental group were found to be significantly less prejudiced in the **posttest** than the control group.

The role of the control group in such an experiment is critical. It serves the function of isolating the experimental stimulus as the single source of change among the experimental subjects. If the pretest and the posttest were separated by a long time span, the experimental subjects might become less prejudiced due to factors lying outside the scope of the experiment. The film might then be irrelevant to the observed reduction in prejudice. If this occurs, then the control group should also decline in prejudice. The hypothesis would be confirmed only if the experimental group declined in prejudice more than the control group.

Similarly, the control group helps the researcher guard against the effect of the experiment per se. It is possible that the act of testing and retesting would make the subjects more sensitive to the purposes of the study. While they might appear relatively prejudiced in the pretest, the testing itself might alert them to the fact that the researcher wanted to discover how prejudiced they were. Since few people want to be identified as prejudiced, the subjects might be more wary in responding to the posttest questionnaire, attempting to answer in such a way as to avoid appearing prejudiced. Again, however, this factor should operate equally for the control and experimental groups, and the *differential* decline in observed prejudice would be the test of the hypothesis.

The isolation of experimental variables is the key advantage of the controlled experiment. It has several disadvantages as well, however. First, the controlled experiment typically provides no useful descriptive data. If 20 percent of both the control and experimental groups agree with a certain prejudiced statement, we learn nothing about the percentage of the larger population who would agree with it, since the subjects are not typically selected through random sampling methods from the general population. And to the extent that they are drawn from a special subpopulation, such as students, their descriptive value is reduced further.

Second, the controlled experiment represents an artificial test of the hypothesis. The relevance of the experiment to the real world is always subject to question. Using the preceding example, assume that the documentary film appears to reduce antiblack prejudice significantly when administered as part of a scientific experiment in a special laboratory, with the subjects aware that they are participating in an experiment. The film would not necessarily have a similar effect if administered to a mass audience over television or in local theaters.

Finally, the findings may not have generalizable applicability to other segments of the population. Conceivably, the film would reduce prejudice among college students—if subjects were drawn from among students—but would have no impact on nonstudents. A nonexperimental example will illustrate this possibility. For years, it was believed that while the working class was more prejudiced against blacks than was the middle class or the upper class, anti-Semitism was believed to increase with increasing social class. The basis for this conclusion was largely a series of prejudice studies conducted among college students. Students from higher-class families were consistently found to be somewhat more anti-Semitic than those from relatively lower-class families. This finding, however, was due to the fact that the students in the studies all came from a relatively narrow range of higher-class families. Thus, the true finding was that upper-class students were slightly more anti-Semitic than were upper-middle-class students. Subsequent studies among the general population, however, indicated that the working-class respondents were more anti-Semitic, just as they were more antiblack.

The shortcomings of the controlled experiment can be reduced through research sophistication, variation of the experimental design, and replication among greatly differing groups of subjects. Moreover, the controlled experiment can be especially valuable when it is combined with other research methods directed at a single research topic.

Content Analysis

Some research topics may be susceptible to the systematic examination of documents such as novels, poems, government publications, songs, and so forth. This research method is called **content analysis**.

For purposes of illustration, suppose that you were interested in examining the changes in official Soviet attitudes toward the United States. You might limit the time span of the study to the years 1950 to 1990, and you might decide to accept editorials in *Pravda* as your indication of official policy.

You might then either obtain copies of all *Pravda* editorials during the period under question or establish a sampling design that would select, say, every tenth editorial. Each such editorial would then be examined and scored in terms of whether it was favorable to the United States or critical of it—or had no relevance to the United States. This activity would require a systematic scoring method. You would be forced to specify what types of references to the United States would be regarded and scored as favorable and what types of references would be regarded and scored as critical. Conceivably, you might want to weight references differentially in terms of the relative strength of support or criticism. In such a case, you would need to specify the manner in which differential weights were assigned.

Having scored the editorials, you might then aggregate those scores in such a way as to characterize different periods of time. Perhaps you would combine the scores assigned by calendar year. You might then report that 12 percent of the 1950 editorials were generally favorable to the United States, that 8 percent of those in 1951 were favorable, and so forth. The pattern of differences over time would represent the official Soviet attitude toward the United States.

Although these data would serve a useful descriptive purpose, you might want to go beyond description to explain the observed fluctuations in official Soviet attitude toward the United States. Suppose you believed that Soviet attitudes were largely determined by the amount of criticism the U.S.S.R. received from the United States; that is, when the United States criticized the U.S.S.R., the U.S.S.R., in turn, became more critical of the United States.

You might, examine this possible explanation through an additional content analysis of speeches and debates in the United States. You would decide what forms of communication within the United States best reflected the hypothesized stimulus and would then examine and score either all or a sample of them in terms of anti-Soviet criticism. Scoring methods would again have to be specified, and the scores assigned would have to be aggregated for the same periods of time used in the initial study. You would then compare the two patterns of fluctuations to determine whether anti-Soviet criticism was typically followed by anti–United States pronouncements in *Pravda*.

Content analysis has the advantage of providing a systematic examination of materials that are more typically evaluated on an impressionistic basis. A news reporter, for example, might read through *Pravda* editorials over time—making a mental note of those mentioning the United States and perhaps noting those editorials expressing either strong support or strong criticism—and develop a general impression of the fluctuations in official posture. Through systematic content analysis, however, you guard against any inadvertent biases you may build into the examination. You might begin with a suspicion that the U.S.S.R. was relatively pro–United States during a certain period, and you could unconsciously pay more attention to favorable editorial content during that period while generally discounting the negative editorials discovered. By adhering strictly to a preestablished sampling and scoring system, you lessen the influence of your own preconceptions.

Content analysis, like all research methods, has its weak points. First, the type of documents selected for examination may not provide the most appropriate reflection of the variable under study. In the preceding example, the possibility exists that editorials in *Pravda* do not provide the best indication of official Soviet attitude toward the United States. The public speeches of selected government officials or Soviet pronouncements on the floor of the United Nations might be more appropriate. In most such cases, you are not in a position to determine which source represents the best focus of the study. This difficulty could be alleviated somewhat, of course, by examining different sources systematically and determining whether each source leads to the same conclusion.

Second, scoring methods almost always contain an arbitrary element. Some editorial comments on the United States might be sufficiently ambiguous to make scoring difficult if not impossible. Other comments might be regarded as favorable by one observer and critical by another. The researcher can possibly reduce this problem by seeking independent evaluations from other researchers experienced in the particular subject matter. To the extent that several independent observers agree on the scoring of given editorial comments, the utility of that scoring system would seem greater. Ultimately, there is no way to ensure that editorial comments are being correctly scored in any absolute sense. In lieu of this reliability, however, you must be as specific as possible in creating, executing, and reporting the scoring system. At the very least, people reading your research report must know precisely what the scores represent, even though they might disagree with the appropriateness of the scoring system used.

Analysis of Existing Data

Scientific research should not be equated with the collection and analysis of original data. In fact, some research topics can be examined through analysis of data already collected and compiled. The classic example of this can be found in Emile Durkheim's study of suicide.[3] Interested in discovering the primary reasons for suicide, Durkheim conducted his comprehensive inquiry without collecting a single original datum. Rather, he tested a wide range of hypotheses through the examination of published suicide rates for different geographical areas. For example, he examined differential rates of

---

[3] Emile Durkheim, *Suicide: A Study in Sociology,* translated by George Simpson (New York: Free Press, 1951).

suicide among Protestants and Catholics by comparing the rates in predominantly Protestant areas with those in predominantly Catholic areas. He examined the effects of climate by comparing rates in warm areas with those found in cooler areas.

The analysis of existing aggregated data has the great advantage of economy. The researcher does not need to pay the costs of sampling, interviewing, coding, recruiting experimental subjects, and so forth. This form of social research has two important disadvantages, however.

First, the researcher is limited to data that have been collected and compiled, and those data might not adequately represent the variables of interest. A healthy dose of ingenuity frequently can help resolve this problem, however. When Samuel Stouffer wanted to examine the effects of the Depression on the family in America and elsewhere, he considered a variety of possible indicators of a hypothesized breakdown of traditional family norms.[4] Divorce rates provided one indication, but Stouffer went well beyond that to consider rates of interreligious marriages, civil as opposed to religious ceremonies, marriages performed outside the home states of the couples, and so forth. Durkheim exhibited a similar ingenuity in the examination of suicide.

The second problem involves what is called the **ecological fallacy**. Whenever you correlate variables generated from aggregated data, it is difficult to determine whether the same relationship between the variables holds true at the level of individuals. For example, while Durkheim found suicide rates to be consistently higher in predominantly Protestant areas than in predominantly Catholic areas, he had no way of determining whether the Protestants were committing suicide. Conceivably, Catholics living in predominantly Protestant areas might have had the highest suicide rates of all. Also, while Stouffer found more "impulsive" marriages during the Depression years, he had no way of determining whether those involved were those most affected by the Depression. Again, an ingenious and logical examination of the data can help reduce this danger; I encourage you to read the two volumes cited to see some of the ways Durkheim and Stouffer solved the problem.

Case Study

The **case study** is a comprehensive description and explanation of the many components of a given social situation. A community study such as W. Lloyd Warner's examination of "Jonesville" would be an example of the case study approach.[5] In a case study you seek to collect and examine as many data as possible regarding the subject of your study. In a

---

[4] Samuel A. Stouffer, *Social Research to Test Ideas* (New York: Free Press, 1962), pp. 134–153.

[5] W. Lloyd Warner, *Democracy in Jonesville* (New York: Harper & Row, 1949).

community study, you would learn about the history of the community, its religious, political, economic, geographical, and racial makeup, and so forth. You might seek to determine the social-class structure of the community, that is, who are its most prominent and powerful citizens and who are at the bottom of the pile.

In short, you would seek the most comprehensive description possible of the community and would attempt to determine the logical interrelations of its various components. What was the impact of the lumber mill closing in the early 1930s? How did the 1960 reform mayor change the power relations in the community? What caused the invasion of Yuppies in the 1980s?

It is important to realize that the case study approach to social research differs radically from the other methods considered so far in terms of scientific objectives. Whereas most research aims directly at generalized understanding, the case study aims initially at the comprehensive understanding of a single, idiosyncratic case. Whereas most research attempts to limit the number of variables considered, the case study seeks to maximize them. Ultimately, the researcher executing a case study typically seeks insights that will have a more generalized applicability beyond the single case under study, but the case study itself cannot ensure such generalizability.

For example, a given case study might indicate that the influx of un-skilled foreign laborers into the community had the effect of generally moving the indigenous laborers upward occupationally, with many of them assuming supervisory jobs over the new migrants. This discovery might lead you to assume that you had uncovered a general principle of occupational mobility and that a similar change would occur in most communities experiencing an influx of unskilled foreign laborers. The single case study could in no way confirm this hypothesis, however, and further studies in other communities would be required for that purpose.

Participant Observation

The term **participant observation** refers to a method of data collection in which the researcher becomes a participant in the social event or group under study. You might join in a protest march as a method of obtaining data about the other participants, or you might join a religious group you want to study.

In practice, as a participant observer, you might or might not reveal your research role; this decision has important methodological and ethical implications. If you openly admit to other participants that you are conducting a scientific study of the group, your presence may very well affect the phen-omenon you wish to study. Awareness that their actions might be reported in print can affect the actions of participants. On the other hand, if you conceal your research activities and pretend to be a typical member of the group, you will be subject to ethical questions regarding the deception. Moreover, your apparently genuine membership can present scientific problems. What will you do if you are elected president of the group or if you are simply asked for your opinion as to what the group should do next? How you react will affect what you are trying to study. Since research situations and purposes vary so greatly in this regard, a general guideline cannot be provided, but you should be aware of the issues involved.

Like case studies, participant observation aims to collect a mass of detailed information. By immersing yourself in the actual social events in progress, you are in a position to achieve a far greater depth of knowledge than is possible for, say, the content analyst or the experimenter. At the same time, however, the participant observer faces great difficulty in maintaining systematic research procedures. Because it will be humanly impossible to observe and record everything that happens, you must select your data. The attempt to observe and record everything can result in a situation in which unconscious biases form the basis for selection. Thus, for example, if you begin to form the conclusion that women students are becoming more active in the direction of protest-march activities, you may unconsciously become more attuned to noting instances that will support that conclusion. Ultimately, the primary danger is that you will not be able to tell your reader the criteria used in selecting and reporting observations—so your reader will be able to evaluate neither the appropriateness of the criteria used nor the generalizability of your conclusions.

We will return to a brief consideration of these research methods in the conclusion of Chapter 3, which examines survey research. After describing the nature, strengths, and weaknesses of survey methods, we will compare them with the other methods described in this chapter. The conclusions we will reach in that later discussion can be revealed here: (1) Different social research situations call for different methods, and (2) the best format is often one that involves the use of several methods focused on the same topic.

Summary

In this chapter, we have considered the possibility of applying the methods of scientific inquiry to social behavior. While we have noted some special considerations involved in such an application, we have found no fundamental obstacle to social science. Although the characteristics of science in general can be brought to bear on the study of society, this cannot be accomplished through an emulation of the traditional image of science.

We have also seen that social researchers may draw upon a variety of research methods to assist them in observing and understanding social behavior. The following chapter specifically addresses survey research as another method of social research. We will soon see that the basic characteristics of science in general apply equally to survey research in particular.

Additional Readings

Earl Babbie, *Observing Ourselves: Essays in Social Research* (Belmont, CA: Wadsworth, 1986).

Thomas D. Cook and Donald T. Campbell, *Quasi-Experimentation: Design and Analysis Issues for Field Settings* (Chicago: Rand McNally, 1979).

Emile Durkheim, *The Rules of Sociological Method,* translated by Sarah Solovay and John Mueller, edited by George Catlin (New York: Free Press, 1962).

Ole Holsti, *Content Analysis for the Social Sciences and Humanities* (Reading, MA: Addison-Wesley, 1969).

Morton Hunt, *Profiles of Social Research: The Scientific Study of Human Interactions* (New York: Russell Sage Foundation, 1985).

John Lofland, *Analyzing Social Settings* (Belmont, CA: Wadsworth, 1984).

George McCall and J. L. Simmons, eds., *Issues in Participant Observation: A Text and Reader* (Reading, MA: Addison-Wesley, 1969).

William Ray and Richard Ravizza, *Methods Toward a Science of Behavior and Experience* (Belmont, CA: Wadsworth, 1985).

Walter Wallace, *The Logic of Science in Sociology* (Chicago: Aldine-Atherton, Inc., 1971).

3

Survey Research as a

Method of Social Science

This textbook primarily addresses the logic and skills of a particular research method: **survey research.** Before continuing, we will describe briefly the components of a typical survey, although we will see later that survey methods can be applied to a wide variety of topics and designs.

Suppose you are interested in studying certain attitudes among students at a state university. A **sample** of several hundred or more students is selected from the total student body. A **questionnaire** is constructed to elicit information (for example, attitudes) relevant to the researcher's subject of inquiry. The questionnaires are then administered to the sample of students, either in face-to-face interviews, by telephone, or in a mailed, self-administered format. The responses given by each student in the sample are then coded into a standardized form that can be recorded in a **quantitative** manner. Each student's coded responses are then transferred to microcomputer disks. The standardized records for all students are then subjected to an aggregated analysis to provide descriptions of the students in the sample and to determine correlations among different responses. The descriptive and explanatory conclusions reached by this analysis are then generalized to the population from which the sample was selected, in this case to the entire student body.

A Brief History of Survey Research

Survey research is sufficiently similar to other methods of research to give it a rather lengthy history. In particular, surveys are very much like censuses, differing primarily in that a survey typically examines a sample from a population, while a **census** generally implies an enumeration of the entire population. Censuses, of course, date back at least to the ancient Egyptian civilization, when rulers deemed it useful to obtain empirical data describing their subjects.

The political functions of survey research have continued to the present day, with the continuation of censuses, the appearance of political polls conducted on behalf of candidates, and uses made by political sociologists. One of the first political uses of the attitudinal survey appeared in 1880. A German political sociologist mailed questionnaires to some 25,000 French workers to determine the extent of exploitation by employers. The rather lengthy questionnaire included such items as the following:

> Does your employer or his representative resort to trickery in order to defraud you of a part of your earnings?
>
> If you are paid piece rates, is the quality of the article made a pretext for fraudulent deductions from your wages?[1]

The survey researcher in this case was Karl Marx. While 25,000 questionnaires were mailed out, no record exists of any being returned.

Max Weber, one of the founders of modern sociology, is also reported to have employed survey research methods in his research on the Protestant ethic. In addition to his comparative historical examination of economic development, he also studied Protestant and Catholic factory workers in order to provide confirming data at the individual level.[2]

For the most part, however, contemporary survey research is a product of American researchers in this century. The present state of the method has resulted from important developmental work in three separate sectors of American society. First, the continuing work of the U.S. Bureau of the Census has made important contributions to the fields of sampling and data collection. While the Census Bureau is best known for its decennial enumeration of the entire population, the great majority of this agency's activities are devoted to a continuing series of sample surveys that provide up-to-date demographic and economic data between enumerations. The Census Bureau has played a singularly important role in the development of standardized definitions for sampling and methods for implementing those definitions in the field. At the same time, the data generated by the Bureau have constituted an invaluable resource for the development of sampling designs in specific surveys. (Chapter 6 illustrates this point in detail.)

The second source of development is found in the activities of commercial polling firms such as those organized by George Gallup, Elmo Roper, Louis Harris, and others. These firms have generated a continuing source of funds to

---

[1] T. B. Bottomore and Maximilien Rubel, eds., *Karl Marx: Selected Writings in Sociology and Social Philosophy* (New York: McGraw-Hill, 1956). See page 208 for questions cited.
[2] Paul F. Lazarsfeld and Anthony R. Oberschall, "Max Weber and Empirical Research," *American Sociological Review* (April 1965), pp. 185–199.

support the development and use of survey methods, particularly in the areas of product marketing and political polling. Thus, during periods when funds were not available to support academic survey research, the commercial polling firms were able to continue experimentation with sampling methods, question wording, data-collection techniques, and other aspects of survey research. In addition, the commercial polls have served as an invaluable source of data for secondary analysis, and numerous academic books and articles have been published reporting more detailed analysis of data collected initially for descriptive, commercial purposes.[3]

Third, the scientific refinement of survey research, especially sophisticated methods of analysis, has been largely the product of a few American universities. More specifically, it represents the efforts of two men and, later, a scattered group of survey research centers. Samuel A. Stouffer and Paul F. Lazarsfeld must be regarded as the pioneers of survey research as we know it today.

Stouffer's pioneering work largely represented attempts to apply empirical methods of social research to social problems, beginning with analyses of the effects of the Depression in America and with the collection of data regarding the status of black Americans during the 1930s.[4] With the advent of World War II, Stouffer directed the Information and Education Branch of the Army of the United States, bringing together a group of rising social scientists to examine issues relevant to the successful prosecution of the war.[5] Then, during the McCarthy era of the early 1950s, Stouffer conducted national surveys to examine the effects of the anticommunist crusade.[6]

Across these studies and countless others, Stouffer attempted to develop scientific methods of empirical research appropriate to the examination of social phenomena. His efforts flowed from a sound scientific training under the British statisticians Karl Pearson and R. A. Fisher and from a native ingenuity. Stouffer's legacy remains evident in currently used study designs, sampling methods, questionnaire designs, the logic of analysis, and other survey methods.

Paul Lazarsfeld came to America from a European intellectual background. Like Stouffer, he was interested in the study of relevant social phenomena such as leadership, communications, economic behavior, and the professions. In the examination of such topics, Lazarsfeld, like Stouffer, continued the development of rigorous techniques for applying empirical methods to social

---

[3] See, for example, Herbert Hyman, *Secondary Analysis of Sample Surveys* (New York: John Wiley & Sons, 1972).
[4] An excellent overview of Stouffer's work can be found in the posthumously published book by Samuel A. Stouffer, *Social Research to Test Ideas* (New York: Free Press, 1962).
[5] See Samuel A. Stouffer, et al., *The American Soldier: Studies in Social Psychology in World War II,* vol. i–v (Princeton, NJ: Princeton University Press, 1949 and 1950).
[6] Samuel A. Stouffer, *Communism, Conformity, and Civil Liberties* (Garden City, NY: Doubleday & Company, Inc., 1955).

issues. In the area of political behavior, Lazarsfeld felt it important to examine voting as a process rather than as a single event. To do this, he designed and executed **panel studies** involving the reinterviewing of a given set of respondents at different times during the course of a political campaign, thereby following changes in voting intentions over time. Not content with description, however, Lazarsfeld combined this effort with a careful examination of the demographic and social factors associated with observed changes.[7]

While Lazarsfeld's contributions to the development of survey research are too many to enumerate here, three seem especially important. First, Lazarsfeld's career overlapped the technical development of mechanized data-processing equipment, first card punchers and sorters, then computers. He must be given primary credit for recognizing the potential of such equipment for use in analytical social research and for starting social scientists on the road to realizing that potential.

Second, Lazarsfeld used mechanized data-processing equipment in elucidating and formalizing the logic of survey analysis. Whereas Stouffer's ingenuity led him to suggest reasons for observed relationships in his World War II studies of the U.S. Army, it was Lazarsfeld who formalized those reasons into a logical model of such relationships and showed how the model could be implemented in practice.[8] (This topic is discussed in more detail in the consideration of the **elaboration model** in Chapter 15.)

Lazarsfeld's third major contribution to survey research was the development of the permanent research center in support of survey methods, beginning with the organization of the Bureau for Applied Social Research at Columbia University. Like Stouffer's Army research group, the Bureau brought together and trained a legion of bright young social scientists; the Bureau, however, was able to survive peacetime. Similar organizations subsequently established include the Survey Research Center at the University of California in Berkeley, the National Opinion Research Center at the University of Chicago, the Institute for Social Research at the University of Michigan, the Survey Research Center at the University of California in Los Angeles, and the Survey Research Laboratory at the University of Wisconsin.

In addition to those listed above, similar institutes and centers now exist on university campuses across the country and around the world. They serve a number of functions. First, while academic departments often provide classroom instruction in survey methods, the survey centers typically allow students to receive practical apprenticeship training, often through work as research assistants. Second, countless surveys are conducted by such centers, funded by government and foundation grants or on behalf of commercial clients.

---

[7] Paul F. Lazarsfeld, *The People's Choice* (New York: Columbia University Press, 1948).
[8] See Patricia Kendall and Paul F. Lazarsfeld, "Problems of Survey Analysis," in Robert Merton and Paul Lazarsfeld, eds., *Continuities in Social Research Studies in the Scope and Method of the American Soldier* (New York: Free Press, 1950).

Similarly, the centers provide consultation and services to other researchers using survey methods. Finally, such centers play an important role in methodological development in survey research. While an individual researcher conducting a particular research project might be reluctant to experiment with alternative data-collection techniques, an organized research unit is able to do so over the course of several studies.

Before concluding this short history of survey research, we should make some mention of the role played by professional associations. Over the years, professional associations whose members frequently employ survey methods have provided forums for discussing new techniques and empirical findings through their association meetings and professional journals. Perhaps the three primary associations in this regard are the American Sociological Association, the American Political Science Association, and the American Marketing Association. At the same time, the American Association for Public Opinion Research (AAPOR) has served as the association of primary relevance to survey researchers, bringing together academic, commercial, and government practitioners. The *Public Opinion Quarterly,* published by AAPOR, is the key journal for survey research.

Subsequent chapters of this book trace the history of certain components of survey research. Against the backdrop of this brief overview, however, it is appropriate to move to an examination of the place of survey research within the general context of science.

The Scientific Characteristics of Survey Research

As noted earlier, survey research is but one of many research tools available to social researchers. It bears repeating that survey methods are not appropriate to many research topics, nor do they necessarily provide the best approach to topics to which they might be reasonably applied. Nevertheless, survey research can be used profitably in the examination of many social topics and can be especially effective when combined with other methods.

More important, I firmly believe that survey research provides the best teaching example for instruction in social science methodology. If you fully understand the logic and skills of survey research, you will be excellently equipped to learn and to use other social research methods. My basis for this belief is that survey research is like a crustacean: All the bones are on the outside. In subsequent chapters we will carefully examine all the approximations, compromises, and other shortcomings of survey research. As we will discuss shortly, survey research serves a pedagogical function in that those shortcomings are made clearer in survey research than in other social research methods, thereby permitting more considered evaluations of their implications.

Survey Research Is Logical

Survey research is guided by all the logical constraints discussed in the two preceding chapters. In practice, moreover, survey data facilitate the careful implementation of logical understanding. While this topic will be explored in much greater detail in Chapter 15, an example is appropriate at this point.

In a study of church involvement among Episcopalians,[9] Charles Glock, Benjamin Ringer, and I found that lower-class churchwomen were more actively involved in the church than were women of higher social status; involvement decreased steadily with increasing social class. We sought to explain this phenomenon in terms of a *deprivation theory of church involvement*—people who were denied prestige and status in the secular society would be more likely to involve themselves in church life as an alternative source of gratification. Thus, we suggested, the greater church activity of lower-class churchwomen merely reflected their greater deprivation in the secular society. By extension, we suggested that among those women who *did* enjoy a degree of secular status gratification, social class would not affect church involvement. The data at hand indicated whether the respondents had ever held office in a secular organization. Logically, we would expect that among those women who had held secular office, social class would not affect church involvement. This expectation was tested empirically and was found to be correct.

The format of survey research often permits the rigorous, step-by-step development and testing of such logical explanations. Through the examination of hundreds and even thousands of survey respondents, moreover, it is possible to test complex propositions involving several variables in simultaneous interaction.

Survey Research Is Deterministic

Whenever the survey researcher attempts to explain the reasons for and sources of observed events, characteristics, and correlations, the inquiry must assume a deterministic posture. The fact that the survey format permits a clear and rigorous elaboration of a logical model clarifies the deterministic system of cause and effect.

Moreover, the availability of numerous cases and variables permits the survey analyst to document more elaborate causal processes. We may go beyond the initial observation of a correlation between an independent and a dependent variable to examine the role played by several intervening variables. Thus, returning to the above example, we note that social class does not

---

[9]Charles Y. Glock, Benjamin B. Ringer, and Earl R. Babbie, *To Comfort and to Challenge* (Berkeley: University of California Press, 1967).

directly affect church involvement among Episcopal women but rather has an effect through the intervening variable of secular gratification. Lower-class women are less likely to be elected to offices in secular organizations, with the result that lower-class women, as a group, are more involved in the church, but if they *do* gain secular gratification, their social class has no effect on church involvement.

Survey Research Is General

Sample surveys are almost never conducted for purposes of describing the particular sample under study. Rather, they are conducted for purposes of understanding the larger population from which the sample was initially selected. Thus, the Gallup Poll may interview 1,500 American voters for the purpose of predicting how tens of millions will vote on election day. Sample surveys of consumers aim at learning about the preferences and behaviors of consumers in general.

Similarly, the explanatory analyses in survey research aim at the development of generalized propositions about human behavior. The survey format promotes this general scientific aim in two special ways. First, with a large number of cases studied in a given survey, findings can be replicated among several subsets of the survey sample. If an overall correlation is found between education and purchasing patterns, for example, researchers can easily determine whether this relationship occurs equally among men and women, Protestants and Catholics, whites and blacks, people from different geographical regions, and so forth. The replication of a finding among different subgroups strengthens the assurance that it represents a general phenomenon in society. Second, the careful reporting of the methodology of a given survey promotes later replication by other researchers and/or among other samples and subgroups. In this fashion, the generalizability of the findings can be tested and retested.

Survey Research Is Parsimonious

Because survey researchers have a large number of variables at their disposal, they are in an excellent position to carefully examine the relative importance of each. Like all scientists, survey researchers would like to obtain the greatest amount of understanding from the fewest number of variables. They are not required to guess at the most relevant variables in the initial design of the study, however (or at least not as much as researchers using other methods). Since the survey format lends itself to the collection of many variables that can be quantified and processed by computer, survey researchers can construct a variety of explanatory models and then select the one best suited to their aims.

Survey Research Is Specific

In the context of the scientific characteristic of specificity, the crustacean-like nature of survey research is most relevant. Ironically, this characteristic also exposes survey methods to the greatest amount of criticism.

Suppose a survey concludes that political conservatism and prejudice against women are positively related. Such a conclusion would be based on specific operational definitions of both conservatism and prejudice against women. The measurement of each variable would be constructed from specific responses to specific questionnaire items coded and scored in a specified manner. Since all these details would be spelled out in the research report, the critical reader would be able to discover that respondents described as "very prejudiced" were those who agreed to five specific questionnaire items. The respondents might then object that such a criterion did not fully capture the meaning of "prejudice" as they themselves—and, by extension, others—understand the term. They might then determine that the general conclusion was unfounded.

In one sense, this response would be valid. The conceptualization and measurement of variables lie at the heart of science in practice, and if variables are not conceptualized and measured appropriately, observed correlations among such variables might not be meaningful. Thus, if independent observers disagree with the way in which variables have been measured, they might logically disagree with the general conclusion.

What is too often overlooked in this situation is the ease with which critical readers can arrive at and perhaps document their disagreements. Because survey analysts have described precisely how their measurements have been developed and made, the reader knows precisely what those measurements represent. The superficiality and approximations involved in all scientific research are simply more apparent in surveys.

By way of contrast, suppose you have studied the relationship between conservatism and prejudice against women by immersing yourself as a participant observer in a variety of political groups. You might report that your observations suggest that *liberals* are more prejudiced against women than are conservatives. In your report, you could give a qualitative description of the manner in which you distinguished conservatives from liberals and the prejudiced from the unprejudiced, capturing the depth of meaning these terms have in common language usage. You could avoid the apparent superficiality of questionnaire items and survey measurements.

Unfortunately, the independent observer would have no way of knowing precisely how your descriptions were employed in practice, nor would an observer have any way of judging the extent to which your bias, perhaps unconsciously, might have affected your designations of persons in terms of their politics and their prejudice. This is not to say that your observations and designations were, in fact, biased or that your conclusion was incorrect. Very important, however, is the fact that neither you nor the independent observer would be able to make determinations in these regards. (It should be noted, of course, that participant observation in practice varies in rigor and specificity. The point is that the survey researcher is forced *by the method itself* to be explicit.)

Scientific research aims at ever more sophisticated and more useful conceptualizations and measurements, but at every step along the way the methods used must be made specific. Survey research, by its very nature, lends itself readily to this characteristic.

Conclusion

The other characteristics of science are equally relevant in the context of survey research. Where science must be *empirically verifiable,* survey research provides one method of empirical verification. Also, the comments made in the preceding sections illustrate the manner in which the survey format lends itself to *intersubjectivity*.

Finally, survey research methods facilitate the *openness of science*. Since survey research involves the collection and **quantification** of data, the data gathered become a permanent source of information. A given body of survey data may be analyzed shortly after collection and found to confirm a particular theory of social behavior. If the theory itself undergoes modifications later, it is always possible to return to the set of data and reanalyze the data from the new theoretical perspective. This reanalysis could not be accomplished as easily in the case of less rigorous and less specific research methods.

A Comparison of Survey and Other Methods

The preceding sections have already made a number of comparisons between survey research on the one hand and other social scientific research methods on the other. It has been pointed out that a comprehensive inquiry would profit from the use of different methods focused on a single topic. Three additional points should be made in this respect.

First, the logic of the controlled experiment can provide a useful guide to the logic of survey analysis. Where the experimenter isolates the experimental variable through the use of matched or randomized groups—the experimental and control groups—the survey analyst seeks to accomplish the same end by controlling for variables after the fact. For example, the experimenter may ensure that both the experimental and the control groups have the same sex distribution in order to avoid the possible influence of that variable on the experiment. The survey analyst accomplishes this either by ensuring that subgroups in the sample have the same sex distribution or by testing the observed relationship separately among men and women. The logical goal of isolating relevant variables by ruling out the influence of extraneous variables is the same for both methods, however.

Second, the coding of survey responses is essentially an instance of content analysis. Frequently, the survey researcher will ask **open-ended** questions that solicit a response in the subject's own words. Such responses, however, must always be codified into types of answers. Thus, it might be necessary to code a given set of responses as being either generally supportive of the enactment of a certain law or generally opposed to such enactment. In this respect, survey researchers can learn much from the experiences and methods of content analysts.

Third, survey interviewing can profit from the experience of participant observers. Methods of gaining **rapport,** maintaining neutrality, and making accurate observations are important to both activities.

Is Survey Research Really Scientific?

In view of the criticism sometimes lodged against social scientists, survey researchers, and sociological practitioners of survey research in particular, perhaps I may be permitted to conclude this chapter on a rather chauvinistic note. Several years ago, Allan Mazur published an article entitled "The Littlest Science"[10] in which he addressed the question of whether sociology could call itself scientific. As the title of his article suggests, he concluded that this designation was just barely justifiable. Not surprisingly, the article brought an immediate and often heated response.

Mazur, along with most of his positive and negative reviewers, would agree that no ultimate answer is possible to the question of whether a discipline such as sociology is or is not a science. As we've already seen in Chapter 1, any definition of science must be arbitrary; thus, there can be no absolute answers. At the same time, the question is worth consideration even if it cannot be answered.

It is my view that future researchers may one day look back upon this era and conclude that the use of survey research by sociologists and other social scientists was a critical period *in the development of science in general.* This tentative expectation is not merely the megalomaniacal pipe dream of a survey researcher who is coincidentally a sociologist by training. The comments of two other researchers illustrate the basis for my expectation.

---

[10] Allan Mazur, "The Littlest Science," *American Sociologist* (August 1968), pp. 195–200.

Economist Daniel Suits, addressing a conference of government researchers and planners in Honolulu a few years back, varied from customary academic terminology and spoke not of the "hard sciences" (for example, physics and chemistry) and the "soft sciences" (for example, sociology, political science, and market research) but rather distinguished between the "hard sciences" and the "easy sciences." His point was that physicists can engage in scientific research rather easily by virtue of their subject matter, while social scientists have a much more difficult row to hoe.

Martin Trow has traced some of the implications of this situation in a different context.[11] Addressing the uses of survey methods in the field of education, Trow noted an interesting anomaly. He observed first that the educational setting is nearly ideal for survey research: The prospective subjects are articulate and familiar with questionnaires, they are easily enumerated and sampled, and questionnaires can be administered under controlled conditions in the classroom. Why, then, Trow asked, have most educational surveys been so trivial and unsophisticated? His answer was that conditions have been *too good*. Educational researchers have never been forced to cope with imperfect research conditions, and since they have never been forced to make compromises and approximations in the design and execution of surveys, they have never had to come to grips with the basic logic of scientific survey research. Not having had to vary from the obvious ideal, they have not had to fully understand why it was considered ideal.

Trow's point is, I think, of far more general significance. The very adversities that face the social sciences demand the creation of a more sophisticated logical system of understanding. Moreover, survey research, because of its specificity of operation, may provide the most useful vehicle for dealing with those adversities in a rigorous and systematic manner. Some additional examples should clarify this point further.

When chemists want to study the properties of hydrogen, they are unlikely to concern themselves very much with sampling techniques. Since one hydrogen atom is like any other, any such atom will suffice for the study. Social scientists, on the other hand, cannot study any convenient individual or group of individuals. While chemists can generalize their findings to all hydrogen on the basis of studying only convenient atoms, social scientists must develop a more sophisticated understanding of the logical concept of generalizability as well as operational methods for achieving it. It should be noted, of course, that other scientists face problems of sampling—geneticists, for example—and that social scientists have found guidance in the work of other disciplines. Nevertheless, the problems of sampling and generalizability are greatest in the study of

---

[11] Martin Trow, "Education and Survey Research," in Charles Y. Glock, ed., *Survey Research Methods in the Social Sciences* (New York: Russell Sage Foundation, 1967), pp. 315–375.

social behavior. Survey research provides an excellent vehicle for the development of useful methods and, by extension, fuller understanding.

The act of measurement provides another example of the problems addressed by survey research. Suppose you are attempting to measure something like prejudice. There is no clear ideal conceptualization from which to work, nor is there any easy measurement operation available from which to approximate such an ideal. If you use survey techniques, moreover, the problem is even more basic. You may discover that one questionnaire item elicits a prejudiced response from a respondent while another item elicits an unprejudiced response. Moreover, a slight variation in the wording of a given item will affect the responses obtained.

Similarly, social scientists recognize that the mere presence of researchers can affect their subjects. Asking a respondent to report an opinion may result in a greater crystallization of that opinion than existed prior to the inquiry; some respondents will form opinions on the spot. A comparable situation in the physical sciences would be if a metal bar could stretch or shrink in length whenever the physicist came by to measure its length. Clearly, the situation faced by social scientists demands a more sophisticated understanding of what conceptualization and measurement entail. The multiplicity of relevant variables and the complex and probabilistic nature of causation in social behavior provide further demands for more sophisticated understanding of what science really is. The list of examples could go on.

We should note in this context that the "easy sciences" are not quite as straightforward as many people might imagine. Physicists realize, for example, that they cannot simultaneously measure an object's location and its velocity; measuring one affects the other. Medical scientists are forced to use *placebos* (sugar pills, for example) in tests of new drugs because some patients seem to improve when they believe they are receiving a powerful new drug and/or because medical researchers may see more optimistic developments in patients they believe are receiving the new drug. At the same time, we find nuclear physicists asserting that particles moving in a given direction and experiencing no additional forces will *probably* continue moving in that direction.

In sum, the social sciences constitute a prime example of what physicist Heinz Pagels[12] calls *the sciences of complexity*. Pagels's argument that the development of the computer represents a breakthrough in the possibility of dealing with complex systems is nowhere truer than in the social sciences. Indeed, I suggest that the computer will be for the social sciences what the telescope was for astronomy and the microscope for biology. And that in the bargain the social sciences will radically expand our vision of what *science* can be.

---

[12] Heinz Pagels, *The Dreams of Reason* (New York: Simon & Schuster, 1988).

Additional Readings

William Sims Bainbridge, *Survey Research: A Computer-Assisted Introduction* (Belmont, CA: Wadsworth, 1989).

Don A. Dillman, *Mail and Telephone Surveys: The Total Design Method* (New York: John Wiley & Sons, 1978).

Charles Y. Glock, ed., *Survey Research in the Social Sciences* (New York: Russell Sage Foundation, 1967).

Paul F. Lazarsfeld, "Introduction" to Samuel A. Stouffer, *Social Research to Test Ideas* (New York: Free Press, 1962), pp. xv–xxxi.

Public Opinion Quarterly, vol. 51, no. 4 (Part 2). This fiftieth anniversary issue (Winter 1987) contains several articles reviewing the history of survey research.

PART TWO

Survey Research

Design

The five chapters comprising Part Two of this text address the several aspects of survey design. We will examine some of the different kinds of surveys available to you, the selection of samples for study, and the important issue of measurement.

There is a tendency to regard the analysis of survey data as the most challenging and exciting part of the enterprise and to regard the design of a survey and the actual collection of data as less exciting. I have to confess to feeling this way, and I suspect that most other survey researchers would agree. During analysis you begin to gain an understanding of the subject matter at hand and are in a position to share your findings with colleagues.

It is a mistake, however, to assume that study design is less challenging than analysis or that it requires less brilliance and ingenuity. Quite often, in fact, ingenuity becomes necessary in survey analysis only because of a lack of ingenuity in the design and execution of the study. While you might find satisfaction in being able to overcome design problems through clever analytical manipulations, you should still be embarrassed that the problems even existed; moreover, you will often find that no amount of ingenuity can solve many design problems.

As you become more competent in the various aspects of survey design, you will come to the realization that survey design requires the same logical problem-solving abilities as survey analysis. In this respect, it will become equally exciting and challenging.

Chapter 4 describes the various study designs available to you. The intent is to familiarize you with the range of options and to help you determine which of these options best suits a particular research aim.

Chapters 5 and 6 address survey sampling. Chapter 5 deals with the logic of sampling, while Chapter 6 presents examples to clarify the implementation of that logic in practice.

After covering sampling, we will turn to the issue of *measurement*—beginning with conceptualization and instrument design in Chapter 7. Here we will deal with the process through which survey researchers refine their concepts and develop questionnaire items to measure them. Chapter 8 continues this topic, addressing the construction of composite indexes and scales during data analysis. While Chapter 8 is out of order, chronologically, I've placed it here because it completes the logical flow established in Chapter 7.

4

Types of Study Design

While "survey research" refers to a particular type of empirical social research, many different kinds of surveys exist. We might include under this term censuses of populations, public opinion polls, market research studies of consumer preferences, academic studies of prejudice, epidemiological studies, and so forth. Surveys may differ in their objectives, cost, time, and scope. Moreover, a variety of basic designs may be included under the term "survey."

This chapter begins with a brief discussion of the possible objectives of survey research. Then it examines the concept **unit of analysis.** Finally, it presents an overview of the different strategies available to you in the pursuit of your goals.

Purposes of Survey Research

There are probably as many different reasons for conducting surveys as there are surveys. A politician might commission a survey for the ultimate purpose of getting elected to office. A marketing firm might conduct a survey for the purpose of selling more Brand X soap. A government might conduct a survey as an important basis for the design of a mass transit system or for the modification of a welfare program.

Though the variety of such purposes is too great to begin enumerating here, three general objectives crosscut all these many concerns: description, explanation, and exploration. While a given survey may (and usually does) aim at satisfying more than one of these objectives, examining them as distinct is useful for our present purposes.

Description

Surveys are frequently conducted for the purpose of making descriptive assertions about some population, that is, discovering the distribution of certain traits or attributes. In this regard, the researcher is concerned not with why the observed distribution exists but merely with what that distribution is.

The age and sex distributions reported by the U.S. Bureau of the Census are examples of this type of survey. Similarly, the Department of Labor might seek to describe the extent of unemployment among the American labor force at a given time or at several points over time. Likewise, the Gallup Poll might seek to describe the percentages of the American electorate who will vote for the various candidates for President, the distribution of attitudes regarding the Indochina War, or the distribution of attitudes toward sex education in the schools and toward fluorides in the water. The percentage of a population likely to purchase a new commercial product is still another example of descriptive research.

The sample survey provides a vehicle for discovering such distributions. The distribution of traits among a carefully selected sample of respondents from among the larger population can be measured, and a comparable description of the larger population can be inferred from the sample. (Chapter 5 discusses the logic of such inferences.)

In addition to describing the total sample (and inferring to the total population), survey researchers often describe subsamples and compare them. Thus, the Gallup Poll might begin with a report of the voting intentions of the entire electorate and then separately describe Democrats and Republicans, men and women, voters of different ages, and so forth. While the descriptions of different subsets might be compared, the primary purpose is description rather than explanation of differences. (Such comparisons, however, provide an intermediary step between the logic of description and the logic of explanation, as we will discuss further in Chapter 13.)

Explanation

Although most surveys aim, at least in part, at description, many have the additional objective of making **explanatory** assertions about the population. In studying voter preference, for example, you might want to explain why some voters prefer one candidate while other voters prefer another. In studying unemployment, you might want to explain why part of the labor force is employed while the remainder is not.

An explanatory objective almost always requires **multivariate** analysis—the simultaneous examination of two or more variables. Preferences for different political candidates might be explained in terms of variables such as party affiliation, education, race, sex, and region of the country. By examining the relationships between candidate preferences and the several explanatory variables, a researcher would attempt to "explain" why voters picked one or the other candidate. (The logic of explanation is dealt with at length in Part Four of this book.)

Exploration

Survey methods can also provide a "search device" when you are just beginning your inquiry into a particular topic. During the Free Speech Movement at the University of California at Berkeley in the mid-1960s, for example, I was part of a group of campus researchers planning an exhaustive study of the nature, sources, and consequences of student radicalism. While we had many ideas on these subjects, we were wary that we might have overlooked some additional components of the situation. If a large study were based solely on our preconceptions, we would run the risk of missing some critical elements.

The survey research method provided one technique for resolving this difficulty. A loosely structured questionnaire was constructed, and about fifty students who differed in their political orientations were interviewed in depth. No attempt was made to select a representative sample of students, nor were data collected in a standardized form. Rather, the students interviewed were encouraged to speak freely about their political views and their attitudes toward student radicalism.

These interviews resulted in a largely revised research design for the main study. Respondents did indeed mention factors relevant to student radicalism that we had not initially anticipated. Certain political orientations that had earlier seemed contradictory now made more sense. These additional factors were subsequently taken into account in the main research design.

Before leaving this topic, we should note some of the things that the exploratory study did not accomplish. First, it did not answer the basic research questions that prompted the planning of the research. Also, it did not adequately elucidate the nature of student radicalism nor did it satisfy questions regarding the sources and consequences of student radicalism. The manner in which the data were collected clearly precluded such results. The exploratory study did, however, raise new possibilities, which were later followed up in the more controlled survey.

These, then, are the three basic objectives of survey research. As we noted at the outset, most studies have more than one such objective (sometimes all three), but these objectives provide useful organizing principles in the design of surveys. Before turning to the specific types of study design available for meeting these objectives, we should introduce a basic, but sometimes confusing, term: **units of analysis.**

Units of Analysis

Survey research provides techniques for studying almost anyone. The *ones* under study in a given survey are the units of analysis. Typically, the unit of analysis for a survey is a person, but there is no reason why this need be the case, and it often is not. Whatever the units of analysis, data are collected for purposes of describing each individual unit (e.g., person). The many descriptions are then aggregated and manipulated in order to describe the whole sample studied and, by extension, the population represented by that sample.

In a market survey of consumer preferences for Brands X and Y, each consumer sampled and surveyed is the unit of analysis. Each is described in terms of the brand he or she favors. Then the several preferences are aggregated to describe the population of consumers in terms of the percentages who favor the two brands. In an unemployment survey, members of the labor force are the units of analysis. Each is described as being either employed or unemployed. These individual descriptions are then used to describe the whole labor force in terms of the unemployment rate.

While units of analysis are typically people, they might also be families, cities, states, nations, companies, industries, clubs, governmental agencies, and so forth. In each instance, the individual units of analysis would be described, and those descriptions would be aggregated to describe the population the units represent. For example, you might collect data describing cities in the United States. Each city might be described in terms of its population size, and all American cities might then be described in terms of the mean population. Additional variables describing the cities might be introduced to explain why some cities are larger than others.

A given survey, of course, can involve more than one unit of analysis. A household survey of a particular city might seek to provide the following information: the percentage of residential structures in deteriorating condition, the racial distribution of heads of households, mean annual family income, the unemployment rate, and the age–sex distribution of the resident population. In these examples, the units of analysis would be, respectively, residential structures, households, families, members of the labor force, and residents.

Units of analysis for a given survey can be described on the basis of their components. For example, cities could be described in terms of their unemployment rates or their racial compositions. Surveys might even be conducted for purposes of providing those descriptions. If the object of the study is to describe cities and to aggregate the various descriptions for the purposes of describing all cities, however, then the city is the basic unit of analysis for the study. At the same time, units of analysis can be described in terms of the groups to which they belong. Thus, individuals might be described in terms of the number of people in their families or the condition of their residential structures.

The applicability of survey methods to various units of analysis sometimes confuses the beginning researcher and results in the selection of an inappropriate unit of analysis for a particular line of inquiry. In particular, the dangers of the *ecological fallacy* should again be noted.[1]

---

[1] For a more comprehensive discussion of this topic, see W. S. Robinson, "Ecological Correlations and the Behavior of Individuals," *American Sociological Review* (June 1950), pp. 351–357.

Suppose you are interested in exploring the possible relationship between race and crime: Are blacks or whites more likely to engage in criminal behavior? The appropriate unit of analysis for this inquiry would be the individual person. Samples of black and white respondents might be studied and their respective crime rates computed and compared.

Given the availability of certain municipal data, however, you might be tempted to approach the problem in a different way. You could easily obtain overall crime rates for major American cities, and you could also find data reporting the racial composition of those same cities. In analyzing such data, you might discover that crime rates were higher in cities containing a higher proportion of blacks in their populations. You might then conclude that blacks have higher crime rates than whites. This line of inquiry, however, is subject to the ecological fallacy in that you have no assurance that the crimes committed in predominantly black cities *were committed by blacks*. It is conceivable that the highest crime rates could occur among whites living in predominantly black areas. Such a misinterpretation would not be possible had you employed the correct unit of analysis.[2]

Whatever the nature of the data used to describe the units of analysis, it is important that they be identified in advance. Otherwise, the sample design and the data collection methods might prohibit the analysis appropriate to the study. Whenever more than one unit of analysis is involved in the study, this consideration is even more important. One final example should clarify matters.

Suppose that in a household population study of a given city you decide to collect your data through interviews with the heads of a sample of households, but you want to make assertions about residential structures, households, families, and persons. Data about the structures could be collected both by asking relevant questions of the respondent (How old is this house?) and through observation (How many stories in the structure?). Data about the household might also be collected through interviewing (How many families live in this house?) and observation (What is the race of the head of the household?). The respondent might be asked for data to describe both the families living in the household (What are their annual incomes?) and the individual persons in that household (What is the sex and age of each person?).

In a complex study such as this, with several units of analysis, special care must be taken in the organization of the data for analysis. If you want to

---

[2] It should be recognized that the researcher sometimes is simply not in a position to conduct an inquiry utilizing the most appropriate unit of analysis. In such cases, an ecological analysis might represent the only feasible approach to the topic at the time. Durkheim's famous study of suicide is an example of an excellent case in point; for a discussion of the ways in which Durkheim avoided the ecological fallacy, see Hanan Selvin, "Durkheim's Suicide and Problems of Empirical Research," *American Journal of Sociology* (May 1958), pp. 607–619. Another example is Samuel A. Stouffer, "Effects of the Depression on the Family," in Samuel A. Stouffer, ed., *Social Research to Test Ideas* (New York: Free Press, 1962), pp. 134–153.

determine the percentage of residential structures in deteriorating condition, each structure should be described only once in the computation, regardless of the number of families or persons living there. If you want to determine the percentage of the population living in deteriorating structures, a given structure should be entered in the computation as many times as there are people living in it. If five people live in a given deteriorating structure, for example, each person should be described as living in a deteriorating structure, and all the person descriptions would be aggregated to describe the city's population.

The safest course to pursue in a complex survey such as this would be to create separate data files for each unit of analysis. In the present example, you should create a structure file, a household file, a family file, and a person file. Each file would contain all the data relevant for analysis with that unit of analysis, even though the data might initially have been collected about some other unit of analysis. Thus, each person file might contain an indication of the race of the head of the household to which that person belongs. Once the unit of analysis for a given computation was determined, you could easily make the computation through the use of the corresponding data file.

Basic Survey Designs

After you have specified the objectives of your survey and the units of analysis, you have several different designs from which to choose. In this section, we will discuss **cross-sectional surveys, longitudinal surveys,** and the use of cross-sectional surveys to approximate longitudinal surveys. In the next section, we will discuss variations of these basic designs.

Cross-Sectional Surveys

In a cross-sectional survey, data are collected at one point in time from a sample selected to describe some larger population at that time. Such a survey can be used not only for purposes of description but also for the determination of relationships between variables at the time of the study.

A Gallup Poll to determine voting intentions is an example of a cross-sectional survey. It is worth noting that respondents in such a poll are typically asked "If the election were held today, who would you vote for?" The results are appropriately reported as follows: "If the election were held today, Candidate X would win in a landslide." A survey to determine the unemployment rate would describe a population's unemployment as of the time of the study.

By the same token, a cross-sectional survey of the relationship between religiosity and prejudice would report that relationship as of the time of the study. You might report that religious people were more prejudiced than irreligious people, but you would recognize that the relationship might change later on. A subsequent survey might find quite a different relationship. (Generally speaking, however, such relationships tend to persist longer than simple descriptions.)

Longitudinal Surveys

Some survey designs—either descriptive or explanatory—permit the analysis of data over time. Data are collected at different points in time, and changes in descriptions and explanations are reported. The primary longitudinal designs are **trend studies, cohort studies,** and **panel studies.**

Trend Studies. A given general population may be sampled and studied at different points in time. Though different persons are studied in each survey, each sample represents the same population. The Gallup Polls conducted over the course of a political campaign are a good example of a trend study. At several times during the course of the campaign, samples of voters are selected and asked for whom they will vote. By comparing the results of these several polls, researchers might determine shifts in voting intentions.

In the ongoing study of prejudice in America, surveys have frequently asked respondents whether they felt that black and white children should attend the same schools. Over the years, the percentages favoring integrated schools have consistently increased. These data permit researchers to note trends in attitudes toward integration.

It should be noted that trend studies often involve a rather long period of data collection. Typically, you do not personally collect all the data used in a trend study but instead conduct a secondary analysis of data collected over time by several other researchers.

While our comments so far have been limited to descriptive trend studies, there is no reason why a researcher could not examine trends in relationships among variables. One example is the relationship between religious affiliation and political preference. Traditionally, Catholics and Jews in America have been more likely to vote for the Democratic party than have Protestants, but this relationship might be examined over time.

Cohort Studies. Trend studies are based on descriptions of a *general* population (such as American voters) over time, although the members of that population will change. Persons alive and represented in the first study might be dead at the time of the second, and persons unborn at the time of the first study might be represented in the second. Likewise, a trend study of attitudes among students at State University will reflect a different population of students each time a survey is conducted.

A cohort study focuses on the same specific population each time data are collected, although the samples studied may be different. For example, we might select a sample of students graduating from State University in the class of 1990 to determine their attitudes toward work. Five years later, we might select and study another sample drawn from the same class. While the sample would be different each time, we would still be describing the class of 1990. (If we studied the graduating class of 1995 the second time around, we would have a trend study of graduating classes rather than a cohort study of the class of 1990.)

The following example illustrates a different type of cohort study. At one point in time, you might sample from among all Americans in their twenties. Ten years later, you could sample all persons in their thirties, and so forth. This would constitute a cohort study of a given age group. This same study could also be accomplished through a secondary analysis of previously collected data. At a given point in time, you could analyze the twenty-year-old respondents in a 1970 study, the thirty-year-old respondents in a 1980 study, the forty-year-old respondents in a 1990 study, and so forth.

Panel Studies.   Both trend and cohort studies permit the analysis of process and change over time, which is not easily possible in a cross-sectional survey. These methods do have severe shortcomings, however. While you may determine through a trend study that voters, as a group, are switching from Candidate A to Candidate B, you cannot tell *which* people are switching, thereby hampering your attempts to explain why such switching is occurring.

Panel studies involve the collection of data over time from the same *sample* of respondents. The sample for such a study is called the *panel*. In a political study, you might reinterview all the members of your panel at one-month intervals throughout the campaign. Each time, you would ask them for whom they planned to vote; then, when switching occurred, you would know which persons were switching in which direction. By analyzing other characteristics of the "switchers" and "nonswitchers," you might be able to explain the reasons for switching. (Of course, you could also ask panel members why they had switched their voting intentions.)

Except in certain limiting cases, panel studies would have to be conducted as part of a particular research program. While trend studies and cohort studies could be carried out through a secondary analysis of previously collected data, panel studies could not. As a result, panel studies tend to be expensive and time-consuming, as well as suffering from two additional problems.

Panel attrition, the first problem, refers to the extent of nonresponse that occurs in later waves of the study interviewing. Some persons interviewed in the first survey might be unwilling or unable to be interviewed later on. The strength of panel studies hinges on the ability to examine the same respondents at different points in time; this advantage is lost among respondents who do not participate in the several surveys.

Second, the analysis of panel data can be rather complicated. The chief analytical device is the *turnover table,* which cross-tabulates a given characteristic at more than one point in time. For example, respondents preferring Candidate A in the first survey are divided into those who still do in the second survey and those who now prefer Candidate B. Those who prefer Candidate B the first time are similarly divided. As the number of surveys, the number of variables, and the complexity of the variables increase, the analysis and presentation of the data can become unmanageable (see Chapter 14 on the difficulties of typological analysis).

For all these reasons, panel studies are less frequently conducted in survey research than are other types of survey. Yet you should realize that the panel study is the most sophisticated survey design for most explanatory purposes. (It most closely approximates the classic laboratory experiment.)

Approximating Longitudinal Surveys

The cross-sectional survey is clearly the most frequently used study design, yet many if not most of the questions that a researcher wishes to answer involve some notion of change over time. A number of devices can be employed in a cross-sectional survey in order to approximate the study of process or change.

First, the respondents to a given survey might be able to provide data relevant to questions that involve process. For example, they might be asked to report their family incomes both for the current year and for the previous year. These data might then be used as though they had been collected in a panel study with two waves of interviewing conducted a year apart. Two dangers appear here, however. First is the danger that respondents might not be able to report such information accurately. The farther back they are forced to reach into their memories, the less accurate the information they provide is likely to be. Second, researchers should not be misled into interpreting the earlier-year data as a cross section of the population at that time, since the sample is limited to the present population.

A second way to approximate a study over time is to use age or cohort comparisons within a cross-sectional survey. Young people in a given study might be less religious than old people; this might be interpreted as a decline in religiosity in the population. (On the other hand, the fact that people tend to become more religious as they grow older may completely account for the observed differences.) Another study might find freshmen less intellectually sophisticated than seniors and conclude that college education increases sophistication. (Recognize, however, that the population from which new freshmen are drawn might be less sophisticated, pointing to a trend rather than a process. Or it might be the case that less sophisticated students who entered with the senior class dropped out of school before the survey.)

Finally, cross-sectional data can sometimes be interpreted in logical terms to indicate a process over time. A student drug study, for example, indicated that all students who reported marijuana use also reported past experience with alcohol. Moreover, all students who reported ever using LSD also reported using marijuana (and alcohol). It is reasonable to conclude from these data that the progression of drug use over time is from alcohol to marijuana to LSD. If, for example, some students had used marijuana before using alcohol, then the cross-sectional survey should have uncovered some who had used marijuana but not alcohol; by the same token, some students should have been found who had used LSD but not the other drugs if LSD were used prior to the others. Since no such students were found at the given point in time, while many students were found who had used alcohol only or all drugs studied but LSD, the researchers concluded on logical grounds that the process over time was from alcohol to marijuana to LSD. (*Note:* This conclusion does not warrant the assumption that one drug leads to another physiologically.)

Variations on Basic Designs

The preceding section outlined the basic survey designs; all surveys can be characterized in the terms discussed. At the same time, these basic designs can be modified in a variety of ways in order to meet the particular requirements of a given study. This section briefly discusses some of the more typical modifications.

Parallel Samples

A research problem may sometimes be particularly relevant to more than one population. For example, if you were an educational researcher, you might want to sample student attitudes toward a proposed student conduct code. You might also be interested in knowing how faculty members, and perhaps administrators, felt about the code. In such an instance, you could sample each population separately and administer the same (or slightly modified) questionnaire to each sample. The results produced could then be compared. As another example, you might want to examine the religious beliefs of Methodist church members and compare them with the beliefs held by Methodist clergy. Again, each population would be sampled and studied. Such studies are called **parallel samples.**

In some instances, the sample from one population could be used to generate the sample from the other. For example, university students might be sampled, and questionnaires could be sent to both the students and their parents. The responses given by students as a whole could be compared with the responses given by their parents as a whole.

Contextual Studies

As we noted in our discussion of units of analysis, persons may be described in terms of groups to which they belong. Just as a family can be described as being large, members of that family can be described as belonging to a large family. Collecting data about some portions of a person's environment or milieu and using those data to describe the individual make up a **contextual study**—an examination of the individual's context.

Using the last example under parallel samples, data collected from parents could be used to describe their particular children. A student in the study could be described as having a politically liberal father, an elderly mother, and so forth. These data could then be used in the analysis of the student's own attitudes.

In studying church members, you could collect data about the church each belonged to and, perhaps, about its minister. A given church member, then, could be described as belonging to a rich, large church in the center city that has a minister less than forty years of age.

Such analyses, however, require that the data collected about respondents' contexts be identified with them so as to be included in their data files. This could not be accomplished by mailing anonymous questionnaires to the sample of students and parents. Students would have to be identified on the questionnaire (by name or number), and parents would have to be identified in some fashion that would link them to their particular children.

Sociometric Studies

Typically, surveys study a sample of a given population, collecting data about the individuals in the sample for purposes of describing and explaining the population they represent. The basic survey format, however, can be used for a more comprehensive examination of a given group, with interrelationships among the members of that group noted. A **sociometric study** is a good example of this.

Suppose you want to learn something about the selection of close friends among school children. You might conduct a conventional survey among a sample of students and ask them to provide a variety of information about their closest friends. In a sociometric design, you would study *all* students in a given class and ask each to identify his or her friends by name. In this fashion, you could determine that Jack chose Bill as his best friend, but Bill, in turn, chose Frank. You might also determine that five members of the class chose Mary as their best friend, and nobody chose Ruth. These sorts of analyses could be extended in complexity to provide a comprehensive examination of a whole friendship network. In seeking to explain why certain students were chosen more often than others, you would have a whole body of data about the chosen students (from their self-reporting of characteristics), including whom they chose.

In this fashion, you could examine the possible factors governing friendship formation, including variables such as sex, race, economic status, and intelligence. By collecting data from all the members of the group, you would be spared the considerable task of asking respondents to describe their selections in detail; in addition, you would have access to the friendship network. Variants on this technique could be used with any intact group, and the study could be conducted over time as well as cross-sectionally.

Choosing the Appropriate Design

Given the several options available to the survey researcher, the question becomes "Which design should I select?" Clearly, this question cannot be answered in the abstract, because different research problems call for different designs. Nevertheless, some general guidelines can be provided.

First, if your aim is single-time description, then a cross-sectional survey is probably the most appropriate design. You would identify the population relevant to your interests, select a sample of respondents from that population, and conduct your survey. The same choice would apply for a research interest involving subset description. If you were interested in documenting the differences in the political attitudes of men and women, you could investigate this interest through a cross-sectional survey.

More typically, however, you are interested in examining some type of dynamic process involving change over time. When you address the sources and/or consequences of religiosity, for example, the issue of change over time is implicit, if not explicit. Implicitly, it is assumed that some people become religious and that being religious has subsequent effects on other attitudes and/or behavior. Ideally, you should select a sample of respondents at a point in their life prior to the development of a religious orientation and study them over time, covering the period during which some of the respondents become religious and following them through the period during which religiosity has effects on other aspects of their life.

Clearly, a study such as the one just described would take years to complete. You would probably begin with a sample of preadolescents and follow them through middle age and into their later life. The time required by such a study—and the attendant costs—would in all likelihood preclude it. Because of this fact, such issues are more frequently dealt with in cross-sectional surveys. Rather than noting the effects of various social conditions and experiences as they occur, you examine their possible effects by comparing respondents who have experienced them in the past with those who have not.

For example, married people are generally less religious than the unmarried. The panel study would permit you to note the decline in religiosity following marriage for given respondents. Using a cross-sectional survey instead, you would compare the levels of religiosity among married and unmarried respondents at one point in time, noting that married respondents were less religious in general. Although you could not observe the effects of marriage on religiosity at the time of marriage, you might be willing to infer such an effect on the basis of the single-time difference.[3] (The logic of such inferences is discussed in Part Four.)

Panel surveys are most feasible when the phenomenon under study is relatively short in duration. An election campaign is an example of this situation. Whereas a given campaign should take less than a year, it is feasible to conduct several waves of interviews with the survey panel over the course of the

---

[3] Recognize, of course, that such an inference from cross-sectional data will always be subject to challenge. In this case, it might be argued that religious people are less likely to get married, that is, that the *causal direction* is the opposite of that inferred.

campaign, monitoring changes in voting intentions over the period and collecting data relevant to explaining such changes. With a duration this short, the difficulties of panel attrition should be reduced and you should be able to locate respondents over time. (Over a longer period of time, many willing respondents might simply move out of town and thereby be difficult to reach.)

In summary, whenever the research problem involves an examination of individual change over time, a panel survey would be the most appropriate design in theory. If the process of change occurs during a relatively short period of time, a panel survey might be feasible. More typically, however, you will be forced to rely on cross-sectional data to make inferences about the process of change in individuals over time.

If, on the other hand, the research problem merely deals with broad trends over time—from a descriptive standpoint—you will in most cases have less trouble. In many instances, you will find that other researchers have already collected and reported all the data you require. If you are interested in overall changes of attitudes toward the war in Indochina, for example, you will find that academic and commercial researchers have been collecting relevant information for years. Your task, then, might consist only of locating such studies, comparing the nature of the questionnaire items involved and the sample designs, and then providing a discussion of the observed changes. In other situations, you might find that only one prior survey has collected data relevant to your research interest or that the latest such study is rather dated. In such a case, you might want to conduct a new cross-sectional survey—comparable in sample design and questionnaire items—in order to provide a new measure for purposes of examining trends.

Summary

These, then, are the basic designs and common variations available to the survey researcher. The preceding discussion by no means exhausts the design possibilities, but it should provide you with sufficient stimulus and guidance for constructing the study design most appropriate to your own research needs.

The best studies are often those that combine more than one design, since each design provides a different perspective on the subject under study. At the same time, you should guard against designing a survey so complex that you have neither the time nor the money to execute it. As we stressed in Chapter 3, all survey designs represent compromises. The good researcher reaches the best compromise possible.

Additional Readings

Thomas D. Cook and Donald T. Campbell, *Quasi-Experimentation: Design and Analysis Issues for Field Settings* (Chicago: Rand McNally, 1963).

Charles Y. Glock, "Survey Design and Analysis in Sociology," in Charles Y. Glock, ed., *Survey Research in the Social Sciences* (New York: Russell Sage Foundation, 1967), pp. 1–62.

Morton Hunt, *Profiles of Social Research: The Scientific Study of Human Interactions* (New York: Basic Books, 1985).

Paul F. Lazarsfeld, Ann K. Pasanella, and Morris Rosenberg, eds., *Continuities in the Language of Social Research* (New York: Free Press, 1972), secs. III–IV.

Delbert Miller, *Handbook of Research Design and Social Measurement* (New York: Longman, 1983).

5

The Logic of

Survey Sampling

In this book, as in many others, the term "survey" has been used to mean implicitly "sample survey," as opposed to a study of all members of a given population or group. Typically, survey methods are used in the study of a segment or portion—a sample—of a population for purposes of making estimated assertions about the nature of the total population from which the sample has been selected. Although the practice of using samples in this connection is more or less tacitly accepted, the reasons for sampling are not generally known. We should consider those reasons briefly before turning to the logic and skills of sampling.

Why Sample?

You can probably guess two reasons for sampling: time and cost. The interviewing alone for a comprehensive household interview survey might require one to three hours and $40 to $100 per interview. The savings in studying 2,000 people rather than, say, 500,000 are apparent. Thus, sampling often makes a project possible, whereas a refusal to sample would rule out the study altogether.

Sampling should not be regarded as a necessary evil, however. A point that is not generally recognized, perhaps, is that sample surveys are often more accurate than interviewing every member of a given population. This seemingly bizarre fact is the result of several characteristics of the logistics of survey interviewing.

First, an enormous interviewing project would require a very large staff of interviewers. Researchers typically attempt to limit their staffs to the best available interviewers, and such a project would probably require them to employ everyone in sight, with the result that the overall quality of interviewers would be lower than usually achieved. The quality of data collected would be reduced by the decreased quality of the interviewers. Also, a smaller-scale study would permit more diligent follow-up procedures, increasing the rates of interview completion.

Second, because interviewing all members of a given large population would require a lengthy interviewing period, it would be difficult if not impossible to specify the *time* to which the data refer. If the study were aimed at measuring the level of unemployment in a given large city, the unemployment rate produced by the survey data would refer to the city neither as of the beginning of the interviewing nor as of the end. Rather, the unemployment rate would have to be attributed to some hypothetical date, representing perhaps the midpoint of the interviewing period. (Asking respondents to answer in terms of a specified uniform date introduces the problem of inaccurate recall.) The problem of time attribution is inherent in any interviewing project that is not executed all in one moment, but the seriousness of the problem grows with the duration of interviewing. If the interviewing took ten years to complete—with the unemployment rate presumably changing during that period—the resultant rate would be meaningless.

Finally, the managerial requirements of a very large survey would be far greater than those normally faced by survey researchers. Supervision, record keeping, training, and so forth would all be more difficult in a very large survey. Once again, the quality of data collected in a very large survey might be lower than that obtained in a smaller, more manageable one. (It is worth noting that the Bureau of the Census follows its decennial census with a sample survey for purposes of evaluating the data collected in the total enumeration.)

Are Sample Data Really Accurate?

Despite the foregoing discussion, you may still feel somewhat uneasy about sampling. Because it is clearly possible for a sample to misrepresent the population from which it is drawn, the researcher who utilizes sampling methods faces an inevitable danger. Nevertheless, as we will show in this chapter, established sampling procedures can reduce this danger to an acceptable minimum. Ultimately, sample surveys can provide very accurate estimates about the populations they portray. However, the sample survey researcher must be prepared to tolerate a certain ambiguity, since determination of the degree of accuracy of sample findings is seldom possible.

Political pollsters represent one group of survey researchers who are given an opportunity to check the accuracy of their sample findings. Election day is the final judgment for political pollsters, and their mixed experiences are instructive in relation to the more general question of sample survey accuracy.

Most critics of sample survey methods are familiar with the 1936 *Literary Digest* poll that predicted Alfred M. Landon to win over Franklin D. Roosevelt by a landslide. Polling a sample of more than 2 million voters by mail, the *Digest* predicted that Landon would beat Roosevelt by nearly 15 percentage points. The primary reason for this failure lay in the **sampling frame** used by the pollsters (we will discuss this concept later in more detail). The *Digest* sample was drawn from telephone directories and automobile registration lists, a sampling procedure that had seemed sufficient in the 1920, 1924, 1928, and 1932 elections but that by 1936 did not provide a representative cross section of American voters. In the wake of the Depression and in the midst of the New Deal, unprecedented numbers of poor Americans came to the polls, and these poor people were not adequately represented by telephone directories and automobile registration lists.[1]

In 1936, George Gallup correctly predicted that Roosevelt would win a second term. Gallup's sampling procedures differed from those of the *Literary Digest,* however. Gallup's American Institute of Public Opinion had pioneered the use of **quota sampling** (described later in this chapter), which better ensured that all types of American voters, rich and poor alike, would be adequately represented in the survey sample. Where the *Digest* poll failed to reach and question the poor—and predominantly Democratic—voters, Gallup's quota sampling did.

Twelve years later, Gallup and most other political pollsters suffered the embarrassment of predicting victory for Thomas Dewey over Harry Truman. As Goodman Ace acidly noted, "Everyone believes in public opinion polls. Everyone from the man in the street . . . up to President Thomas E. Dewey."[2] A number of factors conspired to bring about the 1948 polling debacle. For one thing, most pollsters finished their polling too soon despite a steady trend toward Truman over the course of the campaign. The large number of voters who said they did not know for whom they would vote went predominantly to Truman. Most important, however, the failure in 1948 pointed to serious shortcomings inherent in quota sampling, the very method that was such an improvement over the *Literary Digest* sampling methods. In 1948, a number of academic survey researchers had been experimenting with **probability sampling** methods. By and large, these researchers were far more successful than the quota samplers, and probability sampling remains the most respected method used by survey researchers today.

The brief discussion above has presented a partial history of early survey sampling in America but has done so perhaps at the expense of whatever little faith you may have had in sample survey methods. To counterbalance this possible effect, we shall consider the more recent score sheet of political polling accuracy. In November 1988, George Bush received 54 percent of the popular vote for President; Michael Dukakis received 46 percent. American voters didn't have to wait for the results of the vote tallies to know who won, however. Several "exit polls," interviewing samples of voters as they left their

---

[1] For a recent reassessment, see Peverill Squire, "Why the 1936 *Literary Digest* Poll Failed," *Public Opinion Quarterly,* vol. 52 (Spring 1988), pp. 125–133.

[2] Requoted in *Newsweek,* 8 July 1968, p. 24.

Table 5-1   *Exit Poll Results: November 1988*

	Bush	Dukakis
CBS News/*New York Times*	54%	46%
ABC News/*Washington Post*	54%	46%
NBC News/*Wall Street Journal*	54%	46%
Los Angeles Times	54%	46%

SOURCE: Data are taken from *Public Opinion,* Jan./Feb. 1989, pp. 25–26. The figures for CBS have been repercentaged to exclude the "no answers.".

Table 5-2   *Political Polls Prior to the 1988 Election*

	Bush	Dukakis
CBS News/*New York Times* (11/4–5/88)	55%	45%
Gallup Poll (11/3–5/88)	56%	44%
USA Today/CNN (11/3–6/88)	55%	45%
Harris Poll (11/2–5/88)	53%	47%
ABC News/*Washington Post* (11/2–5/88)	55%	45%
NBC News/*Wall Street Journal* (11/1–5/88)	53%	47%

SOURCE: Data are taken from *Public Opinion,* Nov./Dec. 1988, p. 39. All figures have been repercentaged to exclude the "no answers."

polling places, provided the following estimates of how the election would turn out, as shown in Table 5-1.

In place of the 2 million voters polled by the *Literary Digest* in 1936, these four exit polls sampled from 6,000 to 20,000 voters to predict the voting of more than 90 million Americans who went to the polls in 1988. The results of the exit polls were not surprising, however, because they merely carried forward the trend that had been reported in public opinion polls in the week prior to the election. What the major polls had to say, typically based on samples of 2,000 or fewer respondents, is shown in Table 5-2.

Sample surveys can be extremely accurate. At the same time, we should concede that they often are not accurate, even today. The remainder of this chapter presents the reasons and rules for accuracy in sampling.

Two Types of Sampling Methods

It is useful to distinguish two major types of sampling methods: **probability sampling** and **nonprobability sampling.** The bulk of this chapter is devoted to probability sampling, because it is currently the most respected and useful method. A smaller portion of this chapter considers the various types of nonprobability sampling methods.

We will begin with a discussion of the logic of probability sampling, followed by a brief taxonomy of sampling concepts and terminology. Then we will turn to the concept of sampling distribution, which is the basis of estimating the accuracy of sample survey findings. These theoretical discussions will be followed by a consideration of **populations** and **sampling frames**, focusing on practical problems of how to determine the target group of the study and how to begin selecting a sample. Next, we will examine the basic types of survey design: **simple random samples, systematic samples, stratified samples**, and **cluster samples**. Then we will present a short discussion and description of nonprobability sampling.

The chapter closes with a brief consideration of some nonsurvey uses of sampling methods in fields such as content analysis, participant observation, and historical analysis. Hopefully, you will have become so familiar with the *logic* of survey sampling that you will be able to apply that knowledge to a broader variety of situations.

The Logic of Probability Sampling

It should be apparent from the history of political polling that sample surveys can be very accurate. At the same time, it should be equally apparent that samples must be selected in a careful fashion. We consider briefly why this is the case.

The Implications of Homogeneity and Heterogeneity

If all members of a population were identical to one another in all respects, there would be no need for careful sampling procedures; any sample would indeed be sufficient. In this extreme case of **homogeneity**, in fact, *one* case would be sufficient as a sample to study characteristics of the whole population.

Before you dismiss this idea as impossible, recall that much scientific sampling is carried out on this basis. In the physical sciences, it is sometimes safe to make such an assumption and proceed on the basis of this assumption in research. The chemist who wants to test certain properties of carbon, for example, need not undertake a painstaking enumeration of all the carbon in the world and then carefully select a probability sample of carbon molecules for study. By the same token, the medical scientist—or the practicing physician—who wants to examine a person's blood need not draw out all of the person's blood and select a probability sample of blood cells. Again, for most purposes, any sample of blood from the person will suffice.

Faced with variation, or heterogeneity, in the population under study, however, the researcher must use more controlled sampling procedures. The broader applicability of this principle beyond social research is worth noting. The origins of modern sampling theory are found in agricultural research, especially in the work of R. A. Fisher, whose name is still attached to some commonly used survey statistics.

For our purposes, it is more important to note the heterogeneity of social groups. Because people differ in many ways, a given human population is comprised of varied individuals. A sample of individuals from that population, if it is to provide useful descriptions of the total population, must contain essentially the same variation as exists in the population. Probability sampling provides an efficient method for selecting a sample that should adequately reflect the variation that exists in the population as a whole.

Conscious and Unconscious Sampling Bias

Of course, anyone could select a survey sample, even without any special training or care. To select a sample of 100 university students, a person might go to the university campus and begin interviewing students found walking around campus. This kind of sampling method, which is often used by untrained researchers, has very serious problems.

To begin, there is a danger that your own personal biases may affect the sample selected in this manner; hence, the sample would not truly represent the student population. Assume, for example, that you are personally somewhat intimidated by "hippy-looking" students and feel that they would ridicule your research effort. As a result, you might consciously or semiconsciously avoid interviewing such people. On the other hand, you might feel that the attitudes of "straight-looking" students would not be relevant to your research purposes and might avoid interviewing such students. Even if you sought to interview a "balanced" group of students, you probably would not know the proper proportions of different types of students making up such a balance, or you might be unable to identify the different types by just watching them walk by.

Even making a conscientious effort to interview every tenth student entering the university library would not guarantee you a representative sample, since different types of students visit the library with different frequencies. Thus, the sample would overrepresent those students who use the library more often.

Representativeness and Probability of Selection

Survey samples must represent the populations from which they are drawn if they are to provide useful estimates of the characteristics of that population. They need not, however, be representative in all respects; representativeness, as it has any meaning in regard to sampling, is limited to those characteristics that are relevant to the substantive interests of the study. (This point will become more evident in our later discussion of stratification.)

A basic principle of probability sampling is the following: *A sample will be representative of the population from which it is selected if all members of the population have an equal chance of being selected in the sample.*[3] Samples that have this quality are often labeled **EPSEM samples** (equal-probability-of-selection method). While we will discuss variations of this principle later, it is the primary principle providing the basis of probability sampling.

Moving beyond this basic principle, we must realize that samples—even carefully selected EPSEM samples—are seldom, if ever, *perfectly* **representative** of the populations from which they are drawn. Despite this limitation, however, probability sampling offers two special advantages for researchers.

First, probability samples, while never perfectly representative, are typically *more representative* than other types of samples because the biases discussed in the preceding section are avoided. In practice, a greater likelihood exists that a probability sample will be representative of the population from which it is drawn than that a nonprobability sample will be.

Second, and more important, probability theory permits you to estimate the accuracy or representativeness of your sample. Conceivably, you might through wholly haphazard means select a sample that nearly perfectly represents the larger population. The odds are against your doing so, however, and you would be unable to estimate the likelihood that you have achieved representativeness. The probability sample, on the other hand, can provide an accurate estimate of your success or failure. After presenting a brief taxonomy of sampling terminology, we will examine the means whereby you can estimate the representativeness of your sample.

Sampling Concepts and Terminology

The following discussions of sampling theory and practice utilize a number of technical terms. To facilitate your understanding of the discussions, it is important that those terms be defined. For the most part, terms commonly used in other sampling and statistical textbooks will be employed so that you can better understand those other sources.

In presenting this taxonomy of sampling concepts and terminology, I would like to acknowledge a debt to Leslie Kish and his excellent textbook on survey sampling.[4] While I have modified some of the conventions used by Kish, his presentation is easily the most important source of our discussion.

---

[3] We will see shortly that the size of the sample selected, as well as the actual characteristics of the larger population, affects the *degree* of representativeness.

[4] Leslie Kish, *Survey Sampling* (New York: John Wiley & Sons, 1965).

Element

An **element** is that unit about which information is collected and which provides the basis of analysis. Typically, in survey research, elements are people or certain types of people. It should be recognized, however, that other kinds of units—families, social clubs, corporations, and so forth—might constitute the elements of a survey.

Universe

A **universe** is the theoretical and hypothetical aggregation of all elements as defined for a given survey. If the individual American were the element for a survey, then "Americans" would be the universe. A survey universe is wholly unspecified as to time and place, however, and is essentially a useless term.

Population

A **population** is the theoretically specified aggregation of survey elements. While the vague term "Americans" might constitute the universe for a survey, the delineation of the population would include a definition of the element "American" (for example, citizenship and residence) and the time referent for the study (Americans as of when?). Translating the universe "adult New Yorkers" into a workable population would require a specification of the age defining "adult," the boundaries of New York, and so forth. Specifying the term "college student" would include a consideration of full-time and part-time students, degree candidates and nondegree candidates, undergraduate and graduate students, and similar attributes.

Though you must begin with a careful specification of your population, poetic license usually permits you to phrase your report in terms of the hypothetical universe. For ease of presentation, even the most conscientious researcher normally speaks of "Americans" rather than "resident citizens of the United States of America as of November 12, 1989." The primary guide in this matter, as in most others, is that you should not mislead or deceive your readers.

Survey Population

A **survey population** is that aggregation of elements from which the survey sample is actually selected. Recall that a population is a theoretical specification of the universe. As a practical matter, you are seldom in a position to guarantee that every element that meets the theoretical definitions laid down actually has a chance of being selected in the sample. Even lists of elements compiled specifically for sampling purposes are usually somewhat incomplete. Some students are always omitted, inadvertently, from student rosters. Some telephone subscribers request that their names and numbers be unlisted. The survey population, then, is the aggregation of elements from which the sample is selected.

Often, researchers might decide to limit their survey populations more severely than indicated in the preceding examples. National polling firms might limit their "national samples" to the forty-eight adjacent states, omitting Alaska and Hawaii for practical reasons. A researcher who wants to sample psychology professors might limit the survey population to psychology professors serving in psychology departments, omitting those serving in other departments. (In a sense, we might say that these researchers have redefined their universes and populations, providing that they have made the revisions clear to their readers.)

Sampling Unit

A **sampling unit** is that element or set of elements considered for selection in some stage of sampling. In a simple, single-stage sample, the sampling units are the same as the elements. In more complex samples, however, different levels of sampling units might be employed. For example, a researcher might select first a sample of census blocks in a city, then a sample of households on the selected blocks, and finally a sample of adults from the selected households. The sampling units for these three stages of sampling are, respectively, census blocks, households, and adults, only the last of which are the elements. More specifically, the terms "primary sampling units," "secondary sampling units," and "final sampling units" would be used to designate the successive stages.

Sampling Frame

A **sampling frame** is the actual list of sampling units from which the sample, or some stage of the sample, is selected. If a simple sample of students were selected from a student roster, the roster would be the sampling frame. If the primary sampling unit for a complex population sample were the census block, the list of census blocks would comprise the sampling frame—in the form of either a printed booklet, an IBM card file, a magnetic tape, or a floppy disk file.

In a single-stage sample design, the sampling frame is a list of the elements comprising the survey population. In practice, the existing sampling frames often define the survey population rather than the other way around. You often begin with a universe or perhaps a population in mind for your study and then search for possible sampling frames. The frames available for your use are examined and evaluated, and you decide which frame represents a survey population most appropriate to your needs.

The relationship between populations and sampling frames is critical and has not been given sufficient attention. A later section of this chapter pursues this issue in greater detail.

Observation Unit

An **observation unit,** or unit of data collection, is an element or aggregation of elements from which information is collected. Again, the unit of analysis and the unit of observation are often the same—the individual person—but this need not be the case. For example, you might interview heads of households (the observation units) to collect information about every member of the household (the units of analysis).

Your task is simplified when the unit of analysis and the observational unit are the same. Often this is not possible or feasible, however, and in such situations you should be capable of exercising some ingenuity in collecting data relevant to your units of analysis without actually observing those units.

Variable

A **variable** is a set of mutually exclusive characteristics, such as sex, age, employment status, and so forth. The elements of a given population may be described in terms of their individual characteristics on a given variable. Typically, surveys aim at describing the distribution of characteristics comprising a variable in a population. Thus, you might describe the age distribution of a population by examining the relative frequency of different ages among members of the population.

Note that a variable, by definition, must possess *variation*; if all elements in the population have the same characteristic, that characteristic is a constant in the population rather than part of a variable.

Parameter

A **parameter** is the summary description of a given variable in a population. The mean income of all families in a city and the age distribution of the city's population are parameters. An important portion of survey research involves the estimation of population parameters on the basis of sample observations.

Statistic

A **statistic** is the summary description of a given variable in a survey sample. Thus, the mean income computed from a survey sample and the age distribution of that sample are statistics. Sample statistics are used to make estimates of population parameters.

Sampling Error

Sampling error is discussed in more detail in a later section of this chapter. Probability sampling methods seldom, if ever, provide statistics exactly equal to the parameters they are used to estimate. Probability theory, however, permits us to estimate the degree of error to be expected for a given sample design.

Confidence Levels and Confidence Intervals

These terms also are discussed more fully in a later section. The computation of sampling error permits you to express the accuracy of your sample statistics in terms of your level of confidence that the statistics fall within a specified interval from the parameter. For example, you might say that you are "95 percent confident" that your sample statistic (for example, 50 percent favor Candidate X) is within plus or minus (±) 5 percentage points of the population parameter. As the confidence interval is expanded for a given statistic, your "confidence" increases; you might say that you are 99.9 percent confident that your statistic falls within ±7.5 percentage points of the parameter.

Probability Sampling Theory and Sampling Distribution

This section examines the basic theory of probability sampling as it applies to survey sampling; we will consider the logic of sampling distribution and sampling error with regard to a **binomial variable**, that is, a variable comprised of two characteristics.

Probability Sampling Theory

The ultimate purpose of survey sampling is to select a set of elements from a population in such a way that descriptions of those elements (statistics) accurately describe the total population from which they are selected. Probability sampling provides a method for enhancing the likelihood of accomplishing this aim, as well as methods for estimating the degree of probable success.

Random selection is the key to this process. A random selection process is one in which each element has an equal chance of selection independent of any other events in the selection process. Flipping a perfect coin is the most frequently cited example; the "selection" of a head or a tail is independent of previous selections of heads or tails. Rolling a perfect set of dice is another example.

Such images of random selection seldom apply directly to survey sampling methods, however. You are far more likely to utilize tables of random numbers or computer programs that provide a random selection of sampling units. The wide availability of such research aids makes them an adequate beginning point for our discussion of random sampling.

The reasons for using random selection methods, that is, using random-number tables or computer programs, are twofold. First, this procedure serves as a check on conscious or unconscious bias on the part of the researcher. The researcher who undertakes the selection of cases on an intuitive basis might very well select cases that would support preexisting research expectations or hypotheses. Random selection erases this danger. Second, random selection offers you access to the body of probability theory, which provides the basis for your estimates of population parameters and estimates of error. We turn now to an examination of this second aspect.

Binomial Sampling Distribution

The clearest way to discuss the concept of sampling distribution is to utilize a simple survey example. Assume for the moment that we want to study the student population of State University in order to determine approval or disapproval of a student conduct code proposed by the administration. The survey population will be that aggregation of students contained in a student roster (the sampling frame). The elements will be the individual students at State University. The variable under consideration will be attitudes toward the code, a binomial variable—approve and disapprove. We will select a random sample of students for purposes of estimating the entire student body.

Figure 5-1 presents an *x*-axis that represents all possible values of this parameter in the population—from 0 percent approval to 100 percent approval. The midpoint of the axis—50 percent—represents the situation in which half the students approve of the code while the other half disapprove.

Let us assume for the moment that we have given each student on the student roster a number and have selected 100 random numbers from a table of random numbers. The 100 students having the numbers selected are then interviewed and asked for their attitudes toward the student code—whether they approve or disapprove. Let us further assume that this operation provides us with 48 students who approve of the code and 52 who disapprove. We might represent this statistic by placing a dot on the *x*-axis at the point representing 48 percent.

Now let us suppose that we select another sample of 100 students in exactly the same fashion and measure their approval/disapproval of the student code. Perhaps 51 students in the second sample approve of the code; this result might be represented by another dot in the appropriate place on the *x*-axis. Repeating this process once more, we might discover that 52 students in a third sample approve of the code.

Figure 5-2 presents the three different sample statistics representing the percentages of students in each of the three random samples who approved of the student code. The basic rule of random sampling is that such samples drawn from a population provide estimates of the parameter that pertains in the total population. Each random sample, then, gives us an estimate of the percentage of students in the total student body who approve of the student code. Unhappily, however, we have selected three samples and now have three separate estimates.

Figure 5-1

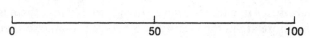

Percent of students approving of the student code

Figure 5-2

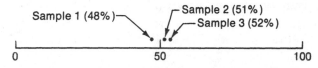

Percent of students approving of the student code

Figure 5-3

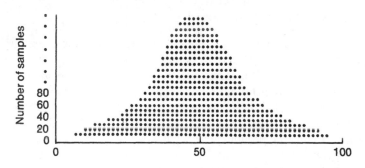

Percent of students approving of the student code

To escape this dilemma, let's go on to draw more and more samples of 100 students each, question each of the samples as to their approval/disapproval of the code, and plot the new sample statistics on our summary graph. In drawing many such samples, we will begin to discover that some of the new samples provide the same estimates given by earlier samples. To account for this situation, we will add a y-axis to the figure representing the number of samples providing a given estimate. Figure 5-3 depicts the product of our new sampling efforts.

The distribution of sample statistics shown in Figure 5-3 is called the **sampling distribution.** We note that by increasing the number of samples selected and interviewed, we have also increased the range of estimates provided by the sampling operation. In one sense we have worsened our dilemma in attempting to guess the parameter in the population. Probability theory, however, provides certain important rules regarding the sampling distribution presented in Figure 5-3.

First, if many independent random samples are selected from a population, the sample statistics provided by those samples will be *distributed around the population parameter* in a known way. Though we see that there is a wide range of estimates, more are in the vicinity of 50 percent than elsewhere in the graph. Probability theory tells us, then, that the true value is in the vicinity of 50 percent.

Second, probability theory provides us with a formula for estimating *how closely* the sample statistics are clustered around the true value. This formula contains three factors: the parameter, the sample size, and the **standard error** (a measure of sampling error).

Formula: $S = \sqrt{\dfrac{PQ}{n}}$

Symbols: P, Q: The population parameters for the binomial; if 60 percent of the student body approve the code and 40 percent disapprove, P and Q are 60 percent and 40 percent, or .6 and .4. Note that $Q = 1 - P$ and $P = 1 - Q$.

   n: The number of cases in each sample.

   S: The standard error.

Assume that the population parameter in the study survey example is 50 percent approval of the code and 50 percent disapproval. Recall that we have been selecting samples of 100 cases each. When these numbers are put into the formula, we find that the standard error equals .05, or 5 percent.

In terms of probability theory, the standard error is a valuable datum because it indicates the extent to which the sample estimates will be distributed around the population parameter. Specifically, probability theory indicates that certain proportions of the sample estimates will fall within specified increments of standard errors from the population parameter. Approximately 34 percent (.3413) of the sample estimates will fall within one standard error above the population parameter, and another 34 percent will fall within one standard error below the parameter. In our example, the standard error is 5 percent, so we know that 34 percent of our samples will give estimates of student approval between 50 percent (parameter) and 55 percent (one standard error above); another 34 percent of the samples will give estimates between 50 percent and 45 percent (one standard error below the parameter). Taken together, then, we know that roughly two-thirds (68 percent) of the samples will give estimates within (plus or minus) 5 percent of the parameter.

Moreover, probability theory dictates that roughly 95 percent of the samples will fall within (plus or minus) two standard errors of the true value and that 99.9 percent of the samples will fall within (plus or minus) three standard errors. In our present example, then, we know that only one sample out of a thousand would give an estimate lower than 35 percent approval or higher than 65 percent approval.

The proportion of samples falling within one, two, or three standard errors of the parameter is constant for any random sampling procedure such as the one just described, providing that a large number of samples are selected. The size of the standard error in any given case, however, is a function of the population parameter and the sample size. If we return to the formula, we note that the standard error will increase as a function of an increase in the quantity PQ. Note further that this quantity reaches its maximum in the situation of an even split in the population. If $P = .5$, $PQ = .25$; if $P = .6$, $PQ = .24$; if $P = .8$, $PQ = .16$; if $P = .99$, $PQ = .0099$. By extension, if P is either 0.0 or 1.0 (that is, if either 0 percent or 100 percent approve of the student code), the standard error will be 0. If everyone in the population has the same attitude (no variation), then every sample will give exactly the same estimate.

The standard error is also an inverse function of the sample size. As the sample size increases, the standard error decreases, and the several samples will be clustered nearer to the true value. Another rule of thumb is evident from the formula: Because of the square root in the formula, the standard error is reduced by half if the sample size is *quadrupled*. In our present example, samples of 100 produce a standard error of 5 percent; to reduce the standard error to 2.5 percent, we would have to increase the sample size to 400. (Appendix B provides a table to assist you in making these estimates.)

The foregoing is provided by established probability theory in reference to the selection of large numbers of random samples. If the population parameter is known and very many random samples are selected, we are able to predict how many of the samples will fall within specified intervals from the parameter. These conditions do not typically pertain in survey sampling, however. Typically, the survey researcher does not know the parameter but conducts a sample survey in order to estimate the parameter. Moreover, survey researchers do not typically select large numbers of samples but select only one sample. Nevertheless, the preceding discussion of probability theory provides the basis for inferences about the typical survey situation.

Whereas probability theory specifies that 68 percent of the samples will fall within one standard error of the parameter, the survey sampler infers that a given random sample has a likelihood of 68 percent of falling within that range. In this regard, we speak of **confidence levels:** You are "68 percent confident" that your sample estimate is within one standard error of the parameter, or you are "95 percent confident" that the sample statistic is within two standard errors of the parameter, and so forth. Quite reasonably, your confidence increases as the margin for error is extended. You are virtually positive (99.9 percent confident) that the sample estimate is within three standard errors of the true value.

Even though you may be confident (at some level) of being within a certain range of the parameter, we have already noted that you seldom know what the parameter is. To resolve this dilemma, you substitute your sample estimate for the parameter into the formula; lacking the true value, you substitute the best available guess.

The result of these inferences and estimations is that the survey researcher is able to estimate both a population parameter and the expected degree of error on the basis of one sample drawn from a population. Beginning with the question "What percentage of the student body approves of the student code?" you could select a random sample of 100 students and interview them. You might then report that your best estimate is that 50 percent of the student body approve of the code and that you are 95 percent confident that between 40 and 60 percent (plus or minus two standard errors) approve. The range from 40 to 60 percent is called the **confidence interval.**

This, then, is the basic logic of probability sampling. Random selection permits you to link your sample survey findings to the body of probability theory for purposes of estimating the accuracy of those findings. All statements of accuracy in sampling must specify both a confidence level and a confidence interval. You can report that you are x percent confident that the population parameter is between two specified values.

The preceding discussion has considered only one type of statistic—the percentages produced by a binomial or dichotomous variable. The same logic, however, applies to the examination of other statistics, such as mean income. Because the computations are somewhat more complicated in such a case, we have considered only binomials in this introduction.

You should be cautioned that the survey uses of probability theory as discussed in this section are not wholly justified on technical grounds. The theory of sampling distribution, for example, makes assumptions that almost never apply in survey conditions. The formula used in estimating the number of samples that would be contained within specified increments of standard errors, for example, assumes an infinitely large population, an infinite number of samples, and sampling with replacement. Moreover, the inferential jump from the distribution of several samples to the probable characteristics of one sample has been grossly oversimplified in our discussion.

The above cautions are offered for reasons of perspective. Researchers often appear to overstate the precision of estimates produced by the use of probability theory in connection with survey research. As we will mention elsewhere in this chapter and throughout the book, variations in sampling techniques and nonsampling factors may further reduce the legitimacy of such estimates. Nevertheless, the calculations discussed in this section can be extremely valuable to you in understanding and evaluating your data. Although the calculations do not provide as precise estimates as you might assume, they can be quite valid for practical purposes, and they are unquestionably more valid than less rigorously derived estimates based on less rigorous sampling methods. Most important, you should be familiar with the basic *logic* underlying the calculations so that you will be able to react sensibly both to your own data and to data reported by others.

Populations and Sampling Frames

The immediately preceding section dealt with a theoretical model for survey sampling. While the survey consumer, student, or researcher must understand that theory, appreciating the less-than-perfect conditions that exist in the field is no less important. This section is devoted to a discussion of one aspect of field conditions that requires a compromise with regard to theoretical conditions and assumptions—the congruence of, or disparity between, populations and sampling frames.

Simply put, a sampling frame is the list, or reasonable facsimile, of elements from which a probability sample is selected. The following section deals with the methods for selecting samples, but we must first consider the sampling frame itself. Properly drawn samples will provide information appropriate for describing the population of elements that comprise the sampling frame—nothing more. It is necessary to make this point in view of the all-too-common tendency for researchers to select samples from a given sampling frame and then make assertions about a population similar to, but not identical to, the survey population defined by the sampling frame. The problem involved here is the broader social-scientific one of generalization and is akin to studying a small Lutheran church in North Dakota for purposes of describing religion in America. In the remainder of this section, we will examine different survey purposes and discuss good and bad sampling frames that might be used to satisfy those purposes.

Surveys of organizations are often the simplest type of survey from a sampling standpoint because organizations typically have membership lists. In such cases, the list of members constitutes an excellent sampling frame. If a random sample is selected from a membership list, the data collected from that sample can be taken as representative of all members—*if all members are included in the list*. If some members have been omitted from the membership list, an effort must be made to sample those nonlisted members; otherwise, the sample survey findings can be taken as representative only of those members on the list.

Populations that can often be sampled from good organizational lists include elementary school, high school, and university students and faculty; church members; factory workers; fraternity or sorority members, members of social, service, or political clubs; and members of professional associations.

The above comments apply primarily to local organizations. Statewide or national organizations often do not have a single membership list that is easily obtainable. There is, for example, no single list of Episcopalian church members. A slightly more complex sample design, however, could take advantage of local church membership lists by first sampling churches and then subsampling the membership lists of the churches selected.

Other lists of individuals can be especially relevant to the research needs of a particular survey. Government agencies maintain lists of registered voters, for example, if you want to conduct a preelection poll or a more detailed examination of voting behavior, but you must make sure that the list is up-to-date. Similar lists contain the names of automobile owners, welfare recipients, taxpayers, business permit holders, licensed professionals, and so forth. While some of these lists might be difficult to obtain, they can provide excellent sampling frames for specialized research purposes. Realizing that the sampling elements in a survey need not be individual persons, we should note the existence of lists of other types of elements: universities, businesses of various types, cities, academic journals, newspapers, unions, political clubs, professional associations, and so forth.

Telephone directories are frequently used for "quick and dirty" public opinion polls. Undeniably, they are easy and inexpensive to use, which no doubt accounts for their popularity, and if you want to make assertions about telephone subscribers the directory is a *fairly good* sampling frame.[5] Unfortunately, telephone directories are all too often taken to be a listing of a city's population or of its voters. Many defects are inherent in this reasoning, but the chief one involves a social-class bias. Poor people are less likely to have telephones; a telephone directory sample, therefore, is likely to have a middle- or upper-class bias. (Chapter 10 examines a special method for sampling in connection with telephone surveys: **random digit dialing.**)

Street directories and tax maps are often used as easily obtained sources for samples of households, but they may suffer from many of the same disadvantages as the telephone directory, such as incompleteness and possible bias. For example, in strictly zoned urban regions, "illegal" housing units are unlikely to appear on official records. As a result, such units would have no chance of selection, and sample findings would not be representative of those units, which are often typically poorer and more overcrowded.

Review of Populations and Sampling Frames

Because survey research literature has given surprisingly little attention to the issues of populations and sampling frames, I have devoted special attention to the topic. To further emphasize the point, I list here, in review, the main guidelines to keep in mind.

1. Sample survey findings can be taken as representative only of the aggregation of elements that comprise the sampling frame.

2. Often, sampling frames do not truly include all the elements that their names might imply. (Student directories do not include all students; telephone directories do not include all telephone subscribers.) Omissions are almost inevitable. Thus, a first concern must be to assess the extent of omissions and correct them if possible. (Realize, of course, that you might feel you can safely ignore a small number of omissions that cannot easily be corrected.)

3. Even to be generalized to the population comprising the sampling frame, it is necessary for all elements to have equal representation in the frame; typically, each element should appear only once. Other-

---

[5] Realize, of course, that a given directory will not include new subscribers or those who have requested unlisted numbers. Sampling is further complicated by the inclusion of nonresidential listings such as stores and offices.

wise, elements that appear more than once will have a greater probability of selection, and the sample will, overall, overrepresent these elements.

Other, more practical, matters relating to populations and sampling frames are treated elsewhere in this book. For example, the form of the sampling frame—a list in a publication, a 3×5 card file, addressograph plates, IBM cards, magnetic tapes, or floppy disks—is very important. Note that such considerations often take priority over scientific considerations; an "easier" list might be chosen over a "harder" one, even though the latter might be more appropriate to the target population. We should not take a dogmatic position in this regard; all researchers should carefully weigh the relative advantages and disadvantages of such alternatives. Most important of all, you must be aware—and must so inform your reader—of the shortcomings of whatever sampling frame you choose.

Types of Sampling Designs

Introduction

You may have reached this point in your reading somewhat aghast at the importance and difficulties of organizing your sampling frame; such a feeling is altogether appropriate and healthy. After you have established your sampling frame, you must then actually select a sample of elements for study.

Up to this point, we have focused on **simple random sampling.** Indeed, the body of statistics typically used by survey researchers assumes such a sample. As we will see shortly, however, you have a number of options in choosing your sampling, and you seldom if ever will choose simple random sampling, for two reasons. First, with all but the simplest sampling frame, simple random sampling is not possible. Second, probably surprising to you, simple random sampling might not be the best (the most accurate) method available. We turn now to a discussion of simple random sampling and the other types of sampling designs available.

Simple Random Sampling

As we noted earlier, simple random sampling (SRS) is the basic sampling method assumed in statistical survey computations. The mathematics of random sampling are especially complex, and we will detour around them in favor of describing the field methods involved in employing this method.

Once a sampling frame has been established in accord with our earlier discussion, you can assign numbers to each element in the list, assigning one and only one number to each element and not skipping any number. A table of random numbers could then be used in the selection of elements for the sample.

If your sampling frame is in a computer-readable form, for example, a floppy disk, a simple random sample could be selected automatically through the use of a fairly simple computer program. In effect, the computer would number the elements in the sampling frame, generate its own series of random numbers, and print out the list of elements selected.

Systematic Sampling

Simple random sampling is seldom used in practice. As we will see later, it is not usually the most efficient sampling method, and as we have already seen it can be rather laborious if done manually. SRS typically requires a list of elements; when such a list is available, researchers usually employ a **systematic sampling** method rather than simple random sampling.

In systematic sampling, every *k*th element in the total list (for example, every 100th) is chosen for inclusion in the sample. If the list contains 10,000 elements and you want a sample of 1,000 elements, you will select every tenth element for your sample. To ensure against any possible human bias in using this method, you select the first element at random. Thus, in the above example, you would begin by selecting a random number between 1 and 10; the element having that number would be included in the sample, plus every tenth element following it. This procedure is technically referred to as a "systematic sample with a random start."

Two terms are frequently used in connection with systematic sampling. The **sampling interval** is the standard distance between elements selected in the sample—10 in the example above. The **sampling ratio** is the proportion of elements in the population that is selected—one-tenth in the example above. (Note that the sampling ratio equals 1/sampling interval.)

In practice, systematic sampling is virtually identical to simple random sampling. If, indeed, the list of elements is "randomized" in advance of sampling, one might argue that a systematic sample drawn from that list is, in fact, a simple random sample. Debates over the relative merits of simple random sampling and systematic sampling have been resolved largely in favor of the simpler method, systematic sampling. Empirically, the results are virtually identical. Also, as we will see in a later section, systematic sampling is slightly more accurate in some instances than simple random sampling.

Systematic sampling involves one danger: The arrangement of elements in the list can make a systematic sample unwise. This danger is usually referred to by the term **periodicity**. If the list of elements is arranged in a cyclical pattern that coincides with the sampling interval, a grossly biased sample might be drawn. Two examples should suffice.

In one study of soldiers during World War II, the researchers selected every tenth soldier from unit rosters. The rosters, however, were arranged in a table of organization listing sergeants first and then corporals and privates squad by squad, and each squad had ten members. As a result, every tenth person on the roster was a squad sergeant, and the systematic sample selected contained only sergeants. It could, of course, have been the case that no sergeants were selected.

As another example, suppose we want to select a sample of apartments in an apartment building. If the sample were drawn from a list of apartments arranged in numerical order (for example, 101, 102, 103, 104, 201, 202, and so on), there would be a danger of the sampling interval coinciding with the number of apartments on a floor or some multiple thereof. In such a case, the samples might include only northwest-corner apartments or only apartments near the elevator. If these types of apartments had some other particular characteristic in common (for example, higher rent), the sample would be biased. The same danger would appear in a systematic sample of houses in a subdivision arranged with the same number of houses on a block.

In considering a systematic sample from a list, then, you should carefully examine the nature of the list. If the elements are arranged in any particular order, you should ascertain whether that order will bias the sample to be selected and should take steps to counteract any possible bias (for example, take a simple random sample from cyclical portions).

In summary, systematic sampling is usually superior to simple random sampling, in convenience if nothing else. Any problems that exist in the ordering of elements in the sampling frame can usually be remedied quite easily.

Stratified Sampling

In the two preceding sections we discussed two alternative methods of sample selection from a list. **Stratified sampling** is not an alternative to these methods but rather represents a possible modification in their use.

Simple random sampling and systematic sampling are important in that they ensure a degree of representativeness and permit an estimate of error. Stratified sampling is a method for obtaining a greater degree of representativeness and thus decreasing the probable sampling error. In order to understand why this is the case, we return briefly to the basic theory of sampling distribution.

Recall that sampling error is reduced by two factors in the sample design. First, a large sample produces a smaller sampling error than does a small sample. Second, a homogeneous population produces samples with smaller sampling errors than does a heterogeneous population. If 99 percent of the population agree with a certain statement, it is extremely unlikely that any probability sample will greatly misrepresent the extent of agreement. If, on the other hand, the population is split fifty-fifty on the statement, then the sampling error will be much greater.

Stratified sampling is based on this second factor in sampling theory. Rather than selecting your sample from the total population at large, you ensure that appropriate numbers of elements are drawn from homogeneous subsets of that population. In a study of university students, for example, you might first organize your population by college class and draw appropriate numbers of freshmen, sophomores, juniors, and seniors. In a nonstratified sample, representation by class would be subject to the same sampling error as other variables. In a sample stratified by class, the sampling error on this variable is reduced to zero.

You might want to utilize an even more complex stratification method. In addition to stratifying by class, you might also stratify by sex, by grade point average, and so forth. In this fashion, you might be able to ensure that your sample would contain the proper numbers of freshman men with a 4.0 average, of freshman women with a 4.0 average, and so forth.

The ultimate function of stratification, then, is to organize the population into homogeneous subsets (with heterogeneity between subsets) and to select the appropriate number of elements from each subset. To the extent that the subsets are homogeneous on the stratification variables, they might also be homogeneous on other variables as well. Since age is related to college class, for example, a sample stratified by class will also be more representative in terms of age. Since occupational aspirations are related to sex, a sample stratified by sex will be more representative in terms of occupational aspirations.

The choice of stratification variables typically depends on what variables are available. Sex can often be determined in a list of names. University lists are typically arranged by class. Lists of faculty members might indicate their departmental affiliation. Governmental agency files might be arranged by geographical region. Voter registration lists are arranged according to precinct.

In selecting stratification variables from among those available, however, you should be concerned primarily with those that are presumably related to variables that you want to represent accurately. Since sex is related to many variables and is often available for stratification, it is often used. Education is related to many variables but is often not available for stratification. Geographical location within a city, state, or nation is related to many things. Within a city, stratification by geographical location usually increases representativeness in areas such as social class and ethnic group. Within a nation, it increases representativeness in a broad range of attitudes as well as in social class and ethnicity.

Methods of stratification in sampling vary. When you are working with a simple list of all elements in the population, two methods are predominant. First, you might group the population elements into discrete groups based on whatever stratification variables are being used. On the basis of the relative proportion of the population represented by a given group, you select— randomly or systematically—a number of elements from that group constituting the same proportion of your desired sample size. For example, if freshman men with a 4.0 average comprise 1 percent of the student population and you want a sample of 1,000 students, you would select 10 students from the group of freshman men with a 4.0 average.

As an alternative method, you might group students as described above and then put those several groups together in a continuous list beginning with all the freshman men with a 4.0 average and ending, say, with all the senior women with a 1.0 average or below. You would then select a systematic sample, with a random start, from the entire list. Given the arrangement of the list, a systematic sample would select proper numbers (within an error range of 1 or 2) from each of the subgroups. (*Note:* A simple random sample drawn from such a composite list would cancel out the stratification.)

The effect of stratification is to ensure the proper representation of the stratification variables in order to enhance representation of other variables related to them. Taken as a whole, then, a stratified sample is likely to be more representative on a number of variables than would be the case for a simple random sample. Although the simple random sample is still regarded as some-what sacred, it should now be clear that you can often do better.

Implicit Stratification in Systematic Sampling

I mentioned earlier that systematic sampling can, under certain conditions, be more accurate than simple random sampling. This is the case whenever the arrangement of the list creates an "implicit stratification." As already noted, if a list of university students is arranged by class, then a systematic sample will provide a stratification by class while a simple random sample will not. Other typical arrangements of elements in lists can have the same effect.

If a list of names comprising the sampling frame for a study is arranged alphabetically, then the list is somewhat stratified by ethnic origins. All the McTavishes are collected together, for example, as are the Lees, Wongs, Yamamuras, Schmidts, Whitehalls, Weinsteins, Gonzaleses, and so forth. To the extent that any one of these groups comprises a substantial subset of the total population, that group will be properly represented in a systematic sample drawn from an alphabetical list.

In a study of students at the University of Hawaii, after stratification by school class, the students were arranged by their student identification num-bers. These numbers, however, were their Social Security numbers. The first three digits of the Social Security number indicate the state in which the number was issued. As a result, within a class, students were arranged by the state in which they had been issued a Social Security number, providing a rough stratification by geographical origins.

You should be aware that an ordered list of elements can be more useful to you than an unordered, randomized list. This point has been stressed here in view of an unfortunate belief that lists should be randomized before systematic sampling takes place. Only if the arrangement presents the problem of *periodicity* discussed earlier should the list be rearranged.

Multistage Cluster Sampling, General

The four preceding sections have dealt with rea-sonably simple procedures for sampling from lists of elements that represent the ideal situation. Unfortunately, however, much interesting social research requires the selection of samples from populations that cannot be easily listed for sampling purposes, such as the population of a city, a state, or a nation; all university students in the United States; and so forth. In such cases, it is necessary to create and execute a more complex sample design. Such a design typically involves the initial sampling of *groups* of elements, known as **clusters**, followed by the selection of elements within each selected cluster. The varieties and procedures of **multistage cluster sampling** are spelled out in some detail in the sampling examples of Chapter 6. Nevertheless, it is appropriate here to outline the method.

Cluster sampling can be used when it is either impossible or impractical to compile an exhaustive list of the elements comprising the target population. All church members in the United States would be an example of such a population. Often, however, the population elements are already grouped into subpopulations, and a list of those subpopulations either exists or can be created practically. For example, church members in the United States belong to discrete churches, and it is possible to discover or create a list of those churches. Following a cluster sample format, then, the list of churches would be sampled in one of the manners discussed in the preceding sections (for example, a stratified systematic sample). Next, you would obtain lists of members from each of the selected churches and sample each of the lists obtained to acquire samples of church members for study.[6]

Another typical situation concerns sampling among population areas such as a city. While no single list of a city's population exists, citizens reside on discrete city blocks or census blocks. It is possible, therefore, to begin by selecting a sample of blocks, then create a list of persons living on each of the selected blocks, and finally, subsample persons on each block.

In a more complex design, you might sample blocks, list the households on each selected block, sample the households, list the persons residing in each household, and finally sample persons within each selected household. This multistage sample design would lead to the ultimate selection of a sample of individuals without requiring the initial listing of all individuals comprising the city's population.

Multistage cluster sampling, then, involves the repetition of two basic steps: listing and sampling. The list of primary sampling units (churches or blocks, for example) is compiled and perhaps stratified for sampling. Then a sample of those units is selected. The selected primary sampling units are then listed and perhaps stratified. The list of secondary sampling units is then sampled, and so forth. The actual methods of listing and sampling are spelled out in considerable detail in the examples in Chapter 6.

Cluster sampling is highly recommended by its efficiency, but that efficiency comes at the price of accuracy. Whereas a simple random sample

---

[6] For example, see Charles Y. Glock, Benjamin B. Ringer, and Earl R. Babbie, *To Comfort and to Challenge* (Berkeley: University of California Press, 1967), app. A.

drawn from a population list is subject to a single sampling error, a two-stage cluster sample is subject to two sampling errors. First, the initial sample of clusters will represent the population of clusters only within a range of sampling error. Second, the sample of elements selected within a given cluster will represent all the elements in that cluster only within a range of sampling error. Thus, for example, you run a certain risk of selecting a sample of disproportionately wealthy city blocks as well as a sample of disproportionately wealthy households within those blocks. The best solution to this problem lies in the number of clusters selected initially and the number of elements selected within each cluster.

Typically, you are restricted to a total sample size; for example, you might be limited to conducting 2,000 interviews in a city. Given this broad limitation, however, you have several options in designing your cluster sample. At the extremes, you might choose one cluster and select 2,000 elements within that cluster, or you might select 2,000 clusters and select one element within each. Of course, neither extreme is advisable, but you are faced with a broad range of choices between these extremes. Fortunately, the logic of sampling distributions provides a general guideline to follow.

Recall that sampling error is reduced by two factors: an increase in the sample size and an increased homogeneity of the elements being sampled. These factors operate at each level of a multistage sample design. A sample of clusters will best represent all clusters if a large number are selected and if all clusters are very much alike. A sample of elements will best represent all elements in a given cluster if a large number of elements are selected from the cluster and if all the elements in the cluster are very much alike.

With a given total sample size, however, if the number of clusters is increased, the number of elements within a cluster must be decreased. In this respect, the representativeness of the clusters is increased at the expense of more poorly representing the elements comprising each cluster, and vice versa. Fortunately, the factor of homogeneity can be used to ease this dilemma.

The elements comprising a given natural cluster within a population are typically more homogeneous than all the elements comprising the total population. The members of a given church are more alike than all church members; likewise, the residents of a given city block are more alike than all the residents of a whole city. As a result, relatively fewer elements might be needed in order to adequately represent a given natural cluster, while a larger number of clusters might be needed to adequately represent the diversity found among the clusters. This fact is most clearly seen in the extreme case of very different clusters comprised of exactly identical elements within each. In such a situation, a large number of clusters would adequately represent the variety among clusters, while only one element within each cluster would be needed to adequately represent all its members. Although this extreme situation never exists in reality, it is closer to the truth in most cases than its opposite—identical clusters comprised of grossly divergent elements.

The general guideline for cluster design, then, is to maximize the number of clusters selected while decreasing the number of elements within each cluster. Note, however, that this scientific guideline must be balanced against an administrative constraint. The efficiency of cluster sampling is based on the ability to minimize the listing of population elements. By initially selecting clusters, you seek to list only the elements comprising the selected clusters, rather than all elements in the entire population. Increasing the number of clusters goes directly against this efficiency factor in cluster sampling. A small number of clusters can be listed more quickly and more cheaply than a large number. (Remember that all the elements in a selected cluster must be listed even if only a few are to be chosen in the sample.)

The final sample design will reflect these two constraints. In effect, you will probably select as many clusters as you can afford. Lest this issue be left too open-ended at this point, however, we present one rule of thumb. Population researchers conventionally aim for the selection of five households per census block. If a total of 2,000 households are to be interviewed, you would aim at 400 blocks with five household interviews on each block. We will return to this rule of thumb in later examples of sample designs.

Before we turn to more detailed procedures available to cluster sampling, it bears repeating that this method almost inevitably involves a loss of accuracy. The manner in which this loss of accuracy appears, however, is somewhat complex. First, as noted earlier, a multistage sample design is subject to sampling error at each of its stages. Since the sample size is necessarily smaller at each stage than the total sample size, the sampling error at each stage will be greater than would be the case for a single-stage random sample of elements. Second, sampling error is estimated on the basis of observed variance among the sample elements. When those elements are drawn from relatively homogeneous clusters, the estimates of sampling error will be too optimistic and must be corrected in light of the cluster sample design. (This problem will be discussed in detail in the later consideration of univariate analysis.)

Multistage Cluster Sampling, Stratification

Thus far, we have discussed cluster sampling as though a simple random sample were selected at each stage of the design. In fact, it is possible to employ stratification techniques as discussed earlier to refine and improve the sample being selected.

Later examples will detail possible methods of stratification; for the present, we should note that the basic options available are essentially the same as those available in single-stage sampling from a list. In selecting a national sample of churches, for example, you might initially stratify your list of churches by denomination, geographical region, size, rural–urban location, and perhaps some measure of social class. United States census information might be used by population researchers to stratify census blocks in terms of ethnic composition, social class, property values, quality of structures, nature of property ownership, and size.

Once the primary sampling units (churches or blocks in our examples) have been grouped according to the relevant available stratification variables, either simple random sampling or systematic sampling techniques could be used to select the sample. You might select a specified number of units from each group or **stratum,** or you might arrange the stratified clusters in a continuous list and systematically sample that list. To the extent that clusters are combined into homogeneous strata, the sampling error at this stage will be reduced. The primary goal of stratification, as before, is homogeneity.

There is no reason why stratification could not take place at each level of sampling. The elements listed within a selected cluster could be stratified prior to the next stage of sampling. Typically, however, this is not done. (Recall the assumption of relative homogeneity within clusters.)

Probability Proportionate to Size
(PPS) Sampling

Thus far I have spoken in a general way about the assignment of sample elements to selected clusters in terms of how many clusters should be selected and how many elements should be within each cluster. This section discusses in greater detail some options available to you.

To ensure the overall selection of a representative sample of elements, you should give each element in the total population an equal chance of selection. The simplest way to accomplish this in a cluster sample is to give each cluster the same chance of selection and to select a given proportion of elements from each selected cluster. Thus, with a population of 100,000 elements grouped in 1,000 clusters (of varying sizes) and a total sample target of 1,000 elements, you might select one-tenth of the clusters (100) with equal probability and subselect one-tenth of the elements in each cluster initially chosen. In this fashion, approximately 1,000 elements would be selected and each element in the population would have the same ($1/10 \times 1/10 = 1/100$) probability of selection. This type of sample selection technique, though the clearest and simplest, is not the most efficient.

Most cluster sampling involves clusters of grossly different sizes (in numbers of elements). The religion researcher finds very large churches and very small ones. The population researcher finds city blocks containing many people and blocks containing very few. Moreover, the small clusters often outnumber the large ones, although the large ones might account for a larger proportion of the total population. Thus, a few very large city blocks might contain a large proportion of a city's population, while the larger number of small blocks actually contains only a small proportion of the population.

The selection of clusters with equal probability, with a fixed proportion of elements taken from the selected clusters, would result in the following situation: (1) A relatively small number of large clusters would be selected in the first stage of sampling. (2) The elements selected to represent all elements in large clusters would be drawn from very few such clusters. In the extreme, all of a city's population residing on ten large city blocks might be represented by people living on only one of those blocks.

An earlier section on cluster sampling discussed the greater efficiency inherent in the selection of many clusters with few elements drawn from each cluster. This principle is put into practice through the method of **probability proportionate to size (PPS) sampling.** This method provides for the selection of more clusters, ensures the representation of elements contained in large clusters, and gives each element in the population an equal chance of selection.

In the first stage of sampling, each cluster is given a chance of selection proportionate to its size (in number of elements). Large clusters have a better chance of selection than small ones. In the second stage of sampling, however, the same *number* of elements is chosen from each selected cluster. The effect of these two procedures is to equalize the ultimate probabilities of selection of all elements, since elements in large clusters stand a poorer chance of selection *within* their cluster than elements in small clusters. For example, a city block containing 100 households will have ten times the chance of selection as a block containing only 10 households. If both blocks are selected, however, and the same number of households is selected from each block, households on the large block will have only one-tenth the chance of selection of households on the small one. The following formula indicates a given element's probability of selection in a PPS sample design.

$$\begin{bmatrix} \text{Element} \\ \text{probability} \end{bmatrix} = \begin{bmatrix} \text{Number of} \\ \text{clusters} \\ \text{selected} \end{bmatrix} \times \begin{bmatrix} \dfrac{\text{Cluster size}}{\text{Population size}} \end{bmatrix} \times \begin{bmatrix} \dfrac{\text{Elements} \\ \text{selected} \\ \text{per cluster}}{\text{Cluster size}} \end{bmatrix}$$

If 100 clusters are selected and 10 elements are selected from each out of a total population of 100,000, the overall probability of selection for each element will be 1,000/100,000, or 1/100. A cluster containing 100 elements has a probability of selection equal to 100 (clusters to be selected) times 100/100,000 (cluster size/population size), or 1/10; each element has a chance of 10/100 (elements per cluster/cluster size), or 1/10, of being selected *within* that cluster; the element's overall chance of selection in this case is 1/10 times 1/10, or 1/100. A cluster containing only 10 elements has a probability of selection of 100 times 10/100,000, or 1/100, but each element's chance of selection within the cluster is 10/10, or 1, making the overall chance of selection equal to 1/100.

Regardless of the number of elements in a cluster, then, each element ultimately has the same probability of selection. This fact can be seen more clearly in the formula if we note that cluster size appears in both the numerator and the denominator and can be cancelled out. The probability of selection then becomes the number of clusters to be chosen times the number of elements to be chosen from each selected cluster, divided by the population size. This is, of course, the sample size divided by the population size.

Two modifications are typically made in this PPS sample design. First, you might feel it necessary for *very* large clusters to be represented in the sample; for example, you might want to ensure that all city blocks (or churches) with more than, say, 1,000 elements are included in the sample. In such a case, you could select all such clusters in the beginning (with a probability of 1.0); the elements in those clusters should then be given a probability equal to the overall sampling ratio. In the previous example, 1/100 of the elements residing on each of the large blocks might be selected.

The second modification concerns small clusters. If a standard number of elements is to be selected from each cluster chosen, clusters containing fewer elements than that standard number present a problem. If the target is ten households from each selected city block, what will you do with blocks containing five households? The usual solution to this problem is to combine small clusters so that each combination contains at least the standard number to be selected. (If the clusters are stratified, combinations should be accomplished within strata. Similarly, small clusters could be attached to larger ones if this procedure will ensure a greater homogeneity within the combined cluster.) The example of an area cluster sample in Chapter 6 demonstrates the necessity of taking this step in order to ensure the consideration of blocks believed to have no households on them.

By way of summary, cluster sampling is a difficult though important sampling method, it may be required whenever it is impossible to compile a list of all the elements comprising the population under study. The preceding discussion of cluster sampling has been regrettably abstract, but the example provided in Chapter 6 provides a clearer picture of the actual steps involved in a complex sample selection.

Disproportionate Sampling and Weighting

Ultimately, a probability sample is representative of a population if all elements in the population have an equal chance of selection in that sample. Thus, in each of the preceding discussions we have noted that the various sampling procedures result in an equal chance of selection—even though the ultimate selection probability is the product of several partial probabilities.

More generally, however, a probability sample is one in which each population element has a *known nonzero* probability of selection, even though different elements might have different probabilities. If controlled probability sampling procedures have been used, any such sample will be representative of the population from which it is drawn if each sample element is assigned a weight equal to the inverse of its probability of selection. Thus, in a case in which all sample elements have had the same chance of selection, each is given the same weight: 1. (This is called a "self-weighting" sample.)

Sometimes, disproportionate sampling and **weighting** are appropriate. For example, you can sample subpopulations disproportionately to ensure sufficient numbers of cases from each subpopulation for analysis.

Let's suppose you are planning a sample survey of 1,000 households in a particular city. Suppose further that the city has a depressed area containing one-fourth of its total population, and you are particularly interested in a detailed analysis of that area. A representative sample from the city would give you about 250 households in the depressed area and 750 in the rest of the city. Two hundred fifty cases might not be a sufficient number of cases for the analysis you plan, however. As one solution, you could select the same number of households (500) from the depressed area as from the remainder of the city. Households in the depressed area, then, would be given a disproportionately better chance of selection than households located elsewhere in the city.

As long as you analyze the two area samples separately or comparatively, you need not worry about the differential sampling. If you intend to combine the two samples to create a composite picture of the entire city, however, you must take the disproportionate sampling into account through a weighting procedure.

Here's a simple, intuitive solution. Because the depressed area comprises one-fourth of the households in the city and the nondepressed area comprises the other three-fourths, the 500 households selected in the nondepressed area represent three times as much of the city as the 500 households selected in the depressed area. The simplest weighting solution, then, would be to count each household selected in the nondepressed area as *three* households. This procedure would make it appear as though you had interviewed 1,500 households in the nondepressed area of the city, which is the number you should have selected and interviewed if you wanted 500 households in the depressed area and could afford a total sample of 2,000.

Sometimes it is possible to actually make copies of the data files for certain cases—the households in the nondepressed area, in this example. Some data analysis programs, however, allow you to specify the weights to be assigned to various cases, and the computer will take account of the weights in calculating statistics in the analysis.

Disproportionate sampling and weighting procedures are sometimes used in situations that involve the errors and approximation often inherent in complex, multistage sampling. This situation will be discussed in some detail in the example of an area cluster sample in Chapter 6, but it is appropriate to mention here some of the conditions under which weighting is often required.

1. In stratified cluster sampling, a given number of clusters might be selected from each stratum, although the sizes of the different strata will vary. Differential weighting can be used to adjust for those variations.

2. A given cluster might be selected in a PPS sample design on the basis of its expected or estimated size, but a field investigation might later indicate that the initial estimate was in error. Thus, the cluster was given a disproportionately high or low chance of selection; weighting can be used to adjust for that error.

3. A sample design might call for the selection of one-tenth of the elements in a cluster, but the cluster might contain fifty-two elements, only five of which are selected for study. Weighting can be used to adjust for the .2 element which logically could not be selected.

4. Ten elements might have been selected for study within a cluster, but two of these could not be studied (for example, they refused to be interviewed). Assuming homogeneity within the cluster, you might assign a weight of 1.25 to each of the studied elements to make up for the two that were not studied.

All these uses of weighting are illustrated in the final sampling example of Chapter 6. With the exception of case 4 above, however, you can derive your own weighting procedure by carefully determining the probability of selection, step by step, for each sample element and assigning a weight equal to the inverse of that probability. Three further aspects of weighting should be discussed before we move on from this topic.

Degrees of Precision in Weighting

In any complex sample design, you face a number of options with regard to weighting in connection with purposively and/or inadvertently disproportionate sampling. You might compute weights for each element to several decimal places, or you might assign rough weights to account for only the grossest instances of disproportionate sampling. In the previous case of the city in which the suburban area was oversampled, it is unlikely that the population of that area comprised exactly one-fourth of the city's population; it might actually have comprised .25001, .2600, or .2816 of the total population. In the first instance, you would most likely choose to apply the rough overall weighting of cases described if no other disproportionate sampling were involved at other stages in the sample design. Perhaps you would do the same in the second and third instances, as well. The precision you seek in weighting should be commensurate with the precision you want in your findings. If your research purposes can tolerate errors of a few percentage points, you will probably not waste your time and effort in weighting exactly. In deciding the degree of precision required, moreover, you should take into account the degree of error to be expected from normal sampling distribution, plus all the various types of nonsampling error.

Ultimately, no firm guideline exists for you to follow in determining the precision to be sought in weighting. As in so many other aspects of survey design, you enjoy considerable latitude. At the same time, however, you should bear your weighting decision in mind when you report your findings. You should not employ only a rough weighting procedure and then suggest that your findings are accurate within a minuscule range of error.

Methods for Weighting

Having outlined the scientific concerns with regard to determining the degree of precision desired in weighting, we should note that the choice will often be made on the basis of methods available for weighting. The three basic methods for weighting are listed here.

1. For the rough weighting of samples drawn from subpopulations, weighted tables can be constructed from the unweighted tables for each of the subsamples. In the earlier example, you could create a raw table of distributions for the suburban sample and for the nonsuburban sample separately, triple the number of cases in each cell of the nonsuburban table, add the cases across the two tables, and compute percentages for the composite table.

2. For more extensive and faster, though still rough, weighting, the data sets for certain cases can be cloned. In the previous example, two additional copies of each nonsuburban household could be made (for a total of three each), the enlarged set of nonsuburban cases could then be combined with the suburban cases, and the enlarged data set could be analyzed as though three times as many nonsuburban households had been studied.

3. Many computer programs for data analysis are designed to assign a precise weight to each case in the original data file. Only this last method is appropriate to refined weighting, since it is impossible to clone fractions of cases with any meaning.

As I mentioned at the outset of this section, scientific concerns are usually subjugated to practical concerns in weighting as in other instances. If the analysis is to be conducted through IBM cards only, for example, weighting must of necessity be approximate rather than precise.

Weighting and Statistical Inference

You should be advised that the weighting procedures described in this section have serious effects on most computations related to statistical inference. If your research purposes require precise statistical inferences (for example, population estimates) made on the basis of carefully weighted data, you should consult a special source on this matter[7] or, better yet, should consult a sampling statistician *before the sample is designed.*

Probability Sampling in Overview

The preceding lengthy and detailed discussion has been devoted to the key sampling method utilized in controlled survey research:

---

[7] For example, Kish, *Survey Sampling.*

probability sampling. In each of the variations examined, we have seen that elements are chosen for study from a population on the basis of random selection with known nonzero probabilities.

Depending on the field situation, probability sampling can be very simple, or it can be extremely difficult, time-consuming, and expensive. Whatever the situation, however, it remains the most effective method for the selection of study elements, for two reasons. First, probability sampling avoids conscious or unconscious **bias** in element selection on the part of the researcher. If all elements in the population have an equal (or unequal and subsequently weighted) chance of selection, there is an excellent chance that the sample so selected will closely represent the population of all elements. Second, probability sampling permits estimates of sampling error. While no probability sample will be perfectly representative in all respects, controlled selection methods permit you to estimate the degree of expected error in that regard.

Having discussed probability sampling at some length, we turn now to a briefer examination of some popular methods of nonprobability sampling.

Nonprobability Sampling

Despite the accepted superiority of probability sampling methods in survey research, nonprobability methods are sometimes used instead, usually for situations in which probability sampling would be prohibitively expensive and/or when precise representativeness is not necessary. The primary methods of nonprobability sampling are described briefly below.

Purposive or Judgmental Sampling

Occasionally it might be appropriate for you to select your sample on the basis of your own knowledge of the population, its elements, and the nature of your research aims. This method of sampling is sometimes called **purposive** or **judgmental sampling.** Especially in the initial design of your questionnaire, you might want to select the widest variety of respondents to test the broad applicability of questions. Though the survey findings would not represent any meaningful population, the test run might effectively uncover any peculiar defects in your research instrument. This situation would be referred to as a pretest, however, rather than a survey proper.

In some instances, you might want to study a small subset of a larger population in which many members of the subset are easily identified but the enumeration of all is nearly impossible. For example, you might want to study the leadership of a student protest movement; many of the leaders are easily visible, but defining and sampling all leaders would not be feasible. In studying all or a sample of the most visible leaders, you could collect data sufficient for your purposes.

In a multistage sample design, you might want to compare liberal and conservative students. Because you would probably not be able to enumerate and sample from all such students, you might decide to sample the memberships of the Young Democrats and the Young Republicans. While such a sample design would not provide a good description of either liberal or conservative students as a whole, it might suffice for general comparative purposes.

Selected precinct sampling for political polls is a somewhat refined judgmental process. On the basis of previous voting results in a given area (city, state, nation), you purposively select a group of voting precincts that, in combination, produce results similar to those of the entire area. Then, in subsequent polls, you select your samples solely from those precincts. The theory is, of course, that the selected precincts provide a cross section of the entire electorate.

Each time there is an election that permits you to evaluate the adequacy of your group of precincts, you would consider revisions, additions, or deletions. Your goal is to update your group of precincts to ensure that it provides a good representation of all precincts. To be done effectively, selected precinct sampling requires considerable political expertise. You should be well versed in the political and social history of the area under consideration so that the selection of precincts is based on an *educated* guess as to persistent representativeness. In addition, this system of sampling requires continuing feedback in order to be effective. You must be in a position to conduct frequent polls and must have periodic electoral validations.

Quota Sampling

As I mentioned earlier, quota sampling begins with a matrix describing the characteristics of the target population. You must know what proportion of the population is male and what proportion is female, for example; for each sex, you must know what proportion falls into various age categories, and so forth. In establishing a national quota sample, you must know what proportion of the national population is, say, urban, Eastern, male, under twenty-five, white, working class, and the like, as well as all the other permutations of such a matrix.

After such a matrix has been created and a relative proportion has been assigned to each cell in the matrix, data are collected from persons having all the characteristics of a given cell. All the persons in the given cell are then assigned a weight appropriate to their portion of the total population. When all the sample elements are so weighted, the overall data should provide a reasonable representation of the total population.

A number of problems are inherent in quota sampling. First, the quota frame (the proportions that different cells represent) must be accurate, and it is often difficult to get up-to-date information for this purpose. The Gallup failure to predict Truman as the presidential victor in 1948 was due partly to this problem.

Second, biases may exist in the selection of sample elements within a given cell, even though its proportion of the population is accurately estimated. An interviewer instructed to interview five persons who meet a given complex set of characteristics might still avoid persons living at the top of seven-story walk-ups or persons having particularly run-down homes and/or vicious dogs.

Some variant on quota sampling is often used in connection with **focus groups** in market research. In this technique, groups of approximately a dozen subjects are typically brought together for focused discussions of products or commercials. It is often appropriate to assemble several groups of different kinds of people or particular mixes of people.

Reliance on Available Subjects

University researchers frequently conduct surveys among the students enrolled in large lecture classes. The ease and inexpense of such a method explains its popularity, but such surveys seldom produce data of any general value. Although it may serve the purpose of a pretest of a question- naire, such a sampling method should not be used for a study purportedly describing students as a whole.

Stopping people at a street corner or some other location is almost never an adequate sampling method, although it is employed all too frequently, as in the case of marketing surveys in supermarket parking lots. Such a survey would be justified only if you wanted to study the characteristics of people passing the sampling point at specified times.

The term "mall intercept interview" is sometimes used in reference to a survey in which—you guessed it—shoppers walking through shopping malls are intercepted and asked to participate in the survey. It is conceivable, of course, that such potential respondents would be appropriate to some specific surveys—a study of shoppers at that mall, for example. In such cases, however, another sampling problem is that most of those approached refuse to partici- pate. In a study conducted in a Chicago mall, however, Hornik and Ellis found that response rates increased when the interviewers gazed at and touched the respondents when asking for their cooperation, especially if the interviewer was female.[8]

I have mentioned this research to indicate that while conventional prob- ability sampling methods are almost always to be preferred in survey research, at times other methods might be preferred or required. The selection of a nonprobability sampling model should not be an excuse for sloppiness, how- ever. Survey researchers need to find ways of procuring a sample that will represent the population they are interested in learning about.

---

[8] Jacob Hornik and Shmuel Ellis, "Strategies to Secure Compliance for a Mall Intercept Interview," *Public Opinion Quarterly,* vol. 52 (Winter 1988), pp. 539–551.

Nonsurvey Uses of Sampling Methods

In the preceding discussion of sampling logic and methods, we have focused, appropriately, on survey research—selecting elements from a population for interviewing or self-administered questionnaires. The basic logic of sampling, however, gives the preceding discussion a more generalized value for the social researcher who uses other data collection methods. We will close this chapter with some brief comments on the nonsurvey uses of sampling methods.

Content Analysis

A content analyst codifies and analyzes documents for the purpose of making descriptive or explanatory assertions about the literature comprised of the documents, the author(s) of the documents, and/or the social milieu of which the documents are a part. Such an analyst might analyze a nation's newspapers, a novelist's works of fiction, the language of legislative bills, and so forth.

Often the volume of documents to be analyzed is too great for complete coverage. In such a case, the sampling techniques discussed in this chapter could easily be adapted to the situation. The sampling units might be individual words, sentences, paragraphs, articles, books, and so forth. The sampling units could be stratified in any appropriate manner, and random, systematic, or even cluster samples could be selected.

Laboratory Experiments

Subjects for laboratory experiments are often selected from among volunteers responding to an advertisement. Sometimes the experimental design will call for matching of subjects in the experimental and control groups. In some cases, quotas will be set for different types of subjects. Whenever the number of potential subjects greatly exceeds the number required for the experiment, standard sampling techniques might be used. Stratification methods could be used as an aid in filling quota requirements.

Participant Observation

Unlike survey researchers, participant observers typically attempt to immerse themselves in the totality of the phenomenon under study. They might attempt to observe all major events taking place, speak to as many participants as possible, and so forth. Obviously, however, no one can observe everything; some selectivity is inevitable. To the extent that such selectivity is uncontrolled, you run the risk of amassing a biased set of observations, just as the inept survey researcher might select a biased sample of respondents.

While I do not mean to suggest that all participant observers everywhere in every research setting should attempt to establish rigorous sampling methods, I am convinced that a participant observer well versed in the logic, and perhaps some of the methods, of survey sampling will be more likely to obtain a representative set of observations. In assessing the mood of students assembled for a protest demonstration, a participant observer should be careful not to speak only with the demonstration leaders, should consider the possible difference between students seemingly attending alone and those attending with friends, should perhaps speak with students at different locations in the gathering, and should be careful to speak with early arrivers and late comers. While the observer will probably not be able to stratify the sample of respondents in a rigorous fashion, he or she might be more sensitive to noting any differences related to such variables and would refine the observations accordingly.

Summary

This rather lengthy chapter has attempted to familiarize you with the major considerations in the logic of survey sampling and with the most common sampling techniques. Clearly, this discussion will not be sufficient to equip you for every field condition that you will face, but I hope that an understanding of the logic of sampling will make it possible for you to arrive at reasonable and sound judgments on your own.

The following chapter describes four survey samples with which I am intimately familiar. By presenting the specific details of these rather different sample designs and the many decisions that went into their execution, I hope to give you some concrete experiences in sampling. Again, these examples cannot exhaust the field situations that you might later face, but the particular sample designs have been chosen to represent the most common sampling situations found in survey research.

Additional Readings

Earl Babbie, *The Practice of Social Research* (Belmont, CA: Wadsworth, 1989), ch. 7.

William G. Cochran, *Sampling Techniques* (New York: John Wiley & Sons, 1963).

Martin R. Frankel and Lester R. Frankel, "Fifty Years of Survey Sampling in the United States," *Public Opinion Quarterly,* vol. 51 (Winter 1987), pp. S127–S138.

M. H. Hansen, W. N. Hurwitz, and W. G. Madow, *Sample Survey Methods and Theory,* 2 vols. (New York: John Wiley & Sons, 1953).

Leslie Kish, *Survey Sampling* (New York: John Wiley & Sons, 1965).

Richard L. Scheaffer, William Mendenhall, and Lyman Ott, *Elementary Survey Sampling* (North Scituate, MA: Duxbury Press, 1979).

6

Examples of

Sample Designs

Chapter 5 presented the basic logic of survey sampling and outlined some procedural options available to you. This chapter presents four case studies of sample designs that represent different sampling situations and designs.

The first example is a stratified systematic sample of students attending the University of Hawaii. The second example is a cluster sample of medical school faculty members across the nation, with the primary sampling units selected with equal probability. The third example concerns a cluster sample of Episcopal churchwomen in northern California, using a PPS (probability-proportionate-to-size) design for primary sampling unit selection. The final example is a complex area sample designed for a household survey in Oakland, California.

Sampling University Students

The purpose of this study was to survey, with a self-administered instrument, a representative cross section of students attending the main campus of the University of Hawaii. The following sections describe the steps and decisions involved in selecting that sample.

Survey Population and Sampling Frame

The obvious sampling frame available for use in this sample selection was the magnetic registration tape maintained by the university administration. The tape contained students' names, local and permanent addresses, Social Security numbers, and a variety of other information such as field of study, class, age, sex, and so forth.

The registration tape, however, contained files on all persons who could, by any conceivable definition, be called students. Because many of these students seemed inappropriate to the purposes of the study, it was necessary to define the *survey population* in a somewhat more restricted fashion. The final definition included those 15,225 day-program degree candidates registered on the Manoa campus of the university, including all colleges and departments, both undergraduate and graduate students, and both American and foreign students. The computer program used for sampling limited consideration to students fitting this definition.

Stratification

The sampling program also permitted the stratification of students prior to sample selection. In this instance, it was decided that stratification by college class would be sufficient, although the students could have been further stratified within class by sex, college, major, and so forth.

Sample Selection

Once the students had been arranged by class (by the sampling program), a systematic sample was selected across the entire rearranged list. The sample size for the study was initially set at 1,100. To achieve this sample, the sampling program was set to employ a 1/14 sampling fraction. The program, therefore, generated a random number between 1 and 14; the student having that number and every fourteenth student thereafter were selected in the sample.

After the sample had been selected in this fashion, the computer was instructed to print each student's name and mailing address on six self-adhesive mailing labels. These labels were then simply transferred to envelopes to be used in mailing the questionnaires.

Sample Modification

The preceding description details the initial design of the sample for the study of university students. Prior to the mailing of questionnaires, it was discovered that unexpected expenses in the production of the questionnaires made it impossible to cover the costs of mailing to some 1,100 students. As a result, one-third of the mailing labels were systematically selected (with a random start) for exclusion from the sample. The final sample for the study was thereby reduced to about 770.

This modification to the sample is mentioned here to illustrate the frequent necessity of changing aspects of the study plan in midstream. Because a systematic sample of students was omitted from the initial systematic sample, the resulting 770 students could still be considered as reasonably representing the survey population. The reduction in sample size did, of course, increase the range of sampling error.

Sampling Medical School Faculty

This section reports the sample design employed to select a sample of medical school faculty members for a national survey studying the effects of scientific orientations on humane patient care. The study design called for a national sample of medical school faculty members in the departments of medicine and pediatrics.

Under ideal conditions, a single list of all faculty members in the two departments would have been obtained or constructed, and a sample would have been selected from that list. Unfortunately, no such list appeared to exist, so the decision was made to select a two-stage cluster sample. In the first stage, a sample of medical schools would be selected; then faculty members would be selected from each of those schools.

The sample design was hampered by the unavailability of data from the very beginning. The study design called for an examination of both full-time and part-time faculty members. Although there were approximately 3,700 full-time faculty in the two departments nationally at the time of the study, no good data were available concerning the numbers of part-time faculty. An analysis of existing data, however, suggested that the total number of both full-time and part-time faculty was around 12,000. For the purposes of the study, it was decided that a sample of 2,000 would be sufficient (an overall sampling fraction of 1/6).

The Selection of Medical Schools

At the time of the study, eighty-four four-year medical schools belonged to the Association of American Medical Colleges. These schools comprised the survey population of schools. The schools were arranged into geographical strata and were then arranged by size (number of students) within strata.

The stratified list of schools was numbered from 1 to 84, and a random number was selected between 1 and 6 (the sampling interval). The school having the number so selected and every sixth school thereafter were selected at the first stage of sampling. Letters were then sent to the deans of the selected schools explaining the purpose of the survey and asking their assistance in obtaining a list of the faculty members in their departments of medicine and pediatrics.

Fourteen medical schools were initially selected. Not all deans were willing to cooperate with the study, however. As refusals were received, an alternative school for each was selected from the list in the following manner: A school adjacent to the school sending the refusal was chosen through the toss of a coin.

Faculty Member Selection

As soon as medical school deans agreed to cooperate with the survey, lists were compiled of all the faculty members in the departments of medicine and pediatrics at each school. All such faculty were included in the final sample and were mailed survey questionnaires.

It should be noted that this sample design was not the best one that might have been employed. The entire sample of faculty members was selected from relatively few schools. A better design would have selected more schools, with fewer faculty members selected from each. For example, one-third of the schools might have been selected and half the faculty at each studied.

The actual sample design was prompted by administrative rather than scientific concerns. A pilot study in the project had shown the difficulty of gaining approval and cooperation from deans. Even when deans agreed to cooperate with the study, they were often rather slow in providing lists of faculty members. The main bottleneck in sampling was at this point. Increasing the number of schools would have directly increased the time and problems involved in the overall sample selection. For this reason alone, the decision was made to take one-sixth of the schools and all the appropriate faculty at each school selected.

Sampling Episcopal Churchwomen

The purpose of this study was to examine the attitudes of women members of churches in the California diocese of the Episcopal Church. A representative sample of all churchwomen in the diocese was desired. As you will by now expect, no single list of such women was available, so a multistage sample design was created. In the initial stage of sampling, churches would be selected; and then women would be selected from each selected church. Unlike the medical school sample, the church sample was selected with *probability proportionate to size* (PPS).

Selecting the Churches

The diocese in question publishes an annual report that contains a listing of the 100 or so churches comprising it, with their respective sizes in terms of membership. This listing constituted the sampling frame for the first stage of sampling.

A total of approximately 500 respondents was desired for the study, so a decision was made to select twenty-five churches with probability

Table 6-1 *Form Used in Listing of Churches*

Church	Membership	Cumulative Membership
Church A	3,000	3,000
Church B	5,000	8,000
Church C	1,000	9,000

proportionate to size and to take twenty women from each church selected. To accomplish this, the list of churches was arranged geographically, and then a table was created similar to the partial listing shown in Table 6-1.

Beside each church in the table was entered its membership; that figure was used to compute a cumulative total running through the list. The final total came to approximately 200,000. The object at this point was to select a sample of twenty-five churches in such a way that each church would have a chance of selection proportionate to the number of its members. To accomplish this, the cumulative totals were used to create ranges of numbers for each church equal to the number of members in that church. Church A in Table 6-1 was assigned the numbers 1 through 3,000; Church B was assigned 3,001 through 8,000; Church C was assigned 8,001 through 9,000; and so forth.

By selecting twenty-five numbers ranging between 1 and 200,000, it would be possible to select twenty-five churches for the study. The twenty-five numbers were selected in a systematic sample as follows. The sampling interval was set at 8,000 (200,000/25), and a random start was selected between 1 and 8,000. Suppose the random number chosen was 4,538. Since that number fell within the range of numbers assigned to Church B (3,001–8,000), Church B was selected.

Increments of 8,000 (the sampling interval) were then added to the random start, and every church within whose range one of the resultant numbers appeared was selected into the sample of churches. It should be apparent that, in this fashion, each church in the diocese had a chance of selection directly proportionate to its membership size. A church with 4,000 members had twice the chance of selection as a church with 2,000 members and ten times the chance of selection as one with only 400 members.

Selecting the Churchwomen

After the sample of churches had been selected, arrangements were made to acquire lists of the women members of each church. It is worth noting here that in practice the lists varied greatly in their form and content. In a number of cases, lists of all members (men and women) were provided, and it was necessary to sort out the women before sampling the lists. The form of the lists varied from typed lists to 3×5 cards printed from addressograph plates.

As the list arrived from a selected church, a sampling interval for that church was computed on the basis of the number of women members and the number desired (twenty). If a church contained 2,000 women, the sampling interval, therefore, was set at 100. A random number was selected and incremented by the sampling interval to select the sample of women from that church. This procedure was repeated for each church.

Note that this sample design ultimately gives every woman in the diocese an equal chance of selection only if half the members of each church are women (or if a constant proportion of the members are women), due to the fact that churches were given a chance of selection based on their total membership (numbers of women were not available). Given the aims of this particular study, the slight inequities of selection were considered insignificant.

A more sophisticated sample design for the second stage would have resolved this possible problem. Since each church was given a chance of selection based on an assumed number of women (assuming 1,000 women in a church of 2,000, for example), the sampling interval could have been computed on the basis of that assumption rather than on the actual number of women listed. If it was assumed in the first stage of sampling that a church had 1,000 women (out of a membership of 2,000), the sampling interval could have been set at 50 (1,000/20). This interval could then have been used in the selection of respondents regardless of the actual number of women listed for that church. If 1,000 women were in fact listed, then their church had the proper chance of selection and twenty women would have been selected from it. If 1,200 women were listed, the church had too small a chance of selection, but this imbalance would have been remedied through the selection of twenty-four women using the preestablished sampling interval. If only 800 women were listed, on the other hand, only sixteen would have been selected.

Sampling Oakland Households

This final example represents one of the most complex sample designs typical of survey research: an area cluster sample. The purpose of this study, conducted in 1966, was to collect data relevant to the study of poverty in the poorer areas of Oakland, California, using the remainder of the city for purposes of comparison. Because the findings of the survey were to be used, in part, to support requests for federal funding for Oakland, it was essential that the data collected provide an accurate description of the city.

For purposes of the study, the city was divided into seven areas; four were officially designated poverty areas, and the remaining three were traditionally viewed as distinct sections of the city. The total sample size for the city was set at 3,500 households on the basis of computations whose complexity exceeds the scope of this book. It is worth noting, however, that the determination of the sample size began with policy discussion concerning the "chance" city officials were willing to take that the survey would, through normal sampling error, underestimate poverty and unemployment levels sufficiently to disqualify the city in its request for federal funding. If these levels were, in fact, high enough to warrant the award of funds, then a perfectly accurate sample would demonstrate this fact. With a small sample, however, the range of sampling error opened the possibility of underestimation. As the sample size was increased, of course, the chance of this underestimation was reduced. In this manner, then, the ultimate sample size of 3,500 households was established. (*Note:* This is the way sample sizes should be established.)

Because the study called for the comparison of all seven areas of the city, describing all areas with equal degrees of accuracy was important. Therefore, the sample was designed so as to select 500 households from each area, even though the seven areas differed greatly in their total numbers of households. The remainder of this discussion is devoted to the sample selection procedures used in only one of the seven areas, since the procedures were essentially replicated for each area.

General Considerations

As we have already stated, an area cluster sample was designed for the study. At the first stage of sampling, a stratified sample of *census blocks*[1] was to be selected with probability proportionate to size. Blocks selected in this manner were to be listed as follows: Enumerators would physically visit each selected block and prepare lists of all the households found there; then a systematic sample of five households would be selected from each of those blocks.

As we noted in the preceding chapter, five households per block is a common convention in area cluster sampling. Both sampling theory and survey experience suggest that a sample of five households represents the point of diminishing returns in the description of blocks. Five households will provide a reasonably accurate description of a given block, due to the homogeneity typically found among households on a single block. Although six households, of course, would provide a more accurate description of a block, the corresponding decrease in the accuracy of the sample of blocks selected to describe the population of all blocks would offset this advantage. (This assumes that the total survey sample size is fixed.) To illustrate this point in the extreme, you could limit your sample to all households found on a single block; this sample would provide a perfectly accurate description of that block, but the block in itself would not provide a very accurate description of all blocks or, by extension, of all households in the city.

---

[1] For the most part, a census block is the same as a city block—a more or less rectangular area bounded by four streets, with houses lined up along the streets. In some places, however, streets wander about erratically; and strangely shaped chunks are designated as census blocks.

Selecting 500 households from each area of the city meant that 100 blocks would be selected, with five households taken from each block. Moreover, to further improve the representativeness of the household sample, it was decided to organize the blocks into relatively homogeneous strata (groups of blocks) and to select two blocks from each stratum. Thus, the initial task in each area was to create fifty strata; then two blocks would be selected from each stratum and, finally, five households would be selected from each block—making up the total of 500 households in the area.

The First-Stage Sampling Frame

Two kinds of data were required for describing census blocks. Most important, it was necessary to know approximately how many households were on each block so that a PPS block sample could be selected. For purposes of stratification, however, it was also necessary to know certain relevant characteristics about the blocks, such as their racial composition and their socioeconomic levels.

Both kinds of data were available to the research team in the form of the 1960 United States census block statistics. This file indicated the 1960 size of all blocks (in numbers of housing units) and also provided variable descriptions such as (1) percent nonwhite, (2) percent renter-occupied, (3) percent deteriorating, and (4) value of the structures (either rent or valuation). Unfortunately, however, the study was being conducted six years after the census data had been collected, so it was anticipated that many of the housing counts would be out of date.

Before the sampling began, two months were spent examining city planning maps of each area. Housing units constructed or demolished since the 1960 census were noted, and these data were used to correct the estimated number of households per block. At the conclusion of this process, a data record was prepared for each census block in the city showing its current expected size (in households) and the several characteristics compiled in the 1960 census (which could not be updated).

The following sections describe the sample selection procedures used in Poverty Area D, which contained an estimated 9,938 households at the time of the sampling. Because the sample target for the area was 500 households, the overall sampling fraction was rounded to 1/20.

Large Block Selection

Each area of the city had some census blocks that contained very large numbers of households. Some were large blocks in geographical size; others contained several large apartment houses. Whereas the presence of these blocks might hinder the stratification techniques planned for the study, and because these blocks were considered very important blocks for the purposes of the study, an initial decision was made that every block containing 200 or more households would be automatically included in the sample. Each such block would be listed, and one-twentieth (overall sampling ratio for the area) of the households listed would be selected for interviewing.

In Poverty Area D, the large blocks contained a total of 702 households. Therefore, thirty-five households were selected from those blocks, leaving 465 to be selected from the remainder of the area.

Handling Small Blocks and Zero Blocks

Several blocks in Poverty Area D (and elsewhere) contained very few households, and some were estimated to contain none. These blocks presented special problems. First, a block with an expected size of zero would have no chance of selection. If the estimates were incorrect and the block did, in fact, have households on it, then those households would have no chance of selection. Second, since the basic sample design called for the selection of five households from each selected block, those with fewer than five households would also present a problem. Moreover, if the size estimates were incorrect, some blocks believed to contain more than five households might in fact contain fewer than five.

To solve these problems, each zero block and each block with an expected size of fewer than ten households was "attached" to an adjacent block. The number of households expected on the small block, if any, was added to the number for the adjacent block, and the pair of blocks was treated as a single block for purposes of sampling. The pair had a chance of selection proportionate to their combined size; if the pair was selected in the sample of blocks, both blocks were listed and sampled as though they were a single block.

Stratification

Because thirty-five households were to be taken from the very large blocks, 465 needed to be selected from the remainder of the area, with its total of 9,236 estimated households. With five households to be taken from each sample block, ninety-three blocks would be required. Since two blocks would be taken from each stratum, the task at this point was to create forty-seven (rounded from 46.5) strata.

Each stratum (group of blocks) had to have two characteristics. First, the blocks in the stratum should be as similar to one another as possible in terms of racial composition, socioeconomic status (SES), and so forth, to ensure that all types of blocks would be selected in their proper proportion for the total sample. Second, each stratum should contain a total of approximately 200 households. Since ten households were to be selected from each stratum (two blocks, five households from each), a stratum size of 200 would produce an ultimate sampling fraction of 1/20, the fraction established for the whole area. Of course, all the blocks contained specified numbers of households, so it was not possible to create strata containing exactly 200 households. (A later discussion will deal with the statistical correction for such variations.)

In creating homogeneity among the blocks grouped in a given stratum, it was possible to employ the block characteristics provided by the 1960 census. To avoid confusion in the following empirical descriptions, note that the creation of homogeneity was undertaken on a largely ad hoc and arbitrary basis. Although one might be tempted to create a stratification format on theoretical grounds at the outset (for example, all blocks with more than 75 percent nonwhite residents, more than 50 percent renter-occupied units, and less than $150 average monthly rent would be combined into a stratum), such an approach would not necessarily be appropriate to the nature of the particular area being sampled. Instead, each group of blocks (comprising an area of the city or a subset thereof) was examined to determine the variations of their characteristics, and a stratification system was developed to suit those particular characteristics. Different areas of the city, therefore, were stratified differently; moreover, it should be recognized that the particular stratification format for a given area was only one of several, perhaps equally appropriate, possibilities.

As a general rule, however, the available stratification variables were considered in a set order for each area of the city: recent growth, racial composition, percent renter/owner occupied, property value, and deterioration. Whenever this ordering of stratification variables was inappropriate in a given area, the inappropriate variables were simply ignored or were considered at a different point in the stratification.[2]

In each of the seven areas of the city, then, we began stratification on the basis of recent growth. When the updated data records were created for each census block, a notation was made as to the number of housing units constructed subsequent to the 1960 census. Since it seemed reasonable to assume that blocks containing new construction would differ from other blocks, this datum was used as the first stratification variable. All the blocks in Poverty Area D containing any new units, for example, were pulled out for separate stratification. Altogether, these blocks contained an estimated 1,254 households. At 200 households per stratum, the "growth" blocks were to be grouped into six strata.

The second stratification variable used was racial composition. All the growth blocks were arranged in terms of the percentage of nonwhites living in each. Some blocks were found to have 80 percent or more nonwhite residents. Taken together, these blocks contained 247 households. Because this figure was relatively close to the target of 200 households per stratum, these blocks were designated as the first stratum.

The blocks containing between 30 and 79 percent nonwhite residents had a total of 385 households; it was decided to create two strata from these blocks. The next stratification variable was the percentage of households

---

[2] For example, one area of the city of Oakland was, at the time of study, virtually all white. As a result, it made little sense to try stratifying the blocks in that area in terms of racial composition, even though it would have been possible to group together those blocks with *any* nonwhite residents. If, on the other hand, one small group of blocks in the area had, say, 20 percent or more nonwhite residents, these might have been put in a separate stratum.

occupied by renters (as opposed to owners). In the strata being considered here, those blocks with 36 percent or more of the households occupied by renters had a total of 214 households; the remaining 171 households were on blocks with less than 36 percent of households occupied by renters. These two groups of blocks were designated as the second and third strata. (*Note:* There is nothing intrinsically meaningful about the cutting point of 36 percent renters. The goal was to create strata of approximately equal size, and the cutting point that accomplished this goal was used.)

Blocks with less than 30 percent nonwhite residents contained 622 households, calling for three strata. In examining these blocks in terms of all the stratification variables, it was discovered that more than 10 percent of the households in one group of blocks were designated as "deteriorating" by the census. These blocks contained a total of 214 households. Whereas 10 percent deterioration is quite high in the context of all blocks, this group was designated as the fourth stratum.

The remaining blocks contained 408 households, calling for two strata. When renter occupation was examined, it was discovered that those blocks with 86 percent or more renters contained 201 households; these blocks were designated as the fifth stratum. The blocks containing the remaining 207 households were designated as the sixth stratum.

In view of the complexity of this procedure, we present in Table 6-2 a schematic summary of the stratification of the growth blocks in Poverty Area D. The groups of blocks designated as strata are indicated by the notation S1, S2, and so forth. Recall that the goal of stratification was to create relatively homogeneous groups of blocks. We might take time to note the characteristics that all the blocks in stratum 5 have in common: (1) All are located in Poverty

Table 6-2   *Stratification of Growth Blocks in Poverty Area D*

All growth blocks (1,254 hh)	
80% or more nonwhite (247 hh)	S1
30–79% nonwhite (385 hh)	
36% or more renters (214 hh)	S2
35% or less renters (171 hh)	S3
29% or less nonwhite (622 hh)	
10% or more deteriorating (214 hh)	S4
9% or less deteriorating (408 hh)	
86% or more renters (201 hh)	S5
85% or less renters (207 hh)	S6

NOTE: hh = household. I have used this more familiar term rather than the technically correct term *dwelling unit*. A dwelling unit is a room or set of rooms intended for the residential use of a person or a family; a household is the person or group of people residing in a dwelling unit.

Area D of Oakland; (2) all have fewer than 200 households; (3) all have experienced a growth in households since 1960; (4) all have less than 30 percent nonwhite residents; (5) all have less than 10 percent of their structures deteriorating; and (6) all have more than 85 percent renters. The extreme homogeneity of this group of blocks should be apparent.

You would do well to recognize that the preceding discussion is limited to the creation of six strata in Poverty Area D. Forty-one more strata were created in this fashion in the same area, and the whole process was replicated six more times for the other areas of the city.

Block Selection within Strata

Two blocks were to be selected from each stratum, with probabilities proportionate to their sizes. We will use stratum 1 from the previous example, which contained 247 households, to illustrate the procedure used in block selection.

To begin, the order of the blocks in stratum 1 was randomized. Then Table 6-3 was created from the household estimates for each block. The first two columns in the table simply identify the blocks by census tract number and census block number (within the tract). The third column presents the estimated block sizes; the fourth column gives the cumulative total across the six blocks. The final column in the table presents the range of numbers assigned to each block on the basis of its size.

Since two blocks were to be selected, the cumulative total for the stratum (247) was divided by 2 (giving 123.5). A random number between 1 and 123.5 (35) was then selected. Block 27/5, with a range of 13–42, contained this random number, so it was selected into the sample of blocks. The random number was then added to 123.5. Block 27/14, with a range of 149–196, contained this new number (158.5), so it was the second block selected into the sample.

Table 6-3   Block Selection Procedure

Census Tract	Census Block	Number of Households	Cumulative Total	Cumulative Range	
27	18	12	12	1–12	
27	5	30	42	13–42	selected
27	23	26	68	43–68	
28	4	80	148	69–148	
27	14	48	196	149–196	selected
28	2	51	247	197–247	

NOTE: Total hh/2 = 123.5; random number between 1 and 123.5 = 35; random number + (total hh/2) = 158.5.

For this particular study, a computer program was designed to carry out most of the steps described earlier. The blocks were grouped by strata. The computer read the data records in a given stratum, computed and printed the cumulative totals for that stratum, divided the total by 2 and printed that number, and finally generated and printed a random number between 1 and the half-total. It then determined which blocks were to be included in the sample.

Household Selection within Blocks

The procedure described above resulted in the selection of approximately 700 census blocks throughout the city of Oakland. Five households (usually) were to be interviewed on each block. To accomplish this goal, maps were prepared in order to clearly identify each of the selected blocks, and enumerators were sent to prepare a list of all the households on each block. The lists, prepared through the use of standard forms, looked something like the example presented in Table 6-4.

To prepare the listing, the enumerator went to a designated corner of the block and proceeded to walk around it in a circle until he or she returned to the starting point, entering each household on the form as it appeared. Each household, whether it was a single-family house, half a duplex, or an apartment, was entered separately and as it was entered on the list was assigned a number in continuous serial order.

The listing process provided a new estimate of the number of households on each block. (Because enumerators could make mistakes in listing, this number should still be regarded as an *estimate*.) However, recall that each

Table 6-4   *Sample List Sheet*

Hh Number	Street Name	Street Number	Apartment Number or Other Identification
01	Walnut St.	2301	
02	Walnut St.	2303A	Duplex
03	Walnut St.	2303B	Duplex
04	Tenth Ave.	(102?)	No number; brown house with hedge
05	Tenth Ave.	104	Apt. 101
06	Tenth Ave.	104	Apt. 102
07	Tenth Ave.	104	Apt. 103
.			
.			
.			
47	Ninth Ave.	103	
48	Ninth Ave.	101	

block was selected on the basis of an earlier estimate of its size. To take into account discrepancies between these two estimates, we employed a technique mentioned earlier in our discussion of sampling Episcopal churchwomen.

The sampling interval to be used in selecting households from a given block was computed on the basis of the earlier estimate of its size and the five households intended to be selected. If the block size had been estimated at fifty households, then the sampling interval was set at ten. This interval was used in household selection even if the number of households actually listed on the block was larger or smaller. If the block contained sixty households instead of the estimated fifty, six households were selected in the sample; if it contained only forty, then four were selected. (A more precise correction was also employed, which we will discuss shortly.)

The selection of households was accomplished by selecting a random number between 1 and the sampling interval. That random number was then incremented by the sampling interval, and the households listed beside the resulting numbers in the list sheet were designated for interviewing.

When the initial estimates of block size were greatly inaccurate and when the procedures just outlined would produce only one or two sample households (or would produce more than ten), a different procedure was used. In such cases, an arbitrary number of households (no more than ten) was selected, and a note was made to separately weight those interviews during the analysis.

This completes our discussion of the set of procedures used in the selection of some 3,500 households throughout the city of Oakland, California, in 1966. Including the updating of census block sizes and the listing of sample blocks, the whole process took about five months and required a staff of approximately twenty people at its peak.

Weighting the Sample Households

In the simple sample design, each element in the population has the same probability of selection. As a result, the aggregated sample can be taken as representative of the population from which it was selected. If 2,000 respondents were selected from a population of 2 million, then each respondent would be taken to "represent" an additional 999 people who were not selected. To estimate the *number* of people in the population who have a given characteristic, we would multiply the number of people having that characteristic in our sample times 1,000. This *weight* is the inverse of respondents' probabilities of being selected into the sample. When all respondents have the same probability of selection, weighting is irrelevant except with respect to estimating numbers in the population.

When respondents have different probabilities of selection, weighting becomes more important and is relevant even in the computation of percentages. The Oakland study provides an example of the need for, and methods of, weighting sample elements in a complex sample design. No matter how complex the sample design, however, the basic principle still holds: A respondent's weight is the inverse of his or her probability of selection into the sample.

In computing a given respondent's overall probability of selection, we must recall that if several stages of sampling are employed with separate probabilities of selection at each stage, these probabilities must be multiplied to determine the overall probability. If respondents belong to a group (church, block) that has a one-tenth chance of selection and they have a one-tenth chance of selection *within* that group, then their overall probability of selection is 1/100.

In the computation of household weights for the Oakland sample, we must take into account two separate probabilities of selection: the probability of a block being selected and the probability of a household being selected within that block. These probabilities would be computed as follows.

Block Probability. Each block had a probability of selection equal to its estimated size (EBS) divided by the size of the stratum (SS) times 2 (two blocks selected per stratum). We will srite this equation as 2EBS/SS. Note that this formula takes account of both the PPS sampling and the variation in stratasizes. Block 27/5 in our earlier example had an estimated size of thirty households in a stratum containing 247; its probability of selection, therefore, was $(30 \times 2)/247 = .2429$.

Household Probability. Within a given block, each selected household had a probability of selection equal to the number selected on the block (n) divided by the actual number (ABS) listed tor the whole block, written as n/ABS. If Block 27/5 was found during listing to have thirty-four households (instead of thirty) and five of those households were selected, each would have a probability of selection equal to $5/34 = .1471$.

Overall Probability. Multiplying the separate formulas to obtain overall probability, we have the following: (2EBS)(n)/(SS)(ABS). For the example used above, the overall probability of selection is $(2 \times 30 \times 5)/(247 \times 34) = .0357$, or about 1/28. Note that this probability is less than the target sampling fraction of 1/20 for Poverty Area D. The reason for this discrepancy is that stratum 1 had 247 households instead of the target of 200 and Block 27/5 had thirty-four households instead of the thirty estimated. As a result, each of the five households selected on Block 27/5 had a smaller probability of selection than had been intended initially.

Note, however, that if the estimated block size and actual size were identical, the probability formula would be reduced from (2EBS)(n)/(SS)(ABS) to $2n$/(SS), or twice the number of households selected divided by the number of households in the stratum. If the stratum size (SS) were 200, the target sampling fraction of 1/20 would be achieved, since five households would have been selected on the block.

Weighting the Households. All the differences in probabilities of selection were taken care of in the assignment of weights equal to the inverse of a household's overall probability of selection. In the first example given earlier, each household interviewed on Block 27/5 would have been assigned a weight of 28. Each of those households would be assumed to represent itself plus another twenty-seven households in the city of Oakland.

Additional Weighting for Nonresponse.   One final comment should be made. Surveys of this sort never succeed in interviewing all the households initially selected in the sample. Some persons in the sample will refuse to be interviewed, and others will be unavailable.

In this particular study, additional weight was assigned to households to take account of such nonresponse. It was assumed that households that could not be interviewed were more like those on the same block that had been interviewed than like any other possible estimate. As a result, each completed interview was assigned a weight equal to the number of households selected on that block divided by the number actually completed. If four out of five interviews were completed, each completed interview received an additional weight of $5/4 = 1.25$. If all selected households were successfully interviewed, of course, the additional weight would be 1 and could be ignored.

Summary

The preceding four examples of sample designs have been presented in order to give you a more realistic picture of the sampling situations you are likely to face in practice. Although these examples do not exhaust the range of variation in field conditions, study objectives, and sampling techniques, they illustrate the most typical situations. I hope that these examples have illustrated the basic logic behind survey sampling and will better equip you to improvise wisely when you are faced with a novel problem.

Additional Readings

The following research reports provide reasonably detailed discussions of the sampling methods employed.

Gabriel Almond and Sidney Verba, *The Civic Culture* (Princeton, NJ: Princeton University Press, 1963).

Earl R. Babbie, *Science and Morality in Medicine* (Berkeley: University of California Press, 1970).

Charles Y. Glock, Benjamin B. Ringer, and Earl R. Babbie, *To Comfort and to Challenge* (Berkeley: University of California Press, 1967)

Charles Y. Glock and Rodney Stark, *Christian Beliefs and Anti-Semitism* (New York: Harper & Row, 1966).

Samuel A. Stouffer, *Communism, Conformity, and Civil Liberties* (New York: John Wiley & Sons, 1966).

7

Conceptualization and

Instrument Design

Scientific research has two primary goals: *description* and *explanation*. Researchers measure the empirical distribution of values on variables (description) and the associations between variables for purposes of explaining the distribution of values. Chapter 13 discusses the logical interconnections between description and explanation. The purpose of the present chapter is to lay the necessary groundwork by examining the logic and skills of descriptive measurement.

Whether working from a rigorously deduced theory or from a set of tentative suspicions or curiosities, you are faced at some point with a set of unspecified, abstract concepts that you believe will assist your understanding of the world around you. In survey research, these concepts must be converted into questions in a questionnaire in order to permit the collection of empirical data relevant to analysis.

This chapter begins with a few general notes regarding the logic of conceptualization and operationalization. Next, we will consider the different types of data that can be collected by a survey. Finally, we will turn to some of the techniques available for constructing good questions.

Logic of Conceptualization

You are frequently interested in studying concepts such as social class, alienation, prejudice, intellectual sophistication, and so forth. You may suspect that alienation decreases with rising social class, or you may feel that intellectual sophistication reduces prejudice. Before you engage in your empirical research, however, these concepts might be only general ideas in your mind. You would undoubtedly be hard pressed to define precisely what any one of these concepts means to you.

The concept "social status," for example, is frequently used in social research, yet its ultimate meaning is by no means clear. Different definitions of social class include the following elements: income, occupational prestige, education, wealth, power, traditional family status, and moral valuation. Probably no specific combination of these elements would produce a definition of social status that would fully satisfy any researcher, let alone all researchers.

Such unspecified concepts are often said to have a "richness of meaning" in that they contain a variety of elements, thereby summarizing a complex phenomenon. Although the term "social status" evokes different images for different researchers, those images are likely to seem important and meaningful. To permit rigorous, empirical research, however, such general concepts must be specified; that is, they must be reduced to specific, empirical indicators. Inevitably, the operationalization of concepts is unsatisfying both to researchers and to their audiences. Ultimately, concepts rich in meaning must be reduced to oversimplified, inevitably superficial, empirical indicators. Given the significance of this problem, a few additional comments are in order.

I would like to suggest that most concepts of interest to social researchers have *no real meanings*, no ultimate definitions. Social status is a good example to use for illustrating this point. The general concept of social status is surely an old one. Although they might not have verbalized the idea in abstract terms, the earliest humans recognized that some of their numbers had greater social status than others. Some were more powerful, some were held in greater respect, and some were granted greater authority over their fellows. The recognition of such differences has persisted throughout human history. Max Weber drew out many of the implications of such differences. Karl Marx derived a theory of society based on such differences. W. Lloyd Warner and other American sociologists of the 1930s and 1940s noted the same differences among residents of small towns and discovered that members of a community could identify those neighbors with greater or lesser social status.

The existence of status differences among members of a society is clear. Moreover, such differences seem important to an understanding of other aspects of society. In view of this, the term "social status" has been accepted as a summary notation for the phenomenon. But what does the term "social status" really mean? Since it is merely a summary term for denoting a general concept, it has no ultimate meaning. Social status, per se, does not exist except as a convenient notation for a variety of empirical observations. From this standpoint, no researcher can measure social status correctly or incorrectly; we can only make more or less useful measurements.

In this sense, then, scientists never collect data, *they create data*. This idea is essentially the same sentiment expressed by the philosopher of science Alfred North Whitehead when he wrote the following:

> Nature gets credit which should in turn be reserved for ourselves; the rose for its scent; the nightingale for his song; and the sun for its radiance. The poets are entirely mistaken. They should address their lyrics to themselves, and should turn them into odes of self-congratulation on the excellency of the human mind.

Nature is a dull affair, soundless, colourless; merely the hurrying of material, endlessly, meaninglessly.[1]

When you ask several questions and combine the answers to those questions into an index that you call social class, you have *created* a measure of social class; in a real sense, you have *created* a social class ranking and grouping among your respondents. You have not simply tapped a grouping and ranking that already existed in a real sense.

Perhaps the measurement you have created will be useful. It might assist you in understanding the data at hand, or it might be useful in the development of social theories involving social class. Asking whether you have *really* measured social class in any ultimately valid sense, however, is senseless, because the concept exists only in our minds.

The notion that researchers create data rather than collect it is even more basic than the social status illustration suggests. In the earlier days of World War II, Hadley Cantril conducted two national surveys of the American people.[2] The design of the samples and the cross-checking of their characteristics indicated that both surveys provided good estimates of the United States population. In one of the surveys, respondents were asked "Do you think that the United States will succeed in staying out of the war?" A majority (55 percent of those with an opinion) said yes. In the other survey, respondents were asked "Do you think the United States will go into the war before it is over?" A majority (59 percent of those with an opinion) said yes.

These items are often cited as an example of the effects of "biased" questions. Whenever the example is presented to students, however, they typically ask "Which question provided the *correct* answer regarding American expectations about our getting into the war?" This question cannot be answered. We can only conclude that in 1939 there was no such thing as "the American attitude on the likelihood of our getting into the war." No percentage represented the proportion who *really* thought we would enter the war. Thus, you could not *collect* such data; you could only create it by asking questions, and the way the questions were asked had an impact on the answers received.

In a very real sense, then, you can never make *accurate* measurements, only *useful* ones. This supposition should not be taken as a justification for scientific anarchy, however. Rigorous research is still possible; it is simply more difficult than might have been imagined. The remainder of this chapter is devoted to the logic and skills of making useful measurements.

---

[1] Alfred North Whitehead, *Science and the Modern World* (New York: The Macmillan Company, 1925), p. 56.
[2] Reported in Claire Selltiz, et al., *Research Methods in Social Relations* (New York: Holt, Rinehart & Winston, 1959), p. 564.

An Operationalization Framework

As the preceding discussion indicates, concepts are general codifications of experience and observations. We observe people living in different types of residential structures and develop the concept of dwelling unit. We note differences in social standing and develop the concept of social status. We note differences in the degree of religious commitment among people and develop the concept of religiosity. It is imperative to recognize, however, that all such concepts are summary notations for experience and observations.

In science, such concepts often take the form of **variables** bringing together a collection of related **attributes**. Thus, the concept "gender" summarizes two distinct attributes: male and female. "College class" is a variable made up of the attributes freshman, sophomore, junior, senior, and graduate student. "Religious affiliation" might include attributes such as Protestant, Catholic, Jew, Buddhist, and so forth.

Many of the more interesting concepts in social science represent **ordinal variables** made up of values arranged along a **dimension**. For example, the concept of social status implies a ranking of values such as "high status," "moderate status," "low status," and so forth. Religiosity, prejudice, alienation, intellectual sophistication, liberalism, and like concepts imply similarly ranked sets of values.

Operationalization is the process whereby researchers specify empirical observations that can be taken as indicators of the attributes contained within a given concept. If the concept is religiosity, operationalization is the process for specifying empirical measurements that will indicate whether respondents are highly religious, moderately religious, nonreligious, and so forth. Typically, several such indicators will be specified and combined during the analysis of data to provide a composite measure (index or scale) representing the concept.

Since virtually all concepts are ad hoc summaries of experience and observations, they do not have real, ultimate meanings. Thus, you cannot make correct or incorrect measurements but can only determine how well the measurements contribute to your understanding of the empirical data at hand and to the development of theories of social behavior. From this perspective, it is possible to provide a set of guidelines for the operationalization process that might enhance the utility of the process in given research activities.

Assume for the moment that you are interested in studying religiosity. Perhaps you want to learn why some people are more religious than others, or perhaps you want to know the consequences of being more or less religious. Although you no doubt will begin with a general notion of what you mean by religiosity, it is equally likely that you will have no specific indicator in mind. As a beginning of the operationalization process, you should enumerate all the different subdimensions of the variable. In doing this, you should pay attention to previous research on the topic as well as to the commonsense conceptions of religiosity.

Charles Y. Glock has devoted considerable theoretical and empirical attention to the different subdimensions of religiosity; his studies can provide an excellent starting point for your considerations.[3] Glock discusses *ritual involvement* as participation in activities such as weekly church services, communion (for Christians), prayers before meals, and so forth. *Ideological involvement* concerns the acceptance of traditional religious beliefs. *Intellectual involvement* refers to the extent of respondents' knowledge about their religion. *Experiential involvement* refers to the extent of religious experiences that respondents have, such as hearing God speak to them, having religious seizures, and so forth. Finally, Glock examines *consequential involvement* as the extent to which social behavior is motivated by religious concerns and in accord with religious teachings.

Though these five subdimensions of religiosity help organize many possible indicators, other indicators might not fit easily into them. Giving money to the church or to church-related activities might be considered an indication of religiosity, as might participation in church social activities. The list could go on. Your first task, then, is to compile as exhaustive a list as possible of all the different indicators that might be included within your general concept of religiosity.

At the same time that you are deciding all the things that religiosity possibly *is*, you must also consider what it is *not*. You must take special care not to design questionnaire items that tap both religiosity and the variables you will relate to it in the analysis. If, for example, you want to determine the relationship between religiosity and attitudes toward war, items measuring commitment to the Christian doctrine of "peace on Earth" would not be good measures of religiosity. Because the responses would reflect respondents' attitudes toward war—aside from religious concerns—such items would be called "contaminated." Even though responses to the items would surely relate to attitudes regarding participation in war, the findings might contribute nothing useful to understanding the general relationship between religiosity and war attitudes.

You should also pay special attention to the *opposite* of the variable you are attempting to measure. If your goal is to measure religiosity, you should be sensitive to the variable of antireligiosity. While some people might be very religious and others not religious, still others are antireligious. Because antireligiosity can vary in intensity, you must determine the conceptual range of variation on your variable. Perhaps you will want to settle for a measure of religiosity from low to high, grouping the antireligious respondents with those who are simply not positively religious. Or you might want to extend the variable from very religious to very antireligious. In either event, the items

---

[3] Charles Y. Glock and Rodney Stark, *Religion and Society in Tension* (Chicago: Rand McNally, 1965), pp. 18–38.

constructed for measurement should be based on your decision regarding the range. Consider the following example of a questionnaire item:

> Please indicate whether you agree or disagree with the following statement: "Organized religion does more harm than good."

Agreement with the item would indicate an antireligious orientation. Disagreement with the statement, on the other hand, would not necessarily indicate religiosity. Although the responses to this item would surely correlate strongly with measures of religiosity, the item itself does not measure religiosity but only antireligiosity.

The above example illustrates a common problem in the measurement of dimensions. Rather than measuring a variable from low to high, we often make measurements between two polar opposites. For example, we seldom measure degrees of political conservatism, but measure instead variations between liberalism and conservatism. In one study of science and medicine, I attempted to measure differing commitments to scientific perspectives among physicians, but I found myself continually involved in considerations of antiscientific perspectives as well.[4]

This problem has no clear-cut solution. Research interests vary too greatly to permit the elucidation of adequate rules of procedure. The only advice at this point is that you should undertake the time-consuming exercise of listing all possible subdimensions of the variable, noting the dimensions to be excluded from the concept, and specifying the conceptual end points of the dimension that describes the concept. Each questionnaire item considered should then be evaluated against those decisions. You should ask yourself the implications of each possible response: How does the particular response reflect on the basic concept? Only through this process can you generate data relevant to a meaningful later analysis.

Types of Data

The survey research format generates many types of data useful to social research. This section provides a suggestive overview of the different data types.

Although this chapter began with the assertion that data do not exist except through the scientific process of generating them, it nonetheless makes sense to regard some kinds of data as "facts." By "facts" we mean items of information that the respondent believes to represent *truth* and that you generally accept as true. Respondents' demographic characteristics fit into this category. When you ask the respondents to indicate their sex as male or female,

---

[4] Earl Babbie, *Science and Morality in Medicine* (Berkeley: University of California Press, 1970).

they believe their answers to represent indisputable facts and you accept them as such. The same might be said for reports of age, race, region of origin, and so forth.

Sometimes respondents are asked to report information that they accept as statements of truth but that you do not necessarily accept as such. For example, respondents might be asked whether or not God exists. In answering yes or no, they indicate what they believe to be the truth of the matter. You, on the other hand, regard the responses only as descriptions of the respondents, not as answers to the question of whether God exists. Asked to agree or disagree with the statement "Women are more emotional than men," the respondents report what they consider to be the truth, but again you take the answers only as descriptions of the respondents.

In other cases, you ask the respondent to provide information recognized by both of you as subjective attitudes. You might ask "Do you personally think that the President is doing a good job or a bad job?" Both the respondent and you understand that the respondent is giving an opinion and not a fact. The distinction between beliefs and attitudes in this sense often is not clear, but in certain research problems this distinction can be important.

Much interesting social research involves measuring orientations, such as prejudice, that are often not even recognized by the respondents. In a survey, you might ask several questions that, in combination, permit you to describe respondents as being more or less prejudiced. In many cases, the respondents do not understand the latent purpose of the questions and might very well dispute the characterizations of themselves in the analysis.

Survey research does not permit the direct measurement of behavior, although social behavior is frequently the ultimate referent of social research. Survey research does permit the *indirect* measurement of behavior, however, and often in useful ways.

Respondents can be asked to report on their past behavior. What soap did they buy most recently? Did they go to church last week? For whom did they vote in the last election? Such questions are subject to problems of recall and honesty, of course. The respondents might not remember for whom they voted, say, in the 1960 presidential election. They might let their memories play tricks on their candor, especially if the behavior in question is currently regarded as either good or bad. (Postelection political polls often show a higher percentage of the electorate voting for the winner than was the case in the actual voting.) Despite these shortcomings, however, reports of past behavior can often be very useful.

Survey research can also examine prospective behavior, either real or hypothetical. Thus, political polls ask "For whom will you vote?" Sometimes it is useful to create hypothetical situations and ask the respondents how they would behave: "If your political party nominated a woman for President, do you think you would vote for her?"

Measures of prospective behavior are less reliable than measures of past behavior in most instances. In any event, you should bear the shortcomings in mind when you are generating such data. In asking whether respondents would vote for a woman nominated for President by their party, you will probably learn little of value to a prospective female candidate. The descriptive data produced by responses to such a question would probably be quite unreliable. At the same time, however, the data generated by the item might be very useful in describing respondents as more or less antifeminist.

Levels of Measurement

We have seen in the preceding section that the survey researcher is able to measure a variety of social variables. In this section, we will look at such variables from a different perspective, considering four levels of measurement: nominal, ordinal, interval, and ratio.[5]

Nominal Measurements

Nominal measurements merely distinguish the categories that comprise a given variable. Sex, for example, is a nominal variable comprised of the categories male and female. Other examples are religious or political affiliation, region of the country, and college major.

The categories comprising a nominal variable are mutually exclusive, but they bear no other relationship to one another. The remaining levels of measurement discussed below reflect additional relationships among the categories.

Ordinal Measurements

Ordinal measurements reflect a rank order among the categories comprising a variable. Social class is an example of an ordinal variable, comprised perhaps of the categories lower class, middle class, and upper class. Other examples are religiosity, alienation, and anti-Semitism.

Ordinal measurements are used quite often in social scientific research. Although such measurements are often represented by numbers on an index or scale, these numbers have no meaning other than the indication of rank order. Thus, a person scored 5 on an index of alienation would be assumed to be *more alienated* than a person scored 4 on that index, but this ordinal measurement would give no indication of *how much more* alienated the first person was.

---

[5] An excellent treatment of this topic is to be found in James A. Davis, *Elementary Survey Analysis* (Englewood Cliffs, NJ: Prentice-Hall, 1971), pp. 9ff. Most treatments of these levels of measurement use the term *scale* (e.g., *nominal scale*), following the example of S. S. Stevens, "On the Theory of Scales of Measurement," *Science*, vol. 103 (1946), pp. 677–680. Because this book uses the term *scale* in a difference sense (see Chapter 8), I have avoided its use in this context.

Interval Measurements

Interval measurements also utilize numbers to describe conditions, but these numbers have more meaning than do ordinal measurements in that the distances between points have a real meaning. The most common example of an interval measurement is the Fahrenheit temperature scale. The difference between 80 degrees and 90 degrees is the same as the difference between 60 degrees and 70 degrees. (Note that the difference between scores of 4 and 5 on an index of alienation is not necessarily the same as the difference between scores of 3 and 4.)

Ratio Measurements

Ratio measurements have the same characteristics as interval measurements, but they have the additional characteristic of a *true zero*. In comparison with the Fahrenheit temperature scale (an interval measurement), the Kelvin temperature scale (based on absolute zero degrees temperature) is a ratio measurement. Thus, while 40 degrees Fahrenheit is not twice as warm as 20 degrees Fahrenheit, 200 kelvins is twice as warm as 100 kelvins.

In the context of social scientific research, age is an example of a ratio measurement. A twenty-year-old person is twice as old as someone ten years of age. Height, weight, and length of residence in a given city are other examples of ratio measurement.

Implications of Levels of Measurement

In the discussion of survey analysis in Part Three of this book, we will see that different analytical techniques may require specific levels of measurement. If you are analyzing the relationship between two nominal variables, for example, some analytical techniques would be inappropriate for use.

At the same time, it is important to realize that a given variable can be treated differently in terms of the levels of measurement discussed above. Recall, for example, that age is a ratio measurement. Thus, you would be justified in computing a regression equation linking age with height (another ratio measurement). Let's assume, however, that you are studying a sample of high school students ranging in age from, say, fourteen to twenty. Within this limited range, you would probably take advantage of the equal intervals between ages but not the ratio character of the variable. In another type of study, you might choose to group ages into the following ordinal categories: under forty years = *Young*, forty to sixty years = *Middle Aged*, and over sixty = *Old*. Finally, age might be converted to a nominal measurement for certain purposes. Survey respondents might be categorized as members of the post–World War II "baby boom" or not; they might be categorized as being born during the Depression of the 1930s or not. (Another nominal measurement—based on birthdate rather than age—would be assignment in terms of astrological signs.)

In designing your questionnaire, you must consider the type of analysis you will conduct after the collection of your data. You must determine whether and how you can measure a given variable in a way that permits the analysis required. If the analysis requires data in the form of ratio measurements, you must not construct your questionnaire items so as to create only nominal variables.

Guides to Question Construction

In the construction of questionnaire items, you have several options at your disposal. At the same time, the past experience of survey researchers provides a wealth of guidelines that can assist you in the generation of data useful to analysis. This section deals with both topics.

Questions and Statements

Survey research is commonly viewed as involving the asking of questions, yet an examination of a typical survey will probably uncover as many statements as questions. This is not without reason. Often you are interested in determining the extent to which respondents hold a particular attitude or perspective. If you are able to summarize the attitude in a fairly brief statement, you will often present that statement and ask respondents whether they agree or disagree with it. Rensis Likert has formalized this procedure through the creation of the **Likert scale,** a format in which respondents are asked to "strongly agree," "agree," "disagree," "strongly disagree" or perhaps "strongly approve," "approve," and so forth.

Both questions and statements can be used profitably in survey research. Using both in a given questionnaire gives you more flexibility in the design of items and can make the questionnaire more interesting as well.

Open- and Closed-Ended Questions

In the realm of asking questions, you have two options. You can ask **open-ended** questions, in which the respondents are asked to provide their own answers to the question. For example, respondents might be asked "What do you feel is the most important issue facing the United States today?" and be provided with a space to write in their answer (or be asked to report their answer verbally to an interviewer).

In **closed-ended** questions, respondents are asked to select their answer from among a list provided. Closed-ended questions are very popular in survey research because they provide a greater uniformity of responses and are more easily processed. Open-ended responses must be coded prior to data entry, and there is a danger that some respondents will give answers that are essentially irrelevant to your intent. Closed-ended responses, on the other hand, can often be entered directly from the questionnaire, and in some cases, can be marked directly on optical-sensing sheets by the respondents for automatic data entry.

The chief shortcoming of closed-ended questions lies in the structuring of responses. In cases where the relevant answers to a given question are relatively clear, structuring the responses might present no problem. In other cases, however, your structuring of responses might overlook some important ones. In asking about "the most important issues facing the United States," for example, you might provide a checklist of issues but in doing so might overlook certain issues that respondents would have said were important. (Recall our earlier assertion that data are created rather than collected.)

Two guidelines should always be followed in the construction of closed-ended questions. First, the response categories provided should be *exhaustive,* that is, they should include all the possible responses that might be expected. Researchers often support this effort by adding a category labeled "Other (please specify)." In doing this, however, you should realize that respondents will attempt to fit their personal answers into one of the categories provided even though the fit might not be perfect.

Second, the answer categories must be *mutually exclusive,* that is, the respondents should not feel compelled to select more than one. (In some cases, the researcher might want to solicit multiple answers, but such answers will create difficulties in processing.) You can ensure mutually exclusive answers by carefully considering each combination of responses and asking whether a person could reasonably give both. Often you ask the respondent to "select the one *best* answer," but this technique should not be used to make up for a poorly thought out set of responses.

Making Items Clear

It should go without saying that questionnaire items must be clear and unambiguous, but the broad proliferation of unclear and ambiguous questions in surveys makes the point worth stressing here. Often you become so deeply involved in the topic under examination that opinions and perspectives are clear to you but will not be clear to your respondents, many of whom will have given little or no attention to the topic.

On the other hand, you might have only a superficial understanding of the topic and fail to specify the intent of your question sufficiently. The question "What do you think about the proposed antiballistic missile system?" may evoke in the respondent a counter-question: "*Which* proposed antiballistic missile system?" Questionnaire items should be precise so that the respondent knows exactly what question he or she is expected to answer.

Avoiding Double-Barreled Questions

Very frequently, researchers ask respondents for a single answer to a combination of questions. This situation seems to occur most often when the researcher has personally identified with a complex position.

For example, you might ask respondents to agree or disagree with the statement "The United States should abandon its space program and spend the money on domestic programs." Though many people would unequivocally agree with the statement and others would unequivocally disagree, still others would be unable to answer. Some would want to abandon the space program and give the money back to taxpayers. Others would want to continue the space program but also put more money into domestic programs. These latter respondents could neither agree nor disagree without misleading you.

As a general rule, whenever the word *and* appears in a question or questionnaire statement, you should check whether you are asking a double-barreled question.

Ensuring Respondents' Competency to Answer

In asking respondents to provide information, you should continually ask yourself whether they are able to do so reliably. In a study of child rearing, you might ask respondents to report the age at which they first "talked back" to their parents. Quite aside from the problem of defining "talking back to parents," it is doubtful whether most respondents would remember with any degree of accuracy.

As another example, student government leaders occasionally ask their constituents to indicate the manner in which students' fees should be spent. Typically, respondents are asked to indicate the percentage of available funds that should be devoted to a long list of activities. Without a fairly good knowledge of the nature of those activities and the costs involved in them, the respondents cannot provide meaningful answers. ("Administrative costs" will receive little support, even though that category might be essential to the program as a whole.)

One group of researchers examining the driving experience of teenagers insisted on asking an open-ended question concerning the number of miles driven since receiving a driver's license. Though consultants argued that few drivers would be able to estimate such information with any accuracy, the question was asked nonetheless. In response, some teenagers reported driving hundreds of thousands of miles.

Asking Relevant Questions

Similarly, questions asked in a survey should be relevant to most respondents. When attitudes are requested on a topic that few respondents have thought about or really care about, the results are not likely to be very useful. Also, the respondents might express attitudes even though they have never given any thought to the issue, and you run the risk of being misled.

This point is illustrated when researchers ask for responses relating to fictitious persons and issues. In one political poll I conducted on behalf of a candidate, respondents were asked whether they were familiar with each of fifteen political figures in the community. As a methodological exercise, I made up a name: Tom Sakumoto. In response, 9 percent of the respondents said they were familiar with him. Of those respondents, about half reported seeing him on television and reading about him in the newspapers.

When responses are obtained with regard to fictitious issues, you can disregard those responses. But when the issue is real, you might have no way of telling which responses genuinely reflect attitudes and which reflect meaningless answers to an irrelevant question.

The extent to which respondents will make up answers is somewhat controllable by the researcher. A study by Bishop, Tuchfarber, and Oldendick demonstrates this.[6] Respondents were asked one of three questions about whether the "1975 Public Affairs Act" should be repealed. One group of respondents was asked "Do you agree or disagree with this idea, or haven't you thought much about this issue?" Only 3.7 percent reported an opinion as to whether the fictitious law should be repealed. The second group of respondents was asked the same question, except that they were not offered the option of saying they hadn't had a chance to think about it; 24.3 percent expressed an opinion. In the final group, respondents were not given the option of ducking the issue, and those who said they had no opinion were pressured by the interviewers to select one of the answers; 31.5 percent expressed an opinion.

Using Short Items

In the interest of being unambiguous and precise and pointing to the relevance of an issue, you are often led into formulating long and complicated items. This practice should be avoided. That the intent of an item is clear when studied carefully is irrelevant for respondents who do not give it the necessary study. The respondent should be able to read an item quickly, understand its intent, and select or provide an answer without difficulty. In general, you should assume that respondents *will* read items quickly and provide quick answers; therefore, you should provide clear, short items that will not be misinterpreted under such conditions.

Avoiding Negative Items

The appearance of a negation in a questionnaire item paves the way for easy misinterpretation. Asked to agree or disagree with the statement "The United States should not reduce its nuclear weapons arsenals," a sizable portion of the respondents will read over the word *not* and

---

[6] George F. Bishop, Alfred J. Tuchfarber, and Robert W. Oldendick, "Opinions on Fictitious Issues: The Pressure to Answer Survey Questions," *Public Opinion Quarterly*, vol. 50 (Summer 1986), pp. 240–250.

answer on that basis. Thus, some will agree with the statement when they are in favor of reducing nuclear weapons, while others will agree when they oppose it. And you might never know which responses reflect this confusion.

In a study of support for civil liberties, respondents were asked whether they felt "the following kinds of people should be prohibited from teaching in public schools" and were presented with a list including such items as a Communist, a Ku Klux Klansman, and so forth. The response categories "yes" and "no" were given beside each entry. A comparison of the responses to this item with responses to other items reflecting support for civil liberties strongly suggested that many respondents gave the answer *yes* to indicate willingness for such a person to teach, rather than to indicate that such a person should be "prohibited." (A subsequent study in the series that gave as answer categories "permit" and "prohibit" produced much clearer results.)

Avoiding "Biased" Items and Terms

I have frequently repeated the contention of this book that survey data are created rather than simply collected. The manner in which data are sought, therefore, determines the nature of the data received. You must be continually sensitive to the effect of question wording on the results that you will obtain.

Most researchers would recognize the likely effect of a question that began "Don't you agree with the President of the United States in the belief that . . . " and no reputable researcher would use such an item. Unhappily, the biasing effect of items and terms is far subtler than this example suggests.

The mere identification of an attitude or position with a prestigious person or agency can bias responses. The item "Do you agree or disagree with the President's proposal to . . . " would have this effect. "Do you agree or disagree with the recent Supreme Court decision that . . . " would have a similar effect. This is not to say that such wording will necessarily produce consensus or even a majority in support of the position identified with the prestigious person or agency, only that support would likely be increased over what would have been obtained without such identification.

Questionnaire items can be biased negatively as well as positively. "Do you agree or disagree with the position of Adolf Hitler when he stated that . . . " is an example of negative bias. Often the use of terms such as liberal, conservative, communist, atheist, and so on will introduce an unintended bias, though such terms are sometimes essential and appropriate to the intent of your question (e.g., "How would you describe yourself politically: liberal, conservative, middle of the road, or something else?"). Similarly, Tom Smith[7] has demonstrated that public opinion polls report a great deal more support for

---

[7] Tom Smith, "That Which We Call Welfare by Any Other Name Would Smell Sweeter," *Public Opinion Quarterly*, vol. 51 (Spring 1987), pp. 75–83.

providing government assistance to "poor people" than to "people on welfare." The term *welfare* has become very negatively charged within American public opinion.

As in all other examples, you must carefully examine the purpose of your inquiry and construct items that will be most useful to it. You should never be misled into thinking that there are ultimately "right" and "wrong" ways of asking the questions.

Measurement Quality

Before turning to specific measurement techniques, we will consider some general criteria for the quality of the measurements we make. To begin, measurements can be made with varying degrees of *precision,* a term that represents the fineness of distinctions made between attributes comprising a variable. The description of a woman as "forty-three years old" is more precise than "in her forties." Saying "11.75 inches long" is a more precise description than saying "about a foot long."

As a general rule, precise measurements are superior to imprecise ones, as common sense would dictate. There are no conditions under which imprecise measurements would be intrinsically superior to precise ones. Precision is not always necessary or desirable, however. If your research purpose is such that knowing a woman to be in her forties is sufficient, then any additional effort invested in learning her precise age is wasted. The operationalization of concepts, then, must be guided partly by an understanding of the degree of precision required. If your needs are not clear, be more precise rather than less.

Don't confuse precision with *accuracy,* however. Describing someone as "born in Stowe, Vermont" is more precise than "born in New England," but suppose the person in question was actually born in Dover, New Hampshire. The less precise description, in this instance, would be more accurate, that is, a better reflection of the real world.

Precision and accuracy are obviously important qualities in research measurement, and they probably need no further explanation. When social scientists construct and evaluate measurements, however, they pay special attention to two technical considerations: reliability and validity.

Reliability

In the abstract, **reliability** is a matter of whether a particular technique, applied repeatedly to the same object, would yield the same result each time. Suppose, for example, that I asked you to estimate how much I weigh. You look me over carefully and guess that I weigh 165 pounds. (Thank you.) Now let's suppose I ask you to estimate the weights of thirty or forty other people, and while you're engrossed in that I slip back into line wearing a clever disguise. When my turn comes again, you guess 180 pounds.

Gotcha! That little exercise would have demonstrated that having you estimate people's weights was not a very reliable technique.

Suppose, however, that I had loaned you my bathroom scale to use in weighing people. No matter how clever my disguise, you would presumably announce the same weight for me both times, indicating that using the scale provided a more reliable measure of weight than guessing. Reliability, however, does not ensure accuracy any more than precision does. Suppose I've set my bathroom scale to shave five pounds off my weight just to make me feel better. Although you would (reliably) report the same weight for me each time, you would always be wrong.

Suppose we are interested in knowing whether people prefer Burpo or Retcho beer. We could, of course, simply ask them which they prefer—probably the best solution. But suppose we wanted more "objective" data. We might ask our respondents to tell us how many Burpos and Retchos they ever drank. If you think about it, though, this is a pretty unreliable technique. No one is likely to have that kind of information available. Similarly, people are not likely to know how many times they've been to church, how many miles they've driven, and so on.

Asking people questions that are irrelevant to them is also likely to produce unreliable answers. For example, most people probably don't know about China's foreign policy toward Albania and don't even care what it is.

Reliability problems crop up in many other places in survey research. As we will see in Chapter 10, different interviewers sometimes get different answers from respondents as a result of their own attitudes and demeanor. By the same token, different coders might code the same open-ended responses differently. For example, if we were to classify a few hundred specific occupations in terms of some standard coding scheme, say, a set of categories created by the Department of Labor or the Bureau of the Census, you and I would not code all those occupations into the same categories.

How do you create reliable measures? A number of techniques are available for measuring the reliability of questionnaire items, but the methods for *maximizing* reliability are pretty straightforward. Ask people only questions they are likely to know the answers to, ask about things relevant to them, and be clear in what you're asking. The danger is that people *will* give you answers—reliable or not. People will tell you how they feel about China's relationship with Albania even if they haven't the foggiest idea what that relationship is.

Validity

In conventional usage, **validity** refers to the extent to which an empirical measure adequately reflects the *real meaning* of the concept under consideration. First to be considered is something called **face validity**. Particular empirical measures may or may not jibe with our common agreements and our individual mental images associated with a particular concept. You and I might quarrel over the adequacy of measuring people's religiosity by reports of their church attendance, but we could agree that going to church is at least somewhat relevant to religiosity.

Second, researchers have reached concrete agreement on the best way to measure some concepts. The Bureau of the Census, for example, has created operational definitions of concepts such as family, household, and employment status that seem to have a workable validity in most studies that use those concepts.

Edward Carmines and Richard Zeller[8] discuss three additional types of validity: criterion-related validity, content validity, and construct validity. **Criterion-related validity** is sometimes called *predictive validity* and is based on some external criterion. For example, the validity of the college board is shown in its ability to predict the college success of students. The validity of a written driver's test is determined, in this sense, by the relationship between the scores people get on the test and how well they drive. In these examples, college success and driving ability are the *criteria*. In general, behavior can serve as a gauge of criterion validity for the many attitudinal measures made in social research (e.g., Do "prejudiced" people actually discriminate against minorities?), although the relationship between attitudes and behavior is also an important subject of study in its own right.

Content validity refers to the degree to which a measure covers the range of meanings included within the concept. For example, a test of mathematical ability cannot be limited to addition alone but would also need to cover subtraction, multiplication, division, and so forth. As another example, if we say we are measuring prejudice in general, do our measurements reflect prejudice against racial and ethnic groups, religious minorities, women, the elderly, and so on?

Finally, **construct validity** is based on the way a measure relates to other variables within a system of theoretical relationships. Suppose, for example, that you are interested in studying the sources and consequences of "marital satisfaction." As part of your research, you develop a measure of marital satisfaction, and you want to assess its validity. In addition to developing your measure, you will also have developed certain theoretical expectations about the way the variable, marital satisfaction, relates to other variables. To take an obvious example, you might have concluded that satisfied husbands will be less likely than dissatisfied husbands to beat their wives. If your measure of marital satisfaction relates to wife beating in the expected fashion, that result constitutes evidence of your measure's construct validity. If "satisfied" and "dissatisfied" husbands were equally likely to beat their wives, however, that result would challenge the validity of your measure.

---

[8] Edward G. Carmines and Richard A. Zeller, *Reliability and Validity Assessment* (Beverly Hills, CA: Sage, 1979).

Tension Between Reliability and Validity

As a footnote to these discussions, it should be pointed out briefly that a certain tension often exists between the criteria of reliability and validity. We often seem to face a trade-off between the two.

Most of the really interesting concepts that we want to study have many subtle nuances, and it's hard to specify precisely what we mean by them. Researchers sometimes speak of such concepts as having a "richness of meaning." If you doubt this, try to come up with satisfactory definitions and measurements for concepts such as prejudice, satisfaction, alienation, religiosity, and liberalism.

Yet science needs to be specific in order to generate reliable measurements. Very often, then, the specification of reliable operational definitions and measurements seems to rob such concepts of their richness of meaning. Religiosity is more than church attendance; prejudice is more than the expression of a prejudiced attitude on a questionnaire. This situation is a persistent and inevitable dilemma for the social researcher, and you will be effectively forearmed against it by being forewarned. Be prepared for it and deal with it. If you arrive at no clear agreement on how to measure a concept, measure it several different ways. If the concept has several different dimensions, measure them all. Above all, know that the concept does not have any meaning other than what you give it and that the only justification for giving any concept a particular meaning is utility. The goal is to measure concepts in ways that help us understand the world around us.

General Questionnaire Format

The format of a questionnaire can be just as important as the nature and wording of the questions asked. An improperly laid out questionnaire can lead respondents to miss questions, confuse them as to the nature of the data desired, and, in the extreme, result in respondents throwing the questionnaire away. Both general and specific guidelines can be suggested.

As a general rule, the questionnaire should be spread out and uncluttered. You should maximize the "white space" in your instrument. Inexperienced researchers tend to fear that their questionnaires will look too long, and as a result they squeeze several questions on a single line, abbreviate questions, and try to use as few pages as possible. All these efforts are ill-advised and even dangerous. Putting more than one question on a line will result in some respondents skipping the second question. Abbreviating questions will result in misinterpretations. More generally, if respondents find they have quickly completed the first several pages of what initially seemed rather long, they will be less demoralized than if they had spent considerable time on the first page of what seemed to be a short questionnaire. Moreover, they will have made fewer errors and will not have been forced to reread confusing, abbreviated questions, nor will they have been forced to write a long answer in a tiny space.

The desirability of spreading out questions in the questionnaire cannot be overemphasized. Squeezed-together questionnaires are disastrous, whether they are used in mail surveys or in interviews.

Formats for Responses

A variety of methods is available for presenting a series of response categories for the respondent to check in answer to a given question. It has been my experience that *boxes* adequately spaced apart are the best. If the questionnaire is to be set in type or formatted on a good word-processing system, boxes can be generated easily and neatly. It is also possible to provide boxes with a typewriter, however.

If the questionnaire is typed on a typewriter with brackets, excellent boxes can be produced by a left bracket, a space, and a right bracket: [ ]. If brackets are not available, parentheses work reasonably well in the same fashion: ( ). You should not use slashes and underscores, however. First, this technique requires considerably more typing effort; second, the result is not very neat, especially if the response categories must be single-spaced. Figure 7-1 provides a comparison of the different methods.

The worst method of all is to provide open blanks for check marks, because respondents will often enter rather large check marks, making it impossible to determine which response was intended. As a general rule, moreover, double-spacing between response categories is recommended in order to avoid ambiguous check marks.

A very different method might also be considered. Rather than providing boxes to be checked, you might consider entering code numbers beside each response and asking the respondent to *circle* the appropriate number. This method has the added advantage of specifying the number to be entered later in the data-processing stage. If this method is used, however, you should provide clear and prominent instructions to the respondent, as many respondents will be tempted to cross out the appropriate number, thereby making data entry even more difficult. (*Note*: The technique can be used safely in interview surveys, because interviewers can be instructed and tested.)

Contingency Questions

Often in survey research, certain questions will clearly be relevant only to a subset of the respondents. You might ask respon-

Figure 7-1 *Answer Formats*

[ ] Yes	( ) Yes	1. Yes
[ ] No	( ) No	② No
[ ] Don't know	( ) Don't know	3. Don't know

dents if they are familiar with a certain issue and then ask those who are what they think about it. You might be interested in the draft status of young people, realizing that only men will have a draft status. Or you might ask if respondents belong to a certain organization and, if so, whether they have ever held an office in that organization.

The second question in each example is called a **contingency question,** meaning that the second question is *contingent* upon the response to the first. Thus, whether you ask for attitudes toward the issue is contingent upon whether the respondent is familiar with it.

The proper use of contingency questions can facilitate the respondent's task in answering the questionnaire and can also improve the quality of the data produced. The alternative to contingency questions should be avoided for the same reasons. Do not ask the question "If you have ever belonged to the PTA, did you ever hold an office in it?" This latter format will force all respondents to read the question, even though it is irrelevant to many. Those for whom it is irrelevant might be forced to decide whether to skip the question, write in "not applicable," answer no, or throw the questionnaire away. (The latter choice is popular.)

There are a number of contingency question formats. I think that the best is one in which the contingency questions are indented on the question- naire, set off in boxes, and connected with the base question by arrows from appropriate responses. Figure 7-2 illustrates such a question. Used properly, complex sets of contingency questions can be constructed without confusing the respondent, as illustrated in Figure 7-3.

While the above examples have referred primarily to self-administered questionnaires, the proper presentation of contingency questions is even more important in interview questionnaires. The mail questionnaire respondent can, with some dissatisfaction, reread a question, but if the interviewer becomes confused and asks improper questions, the whole interview is jeopardized.

Figure 7-2

23. Have you ever belonged to the local PTA?

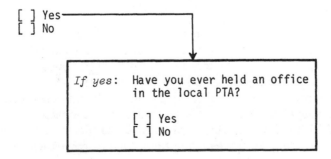

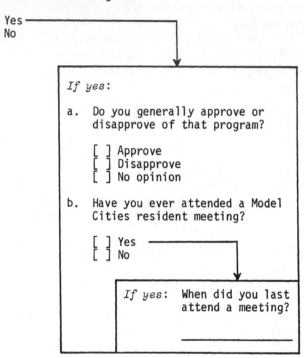

Figure 7-3

14.   Have you ever heard anything about
      the Model Cities Program?

      [ ] Yes
      [ ] No

If yes:

a.   Do you generally approve or
     disapprove of that program?

     [ ] Approve
     [ ] Disapprove
     [ ] No opinion

b.   Have you ever attended a Model
     Cities resident meeting?

     [ ] Yes
     [ ] No

     If yes: When did you last
                 attend a meeting?

     _____

Figure 7-4 shows a portion of an interview questionnaire aimed at determining the person's occupational status. Reading through the question, giving different answers and following the appropriate arrows to the next question, will give you a sense of the researcher's intent as well as enabling you to see how easily the interviewer was able to collect the relevant information from all respondents.

Matrix Questions

Quite often, you will want to ask several questions that have the same set of answer categories. This is typically the case whenever the Likert-type response categories are used. In such cases, it is often possible to construct a **matrix question** as illustrated in Figure 7-5.

This format has a number of advantages. First, it uses space efficiently. Second, respondents will probably find it faster to complete a set of questions presented in this fashion. Third, this format may increase the comparability of responses given to different questions for the respondent as well as the

Figure 7-4

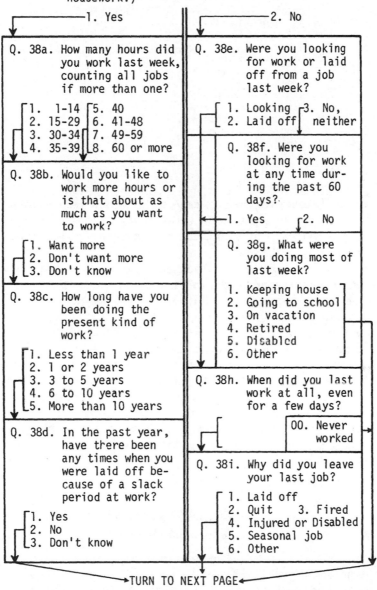

Q. 38. Did you work at any time during the past week?
(Include part-time work, but do not include
housework.)

— 1. Yes   — 2. No

Q. 38a. How many hours did
you work last week,
counting all jobs
if more than one?

1. 1-14   5. 40
2. 15-29   6. 41-48
3. 30-34   7. 49-59
4. 35-39   8. 60 or more

Q. 38b. Would you like to
work more hours or
is that about as
much as you want
to work?

1. Want more
2. Don't want more
3. Don't know

Q. 38c. How long have you
been doing the
present kind of
work?

1. Less than 1 year
2. 1 or 2 years
3. 3 to 5 years
4. 6 to 10 years
5. More than 10 years

Q. 38d. In the past year,
have there been
any times when you
were laid off be-
cause of a slack
period at work?

1. Yes
2. No
3. Don't know

Q. 38e. Were you looking
for work or laid
off from a job
last week?

1. Looking   3. No,
2. Laid off   neither

Q. 38f. Were you
looking for work
at any time dur-
ing the past 60
days?

— 1. Yes   2. No

Q. 38g. What were
you doing most of
last week?

1. Keeping house
2. Going to school
3. On vacation
4. Retired
5. Disabled
6. Other

Q. 38h. When did you last
work at all, even
for a few days?

00. Never
worked

Q. 38i. Why did you leave
your last job?

1. Laid off
2. Quit   3. Fired
4. Injured or Disabled
5. Seasonal job
6. Other

→TURN TO NEXT PAGE←

Figure 7-5 Matrix Question Format

```
17.  Beside each of the statements presented below,
     please indicate whether you Strongly Agree (SA),
     Agree (A), Disagree (D), Strongly Disagree (SD),
     or are Undecided (U).

                          SA    A    D    SD   U
     a. What this country
        needs is more law and
        order. . . . . . . .  [ ]  [ ]  [ ]  [ ]  [ ]
     b. The police should be
        disarmed in America. .  [ ]  [ ]  [ ]  [ ]  [ ]
     c. During riots, looters
        should be shot on
        sight. . . . . . . .  [ ]  [ ]  [ ]  [ ]  [ ]
     etc.
```

researcher. Because respondents can quickly review their answers to earlier items in the set, they might choose between, say, "strongly agree" and "agree" on a given statement by comparing their strength of agreement with their earlier responses in the set.

There are some dangers inherent in using this format, however. Its advantages may encourage you to structure an item so that the responses fit into the matrix format when a different, more idiosyncratic, set of responses might be more appropriate. Also, the matrix question format can foster a *response-set* among some respondents—respondents might develop a pattern of agreeing with all the statements. This would be especially likely if the set of statements begin with several that indicate a particular orientation (for example, a liberal political perspective), with only a few, later statements representing the opposite orientation. Respondents might assume that all the statements represent the same orientation and, reading quickly, misread some of them, thereby giving the wrong answers. You can reduce this problem somewhat by alternating statements representing different orientations and by making all statements short and clear.

Ordering Questions in a Questionnaire

The order in which questions are asked can affect the responses as well as the overall data collection activity. For example, the presence of one question can affect the answers given to subsequent questions. To illustrate, if a number of questions have been asked about the dangers of pollution in the United States and a subsequent question asks respondents to volunteer (open-ended) what they believe to represent dangers to the United

States, "pollution" will receive more citations than would be the case if the initial questions had not been asked.

If respondents are asked to assess their overall religiosity ("How important is your religion to you in general?"), their responses to later questions concerning specific aspects of religiosity will be aimed at achieving consistency with the prior assessment. The converse is true as well. If respondents are first asked specific questions about different aspects of their religiosity, their subsequent overall assessment will reflect their earlier answers.

Some researchers attempt to overcome this effect by "randomizing" the order of questions—usually a futile effort. To begin, a "randomized" set of questions will probably strike respondents as chaotic and worthless. They will have difficulty answering, moreover, since they must continually switch their attention from one topic to another. Finally, even a randomized ordering of questions will affect answers as discussed above, and you will have no control over the effect.

The safest solution is sensitivity to the problem. Although you cannot avoid the effect of question order, you should be in a position to estimate what that effect will be and thus be able to interpret results in a meaningful fashion. If the order of questions seems an especially important issue in a given study, you might construct more than one version of the questionnaire containing different possible orderings of questions. You would then be in a position to determine the effects of the order. At the very least, you should pretest your questionnaire using the different forms.

The desired ordering of questions differs somewhat in self-administered questionnaires and in interviews. In the former, it is usually best to begin the questionnaire with the most interesting set of questions. The potential respondent who glances casually over the first few questions should want to answer them. Perhaps these questions will ask for attitudes that respondents are aching to express. At the same time, however, the initial questions should not be threatening. (It might be a bad idea to begin with questions about sexual behavior or drug use.) Requests for duller demographic data (age, sex, and the like) should generally be placed at the end of the self-administered questionnaire. Placing these questions at the beginning, as many inexperienced researchers are tempted to do, gives the questionnaire the initial appearance of a routine form, and the person receiving it might not be motivated to complete it.

The opposite is generally true for interview surveys. When the potential respondent first opens the door or answers the phone, the interviewer must begin by quickly gaining rapport with the respondent. After presenting a short introduction to the study, the interviewer can best begin by enumerating the members of the household, requesting demographic data about each. Such questions are easily answered and are generally nonthreatening. After the initial rapport has been established, the interviewer can then move into the area of attitudes and more sensitive matters. An interview survey that began with the question "Do you believe in God?" would probably end rather quickly.

Instructions

Every questionnaire, whether it is to be self-administered by the respondent or administered by an interviewer, should contain clear instructions and introductory comments where appropriate.

General Instructions

Every self-administered questionnaire should begin with basic instructions to be followed in completing it. Although many respondents these days are familiar with normal survey techniques, it is useful to begin by instructing the respondents to indicate their answers to certain questions by placing a check mark or an X in the box beside the appropriate answer or by writing in their answers when called for. If many open-ended questions are used, respondents should be given some guide as to whether brief or lengthy answers are expected. If you encourage written-in answers that allow respondents to elaborate on answers given to closed-ended questions, this option should be noted.

Introductions

If a questionnaire is arranged into content subsections—for example, political attitudes, religious attitudes, background data—it is useful to introduce each section with a short statement concerning its content and purpose. The following is an example of a possible introduction: "In this section, we would like to know what people around here consider the most important community problems." Demographic items at the end of a self-administered questionnaire might be introduced as follows: "Finally, we would like to know just a little about you so we can see how different types of people feel about the issues we have been examining."

Short introductions such as these help the respondent make sense out of the questionnaire and make the questionnaire seem less chaotic, especially when it taps a variety of data. They also help put the respondent in the proper frame of mind for answering the questions.

Specific Instructions

Some questions might require special instructions to facilitate proper response, especially if a given question varies from the general instructions pertaining to the whole questionnaire. Some specific cases will illustrate this situation.

Despite the desirability of mutually exclusive answer categories in closed-ended questions, often more than one answer will apply for some respondents. If you want a single answer, you should make this perfectly clear in the question. An example is "From the list below, please check the *primary* reason for your decision to attend college." Often the main question can be followed by a parenthetical note, for example, "Please check the *one* best answer." If, on the other hand, you want the respondent to check as many answers as apply, you should make this clear as well.

When a set of answer categories is to be rank-ordered by the respondent, the instructions should indicate this and a different type of answer format should be used (for example, blanks instead of boxes). The instructions should indicate how many answers are to be ranked (for example, all, first and second, first and last, most important and least important) and what the order of ranking should be (for example, "Place a 1 beside the most important, a 2 beside the next most important, and so forth.").

Rank-ordering a list of items is often difficult for respondents, because they might have to read the list several times. Moreover, you can usually achieve the same results by having respondents *rate* the items (for example, "Very important," "Important," and so on). As Krosnick and Alwin point out,[9] problems occur only when respondents rate most items the same. Of course, as the authors suggest, rating several items the same might indicate how the respondent really feels about those items, and a forced ranking might in fact produce responses that aren't really true. If you find it essential to have respondents rank-order a list of items, however, you should at least make the list as short as possible. As a rule of thumb, don't ask people to rank-order more than five items.

In multiple-part matrix questions, you should give special instructions unless the same format is used throughout the questionnaire. Sometimes respondents will be expected to check one answer in each column of the matrix, while at other times they will be expected to check one answer in each row. Whenever the questionnaire contains both types of response, an instruction should be added clarifying which type is expected in each case.

Interviewer Instruction

A confusing self-administered questionnaire can lower the morale of the respondent, but a confusing interview questionnaire can lower the morale of both the respondent and the interviewer and endanger the efficiency of the latter. Providing clear supplementary instructions where appropriate to interviewers, therefore, is particularly important.

In an interview questionnaire, different formats must be used for those instructions that the interviewer is to read to the respondents and those that are not to be read. For example, the latter might always be typed in parentheses or in capital letters. An interview could be destroyed if an interviewer were to read

---

[9] Jon A. Krosnick and Duane F. Alwin, "A Test of the Form-resistant Correlation Hypothesis: Ratings, Rankings, and the Measurement of Values," *Public Opinion Quarterly*, vol. 52 (Winter 1988), pp. 526–538.

aloud something on the order of "If the respondent is nearly illiterate, then . . . ."

It is equally important that an interview questionnaire contain a *verbatim script* for the interviewer to read in interviewing.[10] Under ideal circumstances, an interviewer should be able to conduct an entire interview from initial introduction ("Hello, my name is . . . ") to final remarks ("That completes the interview. We would like to thank you for . . . ") without ad-libbing a single word. All transitional statements throughout the questionnaire should be included ("Now we would like to turn from community problems to national problems . . . ") so that the verbatim script sounds natural and conversational. The same is true for the demographic enumeration of household members. Rather than instructing the interviewer to obtain the age of each family member, you should provide a standardized question for each ("How old was [he/she] on January 1, 1990?"). Chapter 10 stresses the importance of the interviewer's following the questionnaire wording exactly, but you should recognize in advance that this will be possible only if the questionnaire is properly constructed.

Reproducing the Questionnaire

Having constructed a questionnaire that will collect data relevant to your aims and that will be efficiently processed, you need to produce enough copies for the actual data collection. The method of reproducing the questionnaires is important to the overall success of the study; a neatly reproduced instrument will encourage a higher response rate and thereby provide better data.

Several alternative methods are available for reproducing questionnaires, and your choice will depend on funds available, local facilities, and timing. Ditto or mimeograph reproductions are generally cheaper and more readily available, but they are the least professional looking in quality. Photocopying machines (for example, Xerox) vary in speed, quality, and price, but it is possible to produce excellent questionnaires with them, especially if you use a computer and a laser printer for the original.

Photo-offsetting a copy of the questionnaire provides better quality and, beyond a certain number of copies, might even be cheaper. A single photo-offset master will make countless copies, whereas a ditto or mimeograph master must be recut after a few hundred copies have been run off, and photocopying might retain the same unit cost even on long runs.

---

[10] Of course, a script would not be appropriate for an unstructured interview in an exploratory study. In that case, the interviewer should be given an outline of topics to cover and perhaps a set of possible questions.

The best method of reproduction from the standpoint of professional quality is typesetting. This method is also the most expensive, however, and it might not be feasible for some projects. Also, typesetting generally takes longer than other methods. In any event, you should explore the local possibilities, balancing the relative values of time, money, and quality.

At the present time, the optimum reproduction system consists of a microcomputer with a professional word-processing or desktop publishing program and a laser printer for creating the original, used in conjunction with the photo-offsetting of copies.

The questionnaire can be constructed in several different ways. In some cases, it might be appropriate to print the questionnaire on an oversized single sheet of paper to be folded into a quasi-booklet of unfolding panels. If several pages are required for the questionnaire, connecting them with a corner staple might be appropriate. The most professional-looking long questionnaire is a printed booklet held together with a "saddle stitch" (a staple in the spine of the booklet). Again, this method is the most expensive.

One final concern regarding the reproduction of the questionnaire is how many to order. In arriving at this decision, you must consider your sample size, the number of follow-up mailings, if any, in a mail survey, and the possible need for discussion copies, samples for other researchers, copies for inclusion as appendixes in research reports or code books, and so forth. As a rough rule of thumb, you should estimate the number required for data collection and multiply that figure by a factor of 1.5 to 2.0 to determine the number to be ordered. Bear in mind that additional copies produced in the initial run of the questionnaires will be far cheaper than copies produced in a second run.

Summary

This chapter covered a variety of subjects ranging from the theoretical and philosophical to the technical and even mundane. It began with the general issues of conceptualization and operationalization in the design of questionnaire items appropriate to the creation of data relevant to your aims. Some general and specific guidelines were provided with regard to the writing of questions and the presentation of those questions in the research instrument. The chapter concluded with a brief discussion of the technical side of questionnaire production.

The combination of such topics in a single chapter might seem bizarre, but there has been a purpose in this presentation. The ultimate *scientific* value of a survey can depend as much on the manner in which the questionnaire is reproduced as on the imagination that went into the operationalization of concepts. The results of survey research are a product of many steps, ranging from the theoretical to the mundane, and a weakness in any step threatens the whole.

Additional Readings

Edward G. Carmines and Richard A. Zeller, *Reliability and Validity Assessment* (Beverly Hills, CA: Sage, 1979).

Julius Gould and William Kolb, *A Dictionary of the Social Sciences* (New York: Free Press, 1964).

Abraham Kaplan, *The Conduct of Inquiry* (San Francisco: Chandler Publishing Co., 1964).

Delbert Miller, *Handbook of Research Design and Social Measurement* (New York: Longman, 1983).

A. N. Oppenheim, *Questionnaire Design and Attitude Measurement* (New York: Basic Books, 1966).

Stanley Payne, *The Art of Asking Questions* (Princeton, NJ: Princeton University Press, 1965).

Howard Schuman and Stanley Presser, *Questions and Answers in Attitude Surveys: Experiments on Question Form, Wording, and Context* (New York: Academic Press, 1981).

Tom W. Smith, "The Art of Asking Questions," *Public Opinion Quarterly,* vol. 51 (Winter 1987), pp. S95–S108.

8

Index and Scale

Construction

As we noted in the preceding chapters, much, if not most, of social research is aimed at determining the associations between variables. Typically, we wish to state that X is related to (or causes) Y. We also noted, however, that the measurement of variables is often a difficult undertaking. Normally, it is impossible to arrive at a wholly unambiguous and completely acceptable measure of any variable. Nevertheless, researchers do not give up the attempt to create ever better and more useful measures.

This chapter addresses the problem of measurement. Specifically, it discusses the construction of indexes and scales as measures of variables. These cumulative measures are frequently used in social research, for several reasons. First, despite the care taken in constructing questionnaires, it is seldom possible to arrive at a single question that adequately represents a complex variable. Any single item is likely to misrepresent some respondents in the study. In attempting to measure religiosity, frequency of church attendance is probably not a sufficient indicator in and of itself. Some respondents who attend church frequently might be judged irreligious on other grounds, and some who never attend church might be judged religious. Indexes and scales combine several questionnaire items, thereby avoiding the biases inherent in single items.

Second, you might want to employ a rather refined ordinal measure of your variable, for example, arranging respondents into several ordinal categories from very low to very high on the variable. A single item might not present a sufficient number of answer categories to provide this range of variation, but an index or scale formed from several items would.

Finally, indexes and scales are *efficient* devices for data analysis. If a single questionnaire item gives us only a crude assessment of respondents on a given variable, several items might give us a more comprehensive and more accurate assessment. However, it is normally impractical to consider simultaneously all the particular responses provided by a given respondent. Indexes and scales (especially scales) are *data-reduction* devices; a respondent's many responses can be summarized in a single score, while at the same time the specific details of those responses can be very nearly maintained.

Indexes Versus Scales

The terms *index* and *scale* are typically used imprecisely and interchangeably in social research literature. The technical definitions originally associated with these terms have subsequently lost their meanings. Before we consider the distinctions that this book makes between indexes and scales, we will first examine what they have in common.

Both scales and indexes are typically *ordinal* measures of variables. Scales and indexes are constructed in such a way as to rank-order survey respondents (or other units of analysis) in terms of specific variables such as religiosity, alienation, socioeconomic status, prejudice, intellectual sophistication, and so forth. A respondent's score on a scale or index of authoritarianism, for example, gives an indication of that person's relative authoritarianism vis-à-vis other respondents.

As the terms are used in this book, both scales and indexes are *composite* measures of variables, that is, measurements based on responses to more than one questionnaire item. Thus, respondents' scores on an index or scale of religiosity would be determined by the specific responses given to several questionnaire items, each of which provides some indication of their religiosity.

For the purposes of this book, we will distinguish between indexes and scales with regard to the manner in which scores are assigned to respondents. An **index** is constructed through the simple cumulation of scores assigned to specific responses to the individual items comprising the index. A **scale** is constructed through the assignment of scores to *response patterns* among the several items comprising the scale. A scale differs from an index in that it takes advantage of any *intensity structure* that might exist among the individual items. A simple example should clarify this distinction.

Suppose we want to measure Americans' support for the civil liberties of avowed Communists. We might ask in the questionnaire whether a Communist should be allowed to pursue the following occupations: (1) lawyer, (2) physician, (3) librarian, (4) engineer, (5) newspaper reporter, (6) professor. Some respondents will be willing to allow Communists to pursue all the occupations listed; some will be unwilling to permit Communists to pursue any. Others, however, will feel that some occupations are permissible and others are not. Respondents who give a mixed set of responses will presumably be indicating that they consider some of the occupations to be more important than others. The relative priorities of the different occupations will vary from respondent to respondent, however; no absolute ranking is inherent in the occupations themselves.

Given the responses provided by a survey sample, you might construct an *index* of respondents' relative commitments to civil liberties for Communists on the basis of the number of occupations that respondents would hold open to Communists. A respondent who would permit Communists to hold all the occupations clearly supports a greater degree of civil liberty than one who would close all the occupations to Communists. Moreover, you would assume that a respondent who would permit Communists to hold three of the occupations is more supportive of Communists' civil liberties than a respondent who would hold open only one or two of the occupations, regardless of which occupations are involved. Such an index might provide a useful and accurate ordinal measure of civil libertarianism. Part A of Figure 8-1 illustrates the simple logic of index construction.

Suppose for a moment, however, that the occupations used in the preceding example had been the following: (1) ditch digger, (2) high school teacher, and (3) President of the United States. In this situation, there is every reason to believe that the three items listed have an *intensity structure*. Respondents who would permit a Communist to be President would surely let such a person be a high school teacher or a ditch digger. On the other hand, respondents who would permit Communists to dig ditches might or might not permit them to pursue the other two occupations. In all likelihood, knowing the *number* of occupations approved for Communists by a respondent would tell you *which* occupations were approved. In such a situation, a composite measure comprised of the three items would constitute a scale as we have used that term. Scale construction is illustrated in Part B of Figure 8-1.

It should be apparent that scales are generally superior to indexes, if for no other reason than that scale scores convey more information than index scores. You should still be wary of the common misuses of the term *scale*; clearly, calling a given measure a scale rather than an index does not make it better. Moreover, you should be cautioned against two other misconceptions regarding the nature of scaling. First, whether the combination of several questionnaire items results in a scale almost always depends on the particular sample of respondents being studied. Certain items might form a scale among respondents in one sample but not among those in another; you should not assume that a given set of items comprises a scale simply because those items formed a scale with a given sample. Second, the use of certain scaling techniques (discussed later in this chapter) does not ensure the creation of a scale any more than the use of items that previously formed scales can offer such ensurance.

An examination of the substantive literature based on survey data shows that indexes are used much more frequently than scales. Ironically, however, the methodological literature contains little, if any, discussion of index construction, though discussions of scale construction abound. There are two apparent reasons for this disparity. First, indexes are more frequently used because scales are often difficult or impossible to construct from the data at hand. Second, methods of index construction are not discussed because they seem obvious and straightforward.

Figure 8-1 *Indexes Versus Scales*

Part A. Index-Construction Logic

Here are several occupations Americans might be willing or unwilling to have occupied by an avowed Communist. By and large, the different occupations would probably be regarded as having more or less the same degree of "importance" by respondents.

To create an index of civil liberties support, we might give respondents one point for each occupation they would be willing to have occupied by a Communist.

| Lawyer | Physician | Librarian | Engineer | Reporter | Professor |

Part B. Scale-Construction Logic

Below are shown three occupations that respondents would probably regard as differing in importance with regard to being open to Communists. The size of the boxes represents the different percentages who would be willing to have those occupations open to Communists.

President

High School Teacher

Ditch Digger

Following are the four scale types we would expect from an analysis of responses. A filled box indicates a willingness to have that occupation held by a Communist.

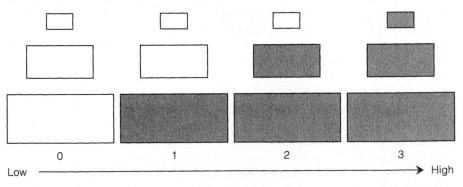

Index construction, however, is not a simple undertaking. Furthermore, I believe that the general failure to develop index-construction techniques has resulted in the creation of many bad indexes in social research. With this in mind, I have devoted most of this chapter to the methods of index construction. Once you fully understand the logic of this activity, you will be better equipped to attempt the construction of scales. Indeed, the carefully constructed index might turn out to be a scale.

Index Construction

Item Selection

A composite index is created for the purpose of measuring some variable. The first criterion for selecting items to be included in the index is **face validity** (logical validity). If you wanted to measure political conservatism, for example, each item considered should appear on its face to indicate respondents' conservatism (or its opposite, liberalism). Political party affiliation would be one such item. If respondents were asked to approve or disapprove of the views of a well-known conservative public figure, their responses to that item might logically provide another indication of the respondents' conservatism. A researcher interested in constructing an index of religiosity might consider items investigating respondents' church attendance, acceptance of certain religious beliefs, frequency of prayer, and so forth. Each item listed would appear to offer some indication of respondents' religiosity.

Typically, the methodological literature on conceptualization and measurement stresses the need for *unidimensionality* in scale and index construction; that is, a composite measure should represent only one dimension. Thus, items reflecting religiosity should not be included in a measure of political conservatism, even though the two variables are empirically related. In this sense, an index or scale should be unidimensional.

At the same time, you should be constantly aware of the subtle nuances that exist within the general dimension you are attempting to measure. In the case of religiosity, the items mentioned above represent different *types* of religiosity. If you want to measure ritual participation in religion, you should limit the items included in the measure to those that specifically indicate ritual participation: church attendance, communion, confession, and the like. If, on the other hand, you want to measure religiosity in a more general way, you should include a balanced set of items that represent each different type of religiosity. Ultimately, the nature of the items included will determine how specific or general your measurement of the variable is.

In selecting items for inclusion in an index, you must also be concerned with the amount of *variance* provided by the items. If an item provides an indication of political conservatism, for example, you should note how many conservatives are indicated by the item. In the extreme, if a given item indicates that no one is a conservative or that everyone is a conservative, the item will not be very useful in the creation of an index. If only 1 percent of a sample approve of a radical right political figure, an item measuring this is not likely to be of much use in the construction of an index.

With regard to variance, two options are available. First, you can select several items that divide respondents about equally in terms of the variable. Thus, you might select several items, each of which indicates about half as conservative and half as liberal. Although none of these items alone would justify the characterization of a respondent as "very conservative," a person who appeared conservative on all the items might be so characterized. The second option is the selection of items differing in variance. One item might indicate about half as conservative, and another might indicate few of the respondents as conservative. (*Note*: This latter option is necessary for scaling, but it is a reasonable method for the construction of an index as well.)

Bivariate Relationships among Items

The second step in index construction is the examination of the bivariate relationships among the items being considered for inclusion. If each item does indeed give an indication of the variable—as suggested on grounds of face validity—then the several items should be related to one another empirically. For example, if several items all reflect a respondent's conservatism or liberalism, then those respondents who appear conservative in terms of one item should appear conservative in terms of others. Recognize, however, that such items will seldom, if ever, be perfectly related to one another; some persons who appear conservative on one item will appear liberal on another. (This disparity creates the need for constructing composite measures in the first place.) Nevertheless, persons who appear conservative on item A should be more likely to appear conservative on item B than persons who appear liberal on item A.

You should examine all the possible bivariate relationships among the several items being considered for inclusion in the index in order to determine the relative strengths of relationships among the several pairs of items. Percentage tables or correlation coefficients, or both, can be used for this purpose. The primary criterion for evaluating these several relationships is the strength of the relationships. The use of this criterion, however, is rather subtle. Clearly, you should be wary of items that are not related to one another empirically. It is unlikely that unrelated items measure the same variable. More to the point, perhaps, a given item that is unrelated to several of the other items probably should be dropped from consideration. At the same time, a very strong relationship between two items is another danger sign. At the extreme, if two items are perfectly related to one another, then only one is necessary for inclusion in the index, because it completely conveys the indications provided by the other. (This problem will become even clearer in the next section.)

An example from the substantive literature of survey research might be useful in illustrating the steps in index construction.[1] A survey of medical school faculty members was concerned with the effect of the "scientific perspective" on the quality of patient care provided by physicians. The primary intent was to determine whether more scientifically inclined doctors were more impersonal in their treatment of patients.

The survey questionnaire offered several possible indicators of respondents' scientific perspectives. Of those, three items appeared—in terms of face validity—to provide especially clear indications of whether or not the doctors were scientifically oriented. The three items were as follows:

1. As a medical school faculty member, in what capacity do you feel you can make your greatest teaching contribution: as a practicing physician or as a medical researcher?

2. As you continue to advance your own medical knowledge, would you say your ultimate medical interests lie primarily in the direction of total patient management or the understanding of basic mechanisms?

3. In the field of therapeutic research, are you generally more interested in articles reporting evaluations of the effectiveness of various treatments or articles exploring the basic rationale underlying the treatments?

In each item, the second answer would indicate a greater scientific orientation than the first answer. Considering the responses to a single item, we might conclude that respondents who chose the second answer are more scientifically oriented than those who chose the first. This *comparative* conclusion is reasonable, but we should not be misled into thinking that respondents who chose the second answer to a given item are "scientists" in any absolute sense. They are simply *more scientific* than those who chose the first answer to the item. This important point will become clearer when we examine the distribution of responses produced by each item.

In terms of the first item, best teaching role, only about one-third of the respondents appeared scientifically oriented; that is, approximately one-third said they could make their greatest teaching contribution as medical researchers. This result does not mean that only one-third of the sample are "scientists," however, for the other two items suggested quite different conclusions. In response to the second item, ultimate medical interests, approximately two-thirds chose the scientific answer, indicating that they were more interested in learning about basic mechanisms than in learning about total patient management. In response to the third item, reading preferences, about 80 percent chose the scientific answer.

---

[1] The example, including the tables presented, is taken from Earl R. Babbie, *Science and Morality in Medicine* (Berkeley: University of California Press, 1970).

To repeat, these three questionnaire items cannot tell us how many "scientists" are in the sample, for none of the items is related to a set of criteria describing what constitutes being a scientist in any absolute sense. Using the items for this purpose would present us with the problem of three quite different estimates of how many scientists were included in the sample.

What these three questionnaire items provide us with are three independent indicators of respondents' relative inclinations toward science in medicine. Each item separates respondents into the more scientific and the less scientific. In view of the different distribution of responses produced by the three items, it is clear that each resulting grouping of more or less scientific respondents will have a somewhat different membership. Respondents who seem scientific in terms of one item will not seem scientific in terms of another. Nevertheless, to the extent that each item measures the same general dimension, we should find some correspondence among the different groupings. Respondents who appear scientific in terms of one item should be more likely to appear scientific in their responses to another item than would respondents who appear nonscientific in their responses to the first. We should find an association or correlation between the responses given to any two items.

The tables in Figure 8-2 provide an examination of the associations among the responses to the three items.[2] Three bivariate (two-variable) tables are presented, showing the conjoint distribution of responses for each pair of items. Although each single item produces a different grouping of "scientific" and "nonscientific" respondents, Figure 8-2 shows that the responses given to each item correspond, to a degree, to the responses given to each other item.

An examination of the three bivariate relationships presented in Figure 8-2 supports the belief that the three items all measure the same variable: scientific orientation. Let's begin by looking at the first bivariate relationship in the table. Faculty assessments of their best teaching roles and faculty expressions of their ultimate medical interests both give indications of scientific orientation. Respondents who answered "researcher" in the first instance would appear to be more scientifically inclined than those who answered "physician." Likewise, those who answered "basic mechanisms" would appear to be more scientifically inclined than those who answered "total patient management" in reply to the question concerning ultimate interests. If both these items do indeed measure the same thing, respondents appearing scientific on the first item ("researchers") should appear more scientific on the second ("basic mechanisms") than respondents who appeared nonscientific on the first ("physicians"). Looking at the data, we see that 87 percent of the "researchers" appear scientific on the second item, as opposed to 51 percent of the "physicians." (*Note:* The fact that the "physicians" are about evenly split in their

---

[2] If you have any difficulty understanding the tables presented in this chapter, you might want to refer to Chapter 14, which deals with the construction and interpretation of such tables.

Figure 8-2 *Bivariate Relationships among Scientific Orientation Items*

A.

	Best Teaching Role	
	Physician	Researcher
Total patient management	49%	13%
Basic mechanisms	51%	87%
	100%	100%
	(268)	(159)

Ultimate Medical Interest

B.

	Reading Preferences	
	Effectiveness	Rationale
Total patient management	68%	30%
Basic mechanisms	32%	70%
	100%	100%
	(78)	(349)

Ultimate Medical Interest

C.

	Reading Preferences	
	Effectiveness	Rationale
Physician	85%	64%
Researcher	15%	36%
	100%	100%
	(78)	(349)

Best Teaching Role

ultimate medical interests is irrelevant. All that is relevant is that they are less scientific than the "researchers" in their medical interests.) The strength of this relationship can be summarized as a 36 percentage point difference.

The same general conclusion can be reached in regard to the other bivariate relationships. The strength of the relationship between reading preferences and ultimate medical interests can be summarized as a 38 percentage point difference; the strength of the relationship between reading preferences and the two teaching roles can be summarized as a 21 percentage point difference.

Initially, the three items were selected on the basis of face validity; each appeared to give some indication of faculty members' orientations to science. By examining the bivariate relationships between the pairs of items, we have found support for the initial belief that the items all measure basically the same thing.

Multivariate Relationships among Items

The discovery of the expected bivariate relationships between pairs of items further suggests their appropriateness for inclusion in a composite index, but it is not a sufficient justification. The next step in index construction is the examination of the multivariate relationships among the items. You must examine the simultaneous relationships among the several variables before combining them in a single index.

Recall that the primary purpose of index construction is the development of a method for classifying respondents in terms of some variable such as political conservatism, religiosity, or scientific orientation. An index of political conservatism, for example, should identify respondents who are very conservative, moderately conservative, not very conservative, and not at all conservative (or moderately liberal and very liberal, respectively, in place of the last two categories). The different gradations in terms of the variable are provided by the combination of responses given to the items included in the index. Thus, a respondent who appeared conservative on all the items would be considered very conservative overall.

An index can provide meaningful gradations in this sense only if each item adds something to the evaluation of each respondent. Recall from the preceding section that two items perfectly related to one another should not be included in the same index. If one item were included, including the other would add nothing to our evaluation of respondents. The examination of multivariate relationships among the items is another way of eliminating "deadwood." It also determines the overall power of the particular collection of items in terms of measuring the variable under consideration.

We return to the earlier example of measuring scientific orientations among a sample of medical school faculty members to help you better understand the purposes of multivariate examination. Figure 8-3 presents the trivariate relationship among the three items. The table in Figure 8-3 is presented somewhat differently from those in Figure 8-2. In this instance, the sample respondents have been categorized in four groups according to (1) their best teaching roles and (2) their reading preferences. The numbers in parentheses indicate the number of respondents in each group. (Thus, sixty-six faculty members said they could best teach as physicians and also said they preferred articles dealing with the effectiveness of treatments.) For each of the four groups, the percentage saying they are ultimately more interested in basic mechanisms than in total patient management is presented. (Of the sixty-six faculty members mentioned above, 27 percent are primarily interested in basic mechanisms.)

Figure 8-3 *Trivariate Relationships among Scientific Orientation Items*

Percentage Interested in Basic Mechanisms

		Best Teaching Role	
		Physician	Researcher
Reading Preferences	Effectiveness	27% (66)	58% (12)
	Rationale	58% (219)	89% (130)

The arrangement of the four groups is based on a previously drawn conclusion regarding scientific orientation. Respondents in the upper left corner of the table are presumably the least scientifically oriented of the four groups in terms of their best teaching roles and their reading preferences. Those in the lower right corner of the table are presumably the most scientifically oriented in terms of those two items.

Recall that expressing a primary interest in "basic mechanisms" was also taken as an indication of scientific orientation. As we should expect, those in the lower right corner are most likely to give this response (89 percent), and those in the upper left corner are least likely to do so (27 percent). Respondents who gave mixed responses in terms of teaching roles and reading preferences hold an intermediate rank in their concern for basic mechanisms (58 percent in both cases).

This table tells us many things. First, we note that the original relationships between pairs of items are not significantly affected by the presence of a third item. Recall, for example, that the relationship between teaching role and ultimate medical interest was summarized as a 36 percentage point difference. Looking at Figure 8-3, we see that among only those respondents most interested in articles dealing with the effectiveness of treatments the relationship between teaching role and ultimate medical interest is 31 percentage points (58 percent minus 27 percent in the first row), and the same is true among those most interested in articles dealing with the rationale for treatments (89 percent minus 58 percent in the second row). The original relationship between teaching role and ultimate medical interest is essentially the same as in Figure 8-2, even among those respondents judged as scientific or nonscientific in terms of reading preferences.

The same conclusion can be drawn as we examine the columns in Figure 8-3. Recall that the original relationship between reading preferences and ultimate medical interests was summarized as a 38 percentage point difference. Looking only at the "physicians" in Figure 8-3, we see that the relationship between the other two items is now 31 percentage points. The same relationship is found among the "researchers" in the second column.

The importance of these observations becomes more evident when we consider what might have happened. The table in Figure 8-4 presents hypothet-

Figure 8-4 *Hypothetical Trivariate Relationship among Scientific Orientation Items*

Percentage Interested in Basic Mechanisms

		Best Teaching Role	
		Physician	Researcher
Reading Preferences	Effectiveness	51% (66)	87% (12)
	Rationale	51% (219)	87% (130)

ical data to illustrate this. These hypothetical data tell a much different story than did the actual data reported in Figure 8-3. In this instance, the original relationship between teaching role and ultimate medical interest evidently persists, even when reading preferences are introduced into the picture. In each row of the table, the "researchers" are more likely than the "physicians" to express an interest in basic mechanisms. Looking down the column, however, we note that no relationship exists between reading preferences and ultimate medical interests. If we know whether respondents think they can best teach as physicians or as researchers, knowing their reading preferences adds nothing to our evaluation of their scientific orientations. If something like Figure 8-4 resulted from the actual data, we would conclude that reading preferences should not be included in the same index as teaching roles, because it will contribute nothing to the composite index.

In the present example, only three questionnaire items were involved. If more items were being considered, then more complex multivariate tables would be in order. In this instance, we have limited our attention to the trivariate analysis of the three items. Recall that the purpose of this step in index construction is to examine the simultaneous interaction of the items in order to determine whether all the items are appropriate for inclusion in the same index.

Index Scoring

After you have determined the best items for inclusion in the index, the next step is to assign scores to specific responses, thereby creating a single composite index out of the several items. Two basic decisions must be made in this regard.

First, you must decide the desirable range of the index scores. Certainly one of the primary advantages of an index over a single item is the range of gradations it offers in the measurement of a variable. As we noted earlier, political conservatism might be measured from "very conservative" to "not at all conservative" (or "very liberal"). How far to the extremes, then, should the index extend? This decision again involves the question of variance. Almost always, as the possible extremes of an index are extended, fewer cases are found at each end. If you attempt to measure political conservatism to its greatest extreme, you might find that almost no one falls into that category.

The first decision, then, concerns the conflicting desires for (1) the greatest possible range of measurement in the index and (2) an adequate number of cases at each point in the index. You will be forced to reach some kind of compromise between these conflicting desires.

The second decision concerns the actual assignment of scores for each specific response. Basically, you must decide whether to give each item an equal weight in the index or to give the items different weights. As we will see later, scale construction is quite different in this regard, but in index construction assignment of scores is an open issue. While there are no firm rules to be followed, I would suggest—and practice tends to confirm—that items should be weighted equally unless compelling reasons exist for differential weighting. That is, the burden of proof should be on differential weighting; equal weighting should be the norm.

Of course, the weighting decision must be related to the balance of the items chosen. If the index is to represent the composite of slightly different aspects of a given variable, then you should give each of those aspects the same weight. In some instances, however, you might feel that two items reflect essentially the same aspect, while the third reflects a different aspect. If you wanted to respect both aspects equally in the index, you might decide to give the different item a weight equal to the combination of the weight of the two similar ones. In such a situation, you might assign a maximum score of 2 to the different item and maximum scores of 1 to each of the two similar items.

Although the rationale for scoring responses should take such concerns as these into account, you will typically experiment with different scoring methods. In your experimentation, you will examine the relative weights given to different aspects and at the same time consider the range and distribution of cases provided. Ultimately, the scoring method chosen will represent a compromise among these different demands. (*Note:* In this activity, as in most survey activities, the decision is open to revision on the basis of later examination. Validating the index, to be discussed shortly, might lead you to recycle your efforts and construct a completely different index.)

In the example taken from the medical school faculty survey, the decision was made to weight each item equally because they had been chosen, in part, on the basis of their representing slightly different aspects of the overall variable, scientific orientation. On each item, the respondents were given a score of 1 for choosing the "scientific" response and a score of 0 for choosing the "nonscientific" response. Each respondent, then, had a chance of receiving a score of 0, 1, 2, or 3, depending on the number of "scientific" responses chosen. This scoring method provided what was considered a useful range of variation—four index categories—and also provided a sufficient number of cases in each category for analysis.

Handling Missing Data

In virtually every survey, some respondents fail to answer some questions (or choose a "don't know" response). Missing data present a problem at all stages of analysis, but especially in index construction. (Again, scaling is different in this regard.) If some respondents fail to provide answers to items being included in a composite index, you face a particular problem in assigning scores to those respondents in constructing the index. However, several methods exist for dealing with this problem.

First, if relatively few respondents have missing data, you might decide to exclude them from the construction and analysis of the index. The primary concerns in this instance are whether the numbers available for analysis will still be sufficient and whether excluding those respondents will result in a biased sample whenever the index is used in the analysis. The possibility of bias can be examined through a comparison, on other relevant variables, of respondents who would be included in the index and those who would be excluded. (In the medical school faculty example discussed earlier, this was the decision made with respect to missing data.)

Second, you might have grounds for treating missing data in the same way as one of the other responses given. For example, if the questionnaire asked respondents to indicate their participation in a number of activities by checking "yes" or "no" for each activity, many respondents might have checked some activities "yes" and left the remainder blank. In such a case, you might decide that a failure to answer meant "no" and score missing data as if the respondents had checked the "no" space.

Third, a careful analysis of missing data might yield an interpretation of their meaning. In constructing a measure of political conservatism, for example, you might discover that respondents who failed to answer a given question were generally as conservative, in terms of other items, as those who gave the conservative answer to the question. As another example, a recent survey measuring religious beliefs found that respondents who chose "don't know" for a given belief were almost identical to the "disbelievers" in their answers regarding other beliefs. (*Note:* You should take these examples not as empirical guides in your own studies but only as suggestive of ways to analyze your own data.) Whenever the analysis of missing data yields such interpretations, you might decide to score such cases accordingly.

Fourth, you might decide to assign an intermediate score for missing data. For example, if a given item is assigned scores of 0, 1, and 2 for its three possible responses, you might assign the intermediate score (1) to the item for those respondents who gave no answer. (This is the same logic whereby the answer "undecided" is often scored as lying between "agree" and "disagree.")

Fifth, you might assign index scores proportionately on the basis of the answers that a respondent does give. For example, assume that six items are being combined in an index, with scores of 0 or 1 assigned to each item. The maximum score a respondent can receive, then, is 6. If a given respondent answers only five items but receives a score of 5 on those items, he or she might be given a proportionate score of $5/5 \times 6 = 6$ on the index. A respondent who received a score of 2 on the four items answered might be given a final score of $2/4 \times 6 = 3$. When these computations result in fractional results, some method for rounding off should be employed to simplify the final index scores.

Finally, you might be unwilling to utilize any of these methods for handling missing data, but your later analysis might require that all respondents in the sample be scored. In such a situation, you could assign scores for missing data on a random basis. For an item assigned the possible scores of 0, 1, and 2, the first respondent failing to answer that item might receive a score of 1, the second a score of 0, the third a score of 2, and so forth. This method is the most conservative method from a research analysis standpoint, because you are "stacking the deck" against yourself. If the resultant index proves to be a powerful tool in your analysis, you might conclude that it would have been even more powerful if all respondents had answered all questions. (Of course, if your purpose is to show that the index is unrelated to other variables, you have stacked the deck in your favor.)

The choice of a particular method for handling missing data depends so much on the research situation that it would be impossible to suggest a single best method or to rank-order the methods described. In general, I would suggest an examination of the respondents who fail to answer both in terms of possible bias in excluding them and in terms of their answers to other items in the index. Understanding your data is the final goal of analysis.

Index Validation

Up to this point, we have discussed the steps involved in the selection and scoring of items that result in a composite index aimed at measuring some variable. If each step discussed earlier is carried out carefully, the likelihood of the index actually measuring the variable is enhanced. The success of creating a useful index is not yet proved, however; *validation* of the index helps accomplish this. The basic logic of validation is as follows. We assume that the composite index measures a variable, that is, that the successive scores on the index group respondents in a rank order in terms of that variable. An index of political conservatism rank-orders groups in terms of their relative conservatism. If the index does this successfully, then persons scored as relatively conservative in terms of the index should appear relatively conservative in terms of all questionnaire items (or other indications) that also reflect political orientation. Several methods are available for validating a composite index.

Item Analysis. The first step in index validation is an internal validation called *item analysis*. You should examine the extent to which the composite index is related to (or predicts responses to) the questionnaire items included in the index itself. If the index has been carefully constructed through the examination of bivariate and multivariate relationships among several items, this step should confirm the validity of the index. For a complex index containing many items, this step provides a more parsimonious test of the independent contribution of each item to the index. If a give item is found to be poorly related to the index, it can be assumed that other items in the index are wiping out the contribution of that item. The item in question, then, contributes nothing to the index's power, and it should be excluded from the index.

Item analysis is an important first test of the index's validity, but it is scarcely a sufficient test. If the index adequately measures a given variable, it should successfully predict other indications of that variable. To test this, we must turn to items not included in the index.

External Validation.   Persons scored as politically conservative on the index should appear conservative in their responses to other items in the questionnaire. It must be realized, of course, that we are talking about relative conservatism; we are unable to formulate a final absolute definition of what constitutes "conservatism" in any ultimate sense. Those respondents scored as the most conservative in terms of the index, however, should be the most conservative in answering other questions. Respondents scored as the least conservative on the index should be the least conservative in answering other items. Indeed, the ranking of groups of respondents on the index should predict the ranking of those groups in their answers to other questions dealing with political orientation.

In our example of the scientific orientation index, several questions in the questionnaire offered the possibility of further validation. Table 8-1 presents some of those items. The items listed provide several lessons regarding index validation. First, we note that the index strongly predicts the responses to the validating items in the sense that the rank order of scientific responses among the four groups is the same as the rank order provided by the index

Table 8-1   *Validation of Scientific Orientation Index*

	Index of Scientific Orientation			
	Low 0	1	2	High 3
Percentage interested in attending scientific lectures at the medical school	34	42	46	65
Percentage who say faculty members should have experience as medical researchers	43	60	65	89
Percentage who would prefer faculty duties involving research activities only	0	8	32	66
Percentage who engaged in research during preceding academic year	61	76	94	99

itself. At the same time, each item gives a *different* description of overall scientific orientation. For example, the last validating item indicates that the great majority of all faculty members were engaged in research during the preceding year. If this were the only indicator of scientific orientation, we would conclude that nearly all faculty were scientific. Nevertheless, faculty scored as more scientific in terms of the index are more likely to have engaged in research than those scored as relatively less scientific. The third validating item provides a different descriptive picture. Only a minority of the faculty overall said they would prefer duties limited exclusively to research. Nevertheless, the percentages giving this answer also correspond to the scores assigned on the index.

Bad Index Versus Bad Validators. A dilemma that must be faced by nearly every index constructor is the apparent failure of external items to validate the index. If the internal item analysis shows inconsistent relationships between the items included in the index and the index itself, something is wrong with the index. If the index fails to strongly predict the external validation items, however, the conclusion to be drawn is more ambiguous. You must choose between two possibilities: (1) The index does not adequately measure the variable in question, or (2) the validation items do not adequately measure the variable and therefore do not provide a sufficient test of the index.

If you have worked long and conscientiously on the construction of the index, you will find the second conclusion very compelling. Typically, you will feel that you have included the best indicators of the variable in the index and that the validating items are, therefore, second-rate indicators. However, you should recognize that the index is purportedly a very powerful measure of the variable; as such, it should be somewhat related to any item that taps the variable even poorly.

When external validation fails, you should reexamine the index before deciding that the validating items are insufficient. One method of doing this involves an examination of the relationships between the validating items and the individual items included in the index. Discovering that some index items relate to the validators but others do not will improve your understanding of the index as it was initially constructed.

No cookbook solution to this dilemma exists; it is an agony that the serious researcher must learn to survive. Ultimately, the wisdom of your decision regarding the index will be determined by the index's utility in your later analyses. Perhaps you will initially decide that the index is a good one and that the validators are defective and later find that the variable in question (as measured by the index) is not related to other variables in the ways expected. At that point, you might return to the composition of the index.

Considerable attention has been given in this text to the construction of simple indexes, for two reasons. First, a review of the empirical research literature points to the popularity of such measures among survey researchers. Second, little, if any, discussion of the techniques for index construction is present in the literature or in methodology textbooks. "Simple" index construction has probably been viewed as too simple to warrant such discussion, and the techniques have remained part of the oral tradition of survey research.

Likert "Scaling"

Earlier in this chapter, a scale was defined as a composite measure constructed on the basis of an intensity structure among items comprising the measure. In scale construction, response patterns across several items are scored, whereas in index construction, individual responses are scored and the independent scores are then summed. By this definition, the measurement method developed by Rensis Likert, called **Likert scaling,** represents a more systematic and refined means of constructing indexes; therefore, this method will be discussed here rather than in the sections on scaling.

The term *Likert scale* is associated with a question format frequently used in contemporary survey questionnaires. Basically, respondents are presented with a statement in the questionnaire and are asked to indicate whether they "strongly agree," "agree," "disagree," "strongly disagree," or are "undecided." Modifications of the wording of the response categories (for example, "approve") can be used, of course.

The particular value of this format is the unambiguous ordinality of the response categories. If respondents were permitted to volunteer or select answers such as "sort of agree," "pretty much agree," "really agree," and so forth, judging the relative strength of agreement intended by the various respondents would be impossible. The Likert format easily resolves this dilemma.

Likert scaling also lends itself to a rather straightforward method of index construction. Because identical response categories will have been used for several items intended to measure a given variable, each such item can be scored in a uniform manner. With five response categories, scores of 0 to 4 or 1 to 5 might be assigned, taking the "direction" of the items into account (for example, assign a score of 5 both to "strongly agree" for positive items and to "strongly disagree" for negative items). Each respondent would then be assigned an overall score representing the summation of the scores received for his or her responses to the individual items.

The Likert method is based on the assumption that the overall score based on responses to the many items that seem to reflect the variable under consideration provides a reasonably good measure of the variable. These overall scores are not the final product of index construction; rather, they are used for purposes of an *item analysis* resulting in the selection of the best items. Essentially, each individual item is correlated with the large, composite measure. Items that correlate most with the composite measure are assumed to provide the best indicators of the variable, and only those items would be included in the index that is ultimately used for analyses of the variable.

Note that the uniform scoring of Likert-item response categories assumes that each item has about the same intensity as the rest. This is the key respect in which the Likert method differs from scaling as the term is used in this book. You should also realize that Likert-type items can be used in a variety of ways; you are by no means bound to the method just described. Such items can be combined with other types of items in the construction of simple indexes; similarly, they can be used in the construction of scales. If all the items being considered for inclusion in a composite measure are in the Likert format, however, then the method described earlier should be considered.

Now we will turn our attention from indexing methods to a selection of scaling techniques. Many methods are available to the survey researcher; we will consider only Bogardus, Thurstone, and Guttman scales.

Scale Construction

Good indexes provide an ordinal ranking of respondents on a given variable. All indexes are based on the assumption that a person with two indications of being scientifically inclined, for example, should be more scientific than a person with only one such indication. What an index might fail to take into account, however, is that not all indications of a variable are equally important. (Of course, you can attempt to resolve this problem by weighting indicators differently.)

Scales offer more ensurance of ordinality by tapping *structures* among the indicators. The items going into a composite measure can have different *intensities* in terms of the variable. The three scaling procedure descriptions that follow illustrate the variety of techniques available.

Bogardus Social Distance Scale

A good example of a scale is the **Borgardus Social Distance Scale.** Suppose you are interested in the extent to which respondents are willing to associate with Albanians. All respondents in the study might be asked the following questions:

1. Are you willing to permit Albanians to live in your country?
2. Are you willing to permit Albanians to live in your community?
3. Are you willing to permit Albanians to live in your neighborhood?
4. Would you be willing to let an Albanian live next door to you?
5. Would you let your child marry an Albanian?

Note that the several questions increase in the closeness of contact that the respondents may or may not want with Albanians. Beginning with the original concern of measuring willingness to associate with Albanians, we have developed several questions indicating differing degrees of intensity on this variable.

The clear differences of intensity suggest a structure among the items. Presumably, if respondents are willing to accept a given kind of association, they would be willing to accept all those associations preceding it in the list, that is, those with lesser intensities. For example, respondents who are willing to permit Albanians to live in their neighborhood will surely accept them in their community and their nation, but they might or might not be willing to accept them as next-door neighbors or as relatives. This, then, is the logical structure of intensity inherent among the items.

Empirically, we would expect to find the largest number of respondents accepting co-citizenship and the fewest accepting intermarriage. In this sense, we speak of "easy items" (co-citizenship) and "hard items" (intermarriage). More respondents agree to the easy items than to the hard ones. With some inevitable exceptions, logic demands that once respondents have refused a certain relationship presented in the scale, they will also refuse all the harder ones that follow it.

The Bogardus Social Distance Scale illustrates the economy of scaling as a data-reduction device. If we know *how many* relationships with Albanians a given respondent will accept, we know *which* relationships he or she will accept. Thus, a single number can accurately summarize five or six survey responses without a loss of information.

Thurstone Scales

Often the inherent structure of the Bogardus Social Distance Scale is not appropriate to the variable being measured. Indeed, such a logical structure among several indicators is seldom apparent. **Thurstone scaling** is an attempt to develop a format for generating groups of indicators of a variable that have at least an *empirical* structure among them. One basic format is that of "equal-appearing intervals."

A group of "judges" is given perhaps a hundred items thought to be indicators of a given variable. Each judge is then asked to estimate how strong an indicator of the variable each item is by assigning scores of, for example, 1 to 13. If the variable were prejudice, for example, the judges would be asked to assign a score of 1 to the very weakest indicators of prejudice, a score of 13 to the strongest indicators, and intermediate scores to indicators that they believed to be somewhere in between.

After all the judges had completed this task, you would examine the scores assigned to each item by all the judges to determine which items produced the greatest agreement among the judges. Those items on which the judges disagreed broadly would be rejected as ambiguous. From among those items producing general agreement in scoring, one or more would be selected to represent each scale score from 1 to 13.

The items selected in this manner would then be included in a survey questionnaire. Respondents who appeared prejudiced on items representing a strength of 5 would be expected to appear prejudiced on items with lesser strengths; if some of those respondents did not appear prejudiced on the item(s) with a strength of 6, they would also not be expected to appear prejudiced on items with greater strengths.

If the Thurstone Scale items had been adequately developed and scored, the economy and effectiveness of data reduction inherent in the Bogardus Social Distance Scale would appear. A single score might be assigned to each respondent (the strength of the hardest item accepted), and that score would adequately represent the responses to several questionnaire items. As is the case with the Bogardus scale, a respondent scored as 6 might be regarded as more prejudiced than one scored as 5 or less.

Thurstone scaling is seldom, if ever, used in survey research today, primarily because of the tremendous expenditure of energies required for the "judging" of items. Several (perhaps ten or fifteen) judges would have to spend a considerable amount of time scoring the many initial items. Since the quality of their judgments would depend on their experience with and knowledge of the variable under consideration, the task might require professional research-ers. Moreover, the meanings conveyed by the several items indicating a given variable tend to change over time. Thus, an item having a given weight at one time might have quite a different weight later on. A Thurstone Scale would have to be periodically updated in order to be effective.

Guttman Scaling

A very popular scaling technique used by survey researchers today is the one developed by Louis Guttman. Like both Bogardus scaling and Thurstone scaling, **Guttman scaling** is based on the fact that some items under consideration might prove to be "harder" indicators of the variable than others. Respondents who accept a given hard item also accept the easier ones. If such a structure appears in the data under examination, we might say that the items form a Guttman Scale. One example should suffice to illustrate Guttman scaling.

In the earlier example describing the measurement of scientific orienta-tion among medical school faculty members, recall that a simple index was constructed. As we will see shortly, however, the three items included in the index essentially form a Guttman Scale. This possibility first appears when we look for relatively "hard" and "easy" indicators of scientific orientation.

That item asking respondents whether they could best serve as practic-ing physicians or as medical researchers is the hardest of the three; only about one-third of the respondents would be judged scientific if this were the single indicator of the variable. If the item concerning ultimate medical interests (total patient management versus basic mechanisms) were used as the only indicator, almost two-thirds would be judged scientific. Reading preference (effectiveness of treatments versus the underlying rationales) is the easiest of the three items; about 80 percent of the respondents would be judged scientific in terms of this item.

Table 8-2 *Scaling Scientific Orientation*

	Reading Preference	Ultimate Interests	Teaching Role	Number of Cases
	+	+	+	116
Scale types:	+	+	−	127
Total = 383	+	−	−	92
	−	−	−	48
	−	+	−	18
Mixed types:	+	−	+	14
Total = 44	−	−	+	5
	−	+	+	7

To determine whether a scalar structure exists among the responses to the three items, we must examine the several possible response patterns given to all three items simultaneously. In Table 8-2, all possible patterns have been presented in a schematic form. For each of the three items, pluses and minuses have been used to indicate the scientific and nonscientific responses, respectively. (A plus indicates a scientific response; a minus indicates a nonscientific response.)

The first four response patterns in the table comprise what we would call the *scale types*—those patterns that form a scalar structure. Following those respondents who selected all three scientific responses (line 1), we see that respondents with only two scientific responses (line 2) have chosen the two easier ones; those with only one scientific response (line 3) chose the easiest of the three. Finally, those respondents who selected none of the scientific responses are shown in line 4.

The second part of the table presents those response patterns that violate the scalar structure of the items. The most radical departures from the scalar structure are the last two response patterns: those who accepted only the hardest item and those who rejected only the easiest one.

The final column in the table indicates the number of survey respondents giving each response pattern. It is immediately apparent that the great majority (90 percent) of the respondents fit into one of the scale types. The presence of mixed types, however, indicates that the items do not form a perfect Guttman Scale.

We should recall at this point that one of the chief functions of scaling is efficient data reduction. Scales provide a technique for presenting data in a summary form while at the same time maintaining as much of the original information as possible.

When the scientific orientation items were formed into an index in our earlier discussion, respondents were given one point on the index for each scientific response they gave. If these same three items were scored as a Guttman Scale, some respondents would receive different scores than they received on the index. Respondents would be assigned scale scores that would permit the most accurate reproduction of their original responses to all three items.

Respondents who fit into the scale types would receive the same scores as they were assigned in the index construction. Persons selecting all three scientific responses, for example, would still be scored 3. Note that if we were told that a given respondent in this group received a score of 3, we could accurately predict that he or she selected all three scientific responses. For persons in the second row of the table, the assignment of the scale score of 2 would lead us to accurately predict scientific responses to the two easier items and a nonscientific response to the hardest. In each of the four scale types, we could accurately predict the actual responses given by all the respondents.

The mixed types in the table present a problem, however. The first mixed type (−+−) was scored 1 on the index to indicate only one scientific response. If 1 were assigned as a scale score, however, we would predict that all respondents in this group had chosen only the easiest item (+−−), thereby making two errors for each such respondent. Scale scores are assigned, therefore, with the aim of minimizing the errors that would be made in reconstructing the original responses. Table 8-3 illustrates the index and scale scores that would be assigned to each response pattern in our example.

As I mentioned earlier, the original index scoring for the four scale types would be maintained in the construction of a Guttman Scale, and no errors would be made in reproducing the responses given to all three items. The mixed types would be scored differently in an attempt to reduce errors. Note, however, that one error is made for each respondent in the mixed types. In the first mixed type, we would erroneously predict a scientific response to the easiest item for each of the eighteen respondents in this group, for a total of eighteen errors.

Table 8-3   *Index and Scale Scores*

	Response Patterns			Number of Cards	Index Scores	Scale Scores*	Total Scale Errors
	+	+	+	116	3	3	0
Scale types:	+	+	−	127	2	2	0
	+	−	−	92	1	1	0
	−	−	−	48	0	0	0
	−	+	−	18	1	2	18
Mixed types:	+	−	+	14	2	3	14
	−	−	+	5	1	0	5
	−	+	+	7	2	3	7

*Note: This table presents one common method for scoring "mixed types," but the reader should be advised that other methods are also used.

The extent to which a set of empirical responses form a Guttman Scale is determined by the accuracy with which the original responses can be reconstructed from the scale scores. For each of the 427 respondents in this example, we will predict three questionnaire responses, for a total of 1,281 predictions. Table 8-3 indicates that we will make forty-four errors using the scale scores assigned. The percentage of correct predictions is called the *coefficient of reproducibility,* the percentage of "reproducible" responses. In the present example, the coefficient of reproducibility is 1,237/1,281, or 96.6 percent.

Except for the case of perfect (100 percent) reproducibility, there is no way of saying that a set of items does or does not form a Guttman Scale in any absolute sense. Virtually all sets of such items *approximate* a scale. As a rule of thumb, however, coefficients of 90 percent or 95 percent are the commonly used standards in determining whether a Guttman Scale exists. If the observed reproducibility exceeds the level you have set, you will probably decide to score and use the items as a scale.[3]

One concluding remark should be made with regard to Guttman scaling. It is based on the structure observed among the *actual data under examination.* This important point is often misunderstood by researchers. It does not make sense to say that a set of questionnaire items (perhaps developed and used by a previous researcher) constitutes a Guttman scale. Rather, we can say only that the items form a scale within a given body of data being analyzed. Scalability, then, is a sample-dependent, empirical question. Although a set of items might form a Guttman Scale among one sample of respondents, there is no guarantee that the items will form such a scale among respondents in another sample. In this sense, then, questionnaire items in and of themselves never form a scale, but a set of empirical observations might.

Typologies

We conclude this chapter with a short discussion of **typology** construction and analysis. Recall that indexes and scales are constructed in order to provide ordinal measures of given variables. We attempt to assign index or scale scores to respondents in such a way as to indicate a rising degree of prejudice, religiosity, conservatism, and so forth. In this regard, we are dealing with single dimensions. Typologies, in contrast, are *multidimensional.*

Often you want to summarize the intersection of two or more dimensions. You might, for example, want to examine political orientations sepa-

---

[3] The decision as to what criteria to use in this regard is, of course, arbitrary. Moreover, a high degree of reproducibility does not ensure that the scale constructed in fact measures the concept under consideration, although it increases confidence that all the component items measure the same thing. Finally, you should be advised that a high coefficient of reproducibility is more likely when few items are involved.

Table 8-4   *A Political Typology*

Domestic Policy Attitudes	Foreign Policy Attitudes	
	Conservative	Liberal
Conservative	A	B
Liberal	C	D

rately in terms of domestic issues and foreign policy. The fourfold presentation in Table 8-4 describes such a typology. Persons in cell A of the table are conservative on both foreign policy and domestic policy; those in cell D are liberal on both issues. Respondents in cells B and C are conservative on one issue and liberal on the other.

Frequently you arrive at a typology in the course of your attempt to construct an index or scale. The items that you thought represented a single variable appear to represent two. In the present example, you might have been attempting to construct a single index of political attitudes, but you discovered—empirically—that foreign and domestic politics had to be kept separate.

In any event, you should be warned against a difficulty inherent in typological analysis. Whenever the typology is used as the *independent variable,* you will probably have no problem. In the example above, you might compute and present the percentages of persons in each cell who say they normally vote for the Democratic party. You could then easily examine the effects of both foreign and domestic policy attitudes on voting behavior.

Analyzing a typology as a *dependent variable,* however, is difficult. If you want to discover why respondents fall into the different cells of the typology, you are in trouble. This becomes apparent when we consider the ways in which you might construct and read your tables. Assume, for example, that you want to examine the effects of sex on political attitudes. With a single dimension you could easily determine the percentages of men and women who have been scored as conservative and liberal on your index or scale. With a typology, however, you would have to present the distribution of the men in your sample among types A, B, C, and D. Then you would have to repeat the procedure for the women in the sample and compare the two distributions. Suppose that 80 percent of the women are scored as type A (conservative on both dimensions), compared with 30 percent of the men. Moreover, suppose that only 5 percent of the women are scored as type B (conservative only on domestic issues), compared with 40 percent of the men. Concluding from an examination of type B that men are more conservative on domestic issues than women would be incorrect, since 85 percent of the women, compared with 70 percent of the men, have this characteristic. The relative sparsity of women scored as type B is due to their concentration in type A. It should be apparent that an interpreta-

Table 8-5 *Nine American Life-styles*

Name	Typical Description
Survivor life-style	Deeply impoverished; poorly educated; many old; many ill
Sustainer life-style	On the edge of poverty; often found in urban slums
Belonger life-style	Traditional middle-class
Emulator life-style	Striving to advance financially; ambitious; typically young
Achiever life-style	Successful in the professions and business; the image of American success
I-Am-Me life-style	Children of Achievers, renouncing traditional success in search of a new life-style
Experiential life-style	Young, educated, successful, somewhat mystical people who place great importance on internal matters
Societaly Conscious life-style	Successful, mature, influential people who are concerned and active in arena of social awareness and responsibility
Integrated life-style	Psychologically mature people who have integrated inner- and outer-directed concerns

SOURCE: Adapted from Mitchell, op. cit., pp. 4–24.

tion of such data is very difficult in areas other than description. Ultimately, you will probably examine the two political dimensions separately, especially if the variables have more categories of responses than in our example.

A more complex typology, developed by Arnold Mitchell and colleagues at the Stanford Research Institute, has attracted quite a bit of attention, especially in marketing circles; it is the *Values and Lifestyle (VALS) typology.*[4] While Americans can be grouped or distinguished in terms of many variables (age, sex, race, education, and so on), Mitchell's research on a variety of demographic and attitudinal variables suggested the possibility of identifying most Americans with one of nine life-styles, presented in Table 8-5.

As Mitchell and others have demonstrated, people identified with the nine major life-styles have very different consumption patterns, suggesting that a particular product might be advertised effectively among some life-styles and not among others. Cadillac sales are not likely to be high among the Survivors,

---

[4] Arnold Mitchell, *The Nine American Lifestyles* (New York: Warner Books, 1983).

nor are rifle sales among the Societally Conscious. Moreover, researchers using the VALS model have been able to identify geographical areas where particular life-styles are concentrated, making it possible, for example, to use direct mail advertising only within those zip codes most likely to produce sales for the particular product being advertised.

Though typologies present special problems for survey analysis, the VALS model demonstrates that they should not be ruled out altogether. Ultimately, of course, you should design and construct composite measures—indexes, scales, or typologies—that are particularly appropriate to your research needs.

Summary

This chapter has addressed the logic and construction of indexes and scales. The techniques described are common in survey analysis and are extremely valuable. Good composite measures such as indexes and scales offer the following advantages:

1. Where single indicators (for example, responses to a questionnaire item) might produce a biased measurement of the variable under examination, a composite measure constructed from several different indicators can eliminate the effects of bias.

2. Composite measures can provide a broader range of variation on the variable. Where a single dichotomous item would provide only two levels of intensity (for example, high and low), the combination of five such items could result in the creation of an index or scale ranging from a low of 0 to a high of 5. If the composite measure is properly constructed, it will provide greater explanatory power in the analysis.

3. Scales can provide an efficient technique for data reduction. An extensive set of questionnaire responses can be summarized in the form of a single scale score without much of the original information being lost.

4. Typologies provide multidimensional classifications, which are often effective as independent variables but are problematic as dependent variables.

Additional Readings

Paul F. Lazarsfeld and Morris Rosenberg, eds., *The Language of Social Research* (New York: Free Press, 1955), sec. 1.

A. N. Oppenheim, *Questionnaire Design and Attitude Measurement* (New York: Basic Books, 1966).

Claire Selltiz et al., *Research Methods in Social Relations* (New York: Holt, Rinehart & Winston, 1959), ch. 10.

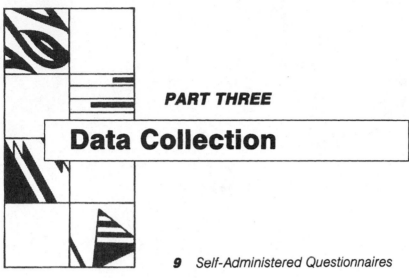

The four chapters comprising Part Three of this text address the various aspects of survey data collection. We will examine the activities that result in the possession of a body of data for analysis and will consider the decisions you must make in terms of the options available to you.

Chapters 9 and 10 focus on the techniques of data collection as they are used in self-administered questionnaires and interview surveys, respectively. The latter chapter considers both face-to-face and telephone interviews.

Chapter 11 deals with the different methods you can use in converting the completed questionnaires into manipulable, quantitative data. We will also look at some new, computerizd methods that create data files as the data are first being collected.

Chapter 12 considers an often overlooked component of professional survey research: pretesting and pilot studies. We will discuss the different techniques you can use to test the various aspects of your study design before you commit your major resources in the field and investigate how you can evaluate the results of such tests.

9

Self-Administered

Questionnaires

Chapters 9 and 10 discuss the manner in which survey researchers actually collect their data for analysis. Having constructed a questionnaire appropriate to your research aims, you must now go about distributing copies of the questionnaire to your sample of respondents. This chapter considers the techniques for accomplishing distribution through the use of self-administered questionnaires; Chapter 10 deals with interview surveys.

Although the mail survey is the typical form of self-administered study, several additional methods are also used. In some cases, it might be appropriate to administer the questionnaire to a group of respondents gathered at the same place at the same time. A survey of students taking Introductory Psychology might be conducted during class, for example. High school students might be surveyed during homeroom period.

Some recent experimentation has been conducted regarding the home delivery of questionnaires. Research workers deliver the questionnaires to the homes of sample respondents and explain the study. The questionnaire is then left for the respondent to complete and is picked up later by the research team.

Home delivery and the mail can be used in combination, as well. In many parts of the country, the United States census is now conducted in this fashion. Questionnaires are mailed to families; then a census enumerator visits the home to pick up the questionnaire and check it for completeness. Using the opposite method, research workers have hand delivered questionnaires with a request that the respondent mail the completed questionnaire to the research office.

On the whole, the appearance of a research worker, whether in delivering the questionnaire, picking it up, or both, seems to produce a higher completion rate than do straightforward mail surveys. Additional experimentation with this method is likely to point out additional techniques for improving completion while reducing costs.

The remainder of this chapter is devoted specifically to mail surveys, the most typical form of self-administered survey.

Mail Distribution and Return

Basic Method

The basic method for data collection through the mail is the transmission of a questionnaire accompanied by a letter of explanation and a return envelope. The respondent completes the questionnaire and returns it to the research office through the mail, using the envelope provided for that purpose.

Alternative Method

In some cases, this process can be further facilitated through the use of a *self-mailing* questionnaire. The questionnaire is constructed in such a manner that the research office's return address and postage are printed on the questionnaire itself. Upon completion, it can be dropped in the mail without an envelope being required.

You should plan this method with caution, however, because the post office has special requirements regarding the form of materials that can be mailed. In particular, questionnaires must be sealed in some manner. Sealing can be accomplished in a number of ways. If the questionnaire is printed in the form of a booklet, it might be possible to obtain a three-panel rather than a two-panel cover. In this form, the back cover has a fold-out panel with an adhesive strip. When questionnaires are mailed out, the fold-out panel is tucked inside the back of the questionnaire. Upon completion of the questionnaire, the respondent can unfold the extra panel, lick the adhesive, and fold the panel around the questionnaire booklet. If the research office return address and postage are already printed on the extra panel, the questionnaire can be placed directly in the mail for return.

This method simplifies the assembly of mailing pieces because including a return envelope is unnecessary. The respondent cannot lose the return envelope without losing the questionnaire itself. Finally, this form of questionnaire has the added appeal of a certain "toy value." To some extent, respondents might want to complete the questionnaire so that they can then play with the cover.

If producing a three-panel cover is not possible, you might be able to have adhesive tabs affixed to the booklet. Rather than folding the entire panel around the completed questionnaire, you can use a smaller tab to seal it.

Finally, you might ask respondents to close and seal the booklet themselves, perhaps with a staple or scotch tape. This method is a little risky, however, for several reasons. First, forcing the respondents to spend extra effort is likely to reduce the response rate. They might not have anything readily available for sealing the booklet at the time of completion, put it off, and eventually forget to return the questionnaire. Second, the variety of sealing methods that respondents devise will probably hinder the processing of returned questionnaires. (1 have received questionnaires sealed with glue, trading stamps, paper clips, and string.) Finally, many respondents will neglect to seal the questionnaires at all, and you might have difficulty getting the post office to deliver them.

Ultimately, self-mailing questionnaires have many advantages in terms of ease, economy, and response rate, but they should be planned and pretested with care. It is vital, moreover, for all experimental models to be cleared with the post office.

Postal Options and Relative Costs

When conducting a mail survey, you have a number of options available for the transmission of questionnaires, both outgoing and incoming. Postal rates change frequently; you should, of course, check current postal rates in planning your study, and you should allow some extra funds for this purpose if the study will be delayed in getting into the field.

Postal-Class Options

First class and bulk rate are the primary postal-class options available. First class is more expensive, but it is also more flexible and gives you better service. Nevertheless, bulk-rate postage can often be used effectively.

In bulk-rate mailing, each mailing piece must be printed with a bulk-rate permit. The permit can be set in type for printing on the envelope or be mimeographed, or a rubber stamp can be created and used. You must obtain a permit number from the post office. (*Note:* Universities or other agencies might already have a number that you can use.) This bulk-rate mailing permit on the envelope takes the place of a stamp.

In order to take advantage of bulk-rate mailing, you must send a minimum of 200 pieces, and the pieces must be arranged in bundles according to zip code. (Check with the post office for specific details.) The post office is then able to transmit bundles of questionnaires to a given zip code without separating and sorting them.

Both first-class and bulk-rate mail seem to move at the same speed through the mails, so bulk-rate mailing presents no disadvantage in terms of speed. Changes of address can present a problem, however. Technically, bulk-rate pieces will be forwarded only within a given city, but in practice some bulk-rate pieces are not forwarded even within a city and others are forwarded between cities. First-class mail is clearly safer in cases of address change.

The primary advantage of bulk rate is cost. Currently, a single mailing piece costs 8.4 cents for bulk rate in contrast to twenty-five cents for first class.[1] When hundreds or thousands of pieces are involved, the savings are considerable.

Bulk-rate mailing can be used only for outgoing questionnaires. They must be returned by first-class mail.

Stamps and Business Reply

Two basic options are available with regard to return postage for questionnaires. You can affix stamps to the envelope or self-mailing questionnaire to cover postage, or you can have the questionnaires imprinted with a business-reply mailing permit. Researchers differ in their assessments of the relative merits of these two methods.

The business-reply permit is similar to the bulk-rate permit in that it is printed on the mailing piece in place of postage stamps. (Check with the post office for additional format requirements.) Business-reply rates are the same as for first-class mail, however, plus a surcharge per piece returned through the mail.[2] As a result, you pay postage only on the questionnaires actually returned, but you pay more per questionnaire than if stamps were used. If stamps are affixed to the envelopes, you are paying postage whether the questionnaire is returned or not. As a general rule, then, you will save money using stamps if you achieve a very high return rate, and you will save money with business-reply postage if the return rate is low.

Other factors are involved in your decision, however. Business-reply postage is easier in that permits can be printed quickly and inexpensively, and you avoid the time and cost of licking and sticking hundreds or thousands of stamps. On the other hand, some researchers believe that the presence of postage stamps on the envelopes will be regarded as a sign of sincerity and that respondents will be more likely to return the questionnaire. (Others fear that respondents will steam off the stamps for use on other mail.)

Methodological studies of this issue do not seem to have resolved the matter. My personal preference is to use business-reply permits for reasons of ease and efficiency.

Monitoring Returns

The mailing of questionnaires sets the stage for an activity that can prove very valuable to the study. As questionnaires are returned to you, you should not sit back idly but should instead undertake a careful recording of methodological data.

---

[1] This bulk-rate price is for pieces weighing up to 3.39 ounces mailed by a nonprofit organization as of June 7, 1989.

[2] As of June 7, 1989, the surcharge ranged from five to forty cents per piece, depending on the type (and cost) of account established with the post office.

An invaluable tool in this activity is a return rate graph. The day on which questionnaires were mailed should be labeled Day 1 on the graph; every day thereafter, the number of returned questionnaires should be logged on the graph. Since this is a rather minor activity, it is usually best to compile two graphs. One should show the number returned each day, rising and then dropping as the number of returns changes. Another should report the *cumulative* number or percentage of returns. In part, this activity provides you with gratification because you get to draw a picture of your successful data collection. More important, however, it is your guide to how the data collection is going. If you plan follow-up mailings, the graph provides a clue as to when such mailings should be launched. (The dates of subsequent mailings should be noted on the graph.)

As completed questionnaires are returned, each should be opened, perused, and assigned an identification number. The numbers should be assigned serially as the questionnaires are returned, even if other identification (ID) numbers have already been assigned. This procedure can offer important advantages, as illustrated by the following two examples.

Assume you are studying attitudes toward a political figure. Further assume that in the middle of the data collection the politician in question is arrested for selling drugs. By knowing the date of that disclosure and the dates when questionnaires were received, you would be in a position to determine the effects of the disclosure.

As a less sensational example, another advantage of serialized ID numbers is their value in helping you estimate nonresponse biases in the survey. Barring more direct tests of bias, you might assume that respondents who failed to answer the questionnaire will be more like those who delayed answering than like those who answered right away. An analysis of questionnaires received at different points in the data collection might then be used for estimates of sampling bias. For example, if grade-point averages (GPAs) reported by students decrease steadily through the data collection, with those replying right away having higher GPAs and those replying later having lower GPAs, then you might tentatively conclude that those who failed to answer at all have lower GPAs yet. Though it would not be advisable to make statistical estimates of bias in this fashion, you could take advantage of approximate estimates.

If respondents have been identified for purposes of follow-up mailing, preparations for those mailings should be made as the questionnaires are returned. The case study that follows later in this chapter will discuss this process in greater detail.

Follow-Up Mailings

The methodological literature on follow-up mailings strongly suggests that it is an effective method for increasing return rates in mail surveys. In general, the longer potential respondents delay replying, the less likely they are to reply at all. Properly timed follow-up mailings, then, provide additional stimuli for responding.

The effects of follow-up mailings will be seen in the response rate curves recorded during data collection. The initial mailing will be followed by a rise and subsequent subsiding of returns, the follow-up mailing will spur a resurgence of returns, and further follow-ups will do the same. In practice, three mailings (an original and two follow-ups) seem to be most efficient.

The timing of follow-up mailings is important. Here the methodological literature offers less precise guides, but it has been my experience that two or three weeks is a reasonable space between mailings. (This period might be increased by a few days if the mailing time—out and in—is more than two or three days.)

Conducting several surveys over time with the same population should help you develop more specific guidelines in regard to follow-up mailings. At one time, the Survey Research Office at the University of Hawaii conducted frequent student surveys, and we were able to refine the mailing/remailing procedure considerably. Indeed, a consistent pattern of returns was found that appeared to transcend differences of survey content. quality of instrument, and so forth. Within two weeks after the first mailing, approximately 40 percent of the questionnaires would be returned; with two weeks after the first follow-up, an additional 20 percent would be received; and within two weeks after the final follow-up, an additional 10 percent would arrive. There are no grounds for assuming that a similar pattern would appear in surveys of different populations, but this illustration should indicate the value of carefully tabulating return rates for every survey conducted.

Follow-up mailings can be administered in a number of ways. The simplest way is to send nonrespondents a letter of additional encouragement to participate. A better method, however, is to send a new copy of the survey questionnaire with the follow-up letter. If potential respondents have not returned their questionnaires after two or three weeks, there is a good likelihood that the questionnaires have been lost or misplaced. Receiving a follow-up letter might encourage respondents to look for the original questionnaire, but if it is not easily found, the letter might go for naught. (The response rates reported previously all involved the sending of additional questionnaires.)

If the individuals in the survey sample are not identified on the questionnaires, remailing only to nonrespondents might not be possible. In such a case, you should send your follow-up mailing to all initial members of the sample, thanking those who have already participated and encouraging those who have not returned the questionnaire to do so. (The case study reported later in this chapter describes another method that can be used in an anonymous mail survey.)

A fair amount of attention has been devoted to ways of increasing response rates in mail surveys. In a recent study, for example, Richard J. Fox and colleagues[3] reviewed eighty-two surveys whose designs included tests of

---

[3] Richard J. Fox, Melvin R. Crask, and Jonghoon Kim, "Mail Survey Response Rates," *Public Opinion Quarterly*, vol. 52 (Winter 1988), pp. 467–491.

different response-increasing techniques, and a few consistent findings emerged. University sponsorship, for example, seems to improve response rates. The number of contacts with respondents is important, too. Also helpful are an advance letter notifying respondents that a questionnaire will follow and a follow-up postcard. Using first-class postage on outgoing envelopes might produce somewhat higher response rates than metered, bulk-rate mail, but the findings regarding this are not consistent. The authors also report that the color of the questionnaire can affect response rates (green does better than white).

Acceptable Response Rates

A question that new survey researchers frequently ask concerns the percentage return rate that should be achieved in a mail survey. It bears repeating here that the body of inferential statistics used in connection with survey analysis assumes that all members of the initial sample complete and return their questionnaires. Because this almost never happens, response bias becomes a concern; you must test for (and hope for) the possibility that the respondents are essentially a random sample of the initial sample and thus a somewhat smaller random sample of the total population.[4]

Nevertheless, overall response rate is one guide to the representativeness of the sample respondents. Achieving a high response rate results in less chance of significant response bias than achieving a low rate. But what is a high response rate?

A quick review of the survey literature will uncover a wide range of response rates. Each rate listed may be accompanied by a statement such as "This is regarded as a relatively high response rate for a survey of this type." (A United States senator made this statement regarding a poll of constituents that achieved a 4 percent return rate.) Despite the great variety of actual return rates and reactions to those rates, some rules of thumb can be followed.

A response rate of at least 50 percent is generally considered *adequate* for analysis and reporting. A response rate of at least 60 percent is considered *good,* and a response rate of 70 percent or more is *very good.* You should bear in mind, however, that these are only rough guides; they have no statistical basis, and a demonstrated lack of response bias is far more important than a high response rate. As a further rule of thumb, the articles contained in the 1988 volumes of the *American Sociological Review* and *Public Opinion Quarterly* report twenty-six survey response rates averaging 69 percent.[5]

---

[4] For more detailed examinations of nonresponse biases, see Marjorie N. Donald, "Implications of Nonresponse for the Interpretation of Mail Questionnaire Data," *Public Opinion Quarterly,* vol. 24, no. 1 (1960), pp. 99–114, and K. A. Brownlee, "A Note on the Effects of Nonresponse on Surveys," *Journal of the American Statistical Association,* vol. 52, no. 277 (1957), pp. 29–32.

[5] Some articles report more than one sample (and response rate), and a couple of articles concerned testing techniques aimed at improving response rates.

In computing response rates, the accepted practice is to omit all questionnaires that could not be delivered. In your methodological report, you should indicate the initial sample size and then subtract the number that could not be delivered due to bad addresses, death, and the like. The number of completed questionnaires is then divided by the *net* sample size to produce the response rate. As a result, the response rate is really a measure of your success in persuading sample members to participate; you do not count against yourself sample members you could not even contact.

While this procedure is the accepted practice, you should be aware of the logical assumption upon which it is based—that nondeliverable questionnaires represent a random sample of the initial sample. Of course, this might not be the case at all. Persons whose questionnaires cannot be delivered are, at the very least, probably more mobile than others in the sample, and mobility can be related to a variety of other variables. Tests for nonresponse bias are still the best guide.

A Case Study

The many steps involved in the administration of a mail survey can best be appreciated in a walk-through of an actual study. We conclude this chapter, then, with a detailed description of a survey conducted among University of Hawaii students. As you will note, the study did not represent the theoretical ideal for such studies, but in that regard it serves our present purpose all the better.

Sample Selection

The sample design and selection for this study were reported as a case study in Chapter 6. By way of general overview, recall that approximately 1,100 students were selected from the university registration tape through a stratified, systematic sampling procedure. For each student selected, the computer printed six self-adhesive mailing labels.

By the time the research team was prepared to distribute the questionnaires, it became apparent that research funds were inadequate to cover several mailings to the entire sample of 1,100 students, because questionnaire printing costs were higher than anticipated. As a result, a systematic two-thirds sample of the mailing labels was chosen, yielding a subsample of 770 students.

Postcards

An earlier decision had been made to keep the survey anonymous in the hope of encouraging more candid responses to some sensitive questions. (Subsequent surveys of the same issues among the same population indicated that anonymity was unnecessary.) Thus, the questionnaires would carry no identification of students. At the same time, it was hoped that follow-up mailing costs could be reduced by remailing only to nonrespondents.

To achieve both of these aims, a special postcard method was devised. Each student was mailed a questionnaire that carried no identifying marks, plus a postcard addressed to the research office with one of the student's mailing labels affixed to the reverse side of the card. The introductory letter asked students to complete and return the questionnaires—assuring them anonymity—and to simultaneously return the postcards. Receipt of the postcards would indicate that they had returned their questionnaires without identifying which questionnaires belonged to whom. This procedure would then facilitate follow-up mailings.

Questionnaire

The thirty-two-page questionnaire was printed in the form of a booklet (photo-offset and saddle-stitched). A three-panel cover, described elsewhere in this chapter, permitted the questionnaire to be returned without an additional envelope being provided.

Introductory Letter

A letter introducing the study and its purpose was printed on the front cover of the booklet. It explained why the study was being conducted (to learn how students felt about a variety of issues), how the students had been selected for the study, the importance of each student responding, and the mechanics of returning the questionnaire.

Students were assured that the study was anonymous, and the postcard method and rationale were explained. A statement of the auspices under which the study was being conducted followed, and a telephone number was provided for those who might want more information about the study. (About five students called for information.)

By printing the introductory letter on the questionnaire, the research team eliminated the necessity of enclosing a separate letter in the outgoing envelope, and the task of assembling mailing pieces was simplified.

Mailing-Piece Assembly

The assembly of materials for the initial mailing involved the following steps: (1) One mailing label for each student was stuck on a postcard; (2) another label was stuck on an outgoing manila envelope; (3) one postcard and one questionnaire were placed in each envelope with a check to ensure that the name on the postcard and on the envelope was the same in each case.

Assembly was accomplished through an assembly-line procedure involving the several members of the research team. Although the procedure was organized somewhat in advance, a certain amount of actual practice was required before the best allocation of tasks and persons was discovered. It is also worth noting that the entire process was delayed several days while the initial batch of manila envelopes was exchanged for larger ones. This delay could have been avoided if a walk-through of the assembly process had been carried out in advance.

Mailing

The distribution of the survey questionnaires had been set up for a bulk-rate mailing. After the questionnaires had been stuffed into the envelopes, they were grouped by zip codes, tied in bundles, and delivered to the post office.

Receiving Returned Questionnaires

Shortly after the initial mailing, questionnaires and postcards began arriving at the research office. Questionnaires were opened, perused, and assigned identification numbers as described earlier in the chapter.

The processing of postcards, however, pointed to an oversight in sample design. Recall from our earlier discussion of the sample design (Chapter 6) that the final arrangement of students in the sampling frame had been by Social Security number, thereby providing a quasi-stratification by region of origin. As a result, the mailing labels were printed in that order (within class strata). Social Security numbers had not been printed on the mailing labels, however, as they were not relevant to the study.

Given a postcard bearing a particular name and address, it was very difficult to locate the corresponding labels among those remaining in the many sheets of computer printout. Thus, it was necessary to cut apart all the labels from the printout sheets and alphabetize them. Then locating a given student's label could be accomplished with a minimum of effort. (*Note:* The labels could have been printed in alphabetical order initially if this problem had been anticipated.) For every postcard received, a search was made for that student's labels, and those labels were destroyed.

Follow-Up Mailings

After a period of two or three weeks, the remaining mailing labels were used to organize a follow-up mailing. The assembly procedures described earlier were repeated, with one exception. A special, separate letter of appeal was prepared and included in the mailing. The new letter indicated that many students had returned their questionnaires already, but that it was very important for all others to do so as well. The letter also indicated that the research office records might be in error, and if students had already returned their questionnaires they should ignore the second appeal and accept thanks for their assistance. If they had not already participated, however, they were encouraged to do so.

The follow-up mailing stimulated a resurgence of returns, as expected, and the same logging procedures were continued. Postcards were used as a basis for destroying the additional mailing labels. Unfortunately, time and financial pressures made it impossible to undertake a third mailing as had initially been planned, but the two mailings resulted in an overall return rate of 62 percent.

Summary

This chapter addressed the nuts and bolts of collecting data through the medium of the self-administered questionnaire. Two summary observations are in order.

First, it is my contention that such mundane matters are vital to high-quality scientific research. The researcher who believes that the quality of a research project is solely a function of analytical and theoretical abilities is sorely misled. A brilliant research design that is improperly executed will result in failure. No detail is so small or so mundane that it can be safely ignored.

Second, the detailed chronicle of one mail survey, with all its problems, should bring home the importance of pretests and pilot studies relating to all aspects of a study design. Only advanced testing can uncover problems that might eventually scuttle an entire study.

Additional Readings

Don A. Dillman, *Mail and Telephone Surveys: The Total Design Method* (New York: John Wiley & Sons, 1978).

Richard J. Fox, Melvin R. Crask, and Jonghoon Kim, "Mail Survey Response Rates," *Public Opinion Quarterly*, vol. 52 (Winter 1988), pp. 467–491.

10

Interview Surveys

Whereas Chapter 9 focused on self-administered questionnaires, this chapter presents alternative methods of data collection. In interview surveys, rather than asking respondents to read questionnaires and enter their own answers, interviewers ask the questions verbally and record the respondents' answers. Interviewing is done either in a face-to-face encounter or over the telephone. This chapter discusses both interview methods.

Importance of Interviewer

Having a questionnaire administered by interviewers rather than by the respondents themselves offers a number of advantages. To begin, interview surveys typically attain higher response rates than do mail surveys.[1] A properly designed and executed interview survey should achieve a completion rate of at least 80 to 85 percent. (Federally funded surveys often require this rate of completion.) It would seem that respondents are more reluctant to turn down an interviewer at their doorstep, or even on the phone, than they are to throw away a mail questionnaire.

Within the context of the questionnaire, the presence of an interviewer generally decreases the number of "don't knows" and "no answers." If minimizing such responses is important to a study, the interviewer can be instructed to probe for answers. ("If you had to pick *one* of the answers, which do you think would come *closest* to your feelings?")

---

[1] In an analysis of response rates to 517 surveys, John Goyder has concluded that the difference in response rates between interview and mail surveys has decreased in recent times and that the earlier differences were primarily a function of more extensive follow-up in the case of interview surveys. See John Goyder, "Face-to-Face Interviews and Mailed Questionnaires: The Net Difference in Response Rate," *Public Opinion Quarterly*, vol. 49 (Summer 1985), pp. 234–252.

Interviewers can also provide a guard against confusion over questionnaire items. If the respondent clearly misunderstands the intent of a question or indicates that he does not understand, the interviewer can clarify matters and thereby obtain relevant responses. (Such clarifications must be strictly controlled, however, through formal specifications, which we will explain later in this chapter.)

Finally, the interviewer can *observe* as well as ask questions. For example, a face-to-face interviewer can note the respondent's race if this is considered too delicate a question to ask. Similar observations can be made regarding the quality of the respondent's dwelling, the presence of various possessions, and other information. Even telephone interviewers can observe a respondent's ability to speak English, general reactions to the study, and so forth.

In one interview survey of students, respondents were given a short self-administered questionnaire to complete, concerning sexual attitudes and behavior, during the course of the interview. While the student completed the questionnaire, the interviewer made detailed notes on the respondent's dress and grooming.

Neutral Role of Interviewer

Survey research is of necessity based on an unrealistic *stimulus-response* theory of cognition and behavior. It must be assumed that a questionnaire item will mean exactly the same thing to every respondent and that every given response must mean the same when given by different respondents. Although this is an impossible goal, survey questions are drafted in such a way as to closely approximate the ideal.

The interviewer must also fit into this ideal situation. The interviewer's presence should affect neither a respondent's perception of a question nor the answer given. The interviewer, then, should be a *neutral* medium through which questions and answers are transmitted.

If this feat is successfully accomplished, different interviewers will obtain exactly the same responses from a given respondent. Interviewer neutrality has a special importance in face-to-face, household surveys. To save time and money, a given interviewer is typically assigned to complete all the interviews in a particular geographical area—a city block or a group of nearby blocks. If the interviewer does anything to affect the responses obtained, the bias thus interjected might be interpreted as a characteristic of the area under study.

Suppose that a face-to-face survey is being done to determine attitudes toward low-cost housing in order to help in the selection of a site for a new government-sponsored development. An interviewer assigned to a given neighborhood might, through word or gesture, communicate his or her own distaste for low-cost housing developments. Respondents might then tend to give responses in general agreement with the interviewer's own position. The results of the survey would indicate that residents of the neighborhood in question strongly resisted construction of the development in their area.

In some surveys, "race matching"—attempting to match the race/ethnicity of respondents and interviewers—might be appropriate. In a secondary analysis of the National Election Studies of 1964 to 1984, Anderson, Silver, and Abramson[2] found that black respondents were more likely to express feelings of warmth and closeness toward whites when they were interviewed by whites than when they were interviewed by blacks. The authors even suggest that an apparent shift in black hostility toward whites between 1976 and 1984 could actually be attributed to an increase in the number of black interviewers. Reese and colleagues[3] report that ethnicity has a similar impact in the case of Hispanic interviewers and respondents.

These examples are intended to highlight the importance of interviewer neutrality in the data-collection process and the difficulty of achieving it completely.

General Rules for Interviewing

The manner in which interviews should be conducted will be affected somewhat both by the survey population and by the nature of the survey content. Nevertheless, it is possible to provide some general guidelines that apply to most, if not all, interviewing situations.

Appearance and Demeanor

As a general rule, interviewers should dress in a fashion fairly similar to that of the people they will be interviewing. A richly dressed interviewer will probably have difficulty getting good cooperation and responses from poorer respondents, and a poorly dressed interviewer will have similar difficulties with richer respondents.

To the extent that the interviewers' dress and grooming differ from those of their respondents, it should be in the direction of cleanliness and neatness in modest apparel. If cleanliness is not next to godliness, it appears to be next to neutrality. Even though middle-class neatness and cleanliness might not be accepted by all sectors of American society, they remain the primary norm and are more likely to be acceptable to the largest number of respondents.

Dress and grooming are typically regarded as signals of a person's attitudes and orientations. At the time this is being written, a young woman wearing a black leather jacket, engineering boots, and a skull-and-crossbones tattoo on her forehead is likely to communicate—correctly or incorrectly—

[2] Barbara A. Anderson, Brian D. Silver, and Paul R. Abramson, "The Effects of the Race of the Interviewer on Race-Related Attitudes of Black Respondents in SRC/CPS National Election Studies," *Public Opinion Quarterly*, vol. 52 (Fall 1988), pp. 289–324.

[3] Stephen D. Reese, Wayne A. Danielson, Pamela J. Shoemaker, Tsan-kuo Chang, and Huei-ling Hsu, "Ethnicity-of-Interviewer Effects Among Mexican-Americans and Anglos," *Public Opinion Quarterly*, vol. 50 (Winter 1986), pp. 563–572.

that she is antiauthority in political outlook, sexually permissive, physically aggressive, prodrugs, and so forth. She is unlikely to obtain candid responses regarding Hell's Angels or punk rockers.

Though the issue of appearance is most obvious in the case of face-to-face interviews, you should realize that we communicate an identity through our speaking as well. Telephone interviewers are not exempt from the concerns expressed above. Consider the following (not recommended) openings to telephone interviews:

- Yo! Just listen to these questions and answer them. You got that? . . . Hey, I'm talking to you.
- So like, I gotta ask you some questions, see, and I need to, like fer sher, get your answers, y'know? Isn't that, like, rully awesome?
- So I'm doin' a survey, and that's a fact.
  And I need your answers, so I'll make some jack.

In demeanor, interviewers should be pleasant, if nothing else. Since they will be prying into the respondent's personal life and attitudes, they must communicate a genuine interest in getting to know the respondent without appearing to be a spy. Interviewers must be relaxed and friendly without being too casual or clinging. One of the most important natural abilities for interviewers to have is the ability to determine very quickly the kind of person the respondent will feel most comfortable with, in other words, the kind of person the respondent would most enjoy talking to. There are two aspects to the importance of a pleasant demeanor. Clearly, the interview will be more successful if the interviewer can become the kind of person the respondent is comfortable with. At the same time, however, there seems to be an ethical issue involved. Because the respondents are asked to volunteer a portion of their time and to divulge personal information about themselves, they deserve the most enjoyable experience the research designer and the interviewer can provide.

Familiarity with the Questionnaire

If interviewers are unfamiliar with the questionnaire, the study suffers and an unfair burden is placed on the respondent. In the latter respect, the interview is likely to take more time than necessary and be generally unpleasant. Moreover, familiarity cannot be accomplished by having the interviewer skim through the questionnaire two or three times. The questionnaire must be studied carefully, question by question, and the interviewer must practice reading it aloud. (Interviewer training is discussed in detail later in this chapter.)

Ultimately, the interviewer must be able to read the questionnaire items to respondents without error and without stumbling over words and phrases. A good guide for interviewers is the style of an actor reading lines in a play or motion picture. The lines must be read as naturally as if they constituted a natural conversation, but the conversation must exactly follow the language set down in the questionnaire. The interviewer should not attempt to memorize the questionnaire.

By the same token, the interviewer must be familiar with the specifications prepared in conjunction with the questionnaire. Inevitably, some items will not exactly fit a given respondent's situation, and a question will arise as to how such questions should be interpreted in the given situation. The specifications provided to the interviewer should give adequate guidance in such cases, but the interviewer must be sufficiently familiar with the organization and contents of the specifications to permit efficient reference to them. It would be better for the interviewer to leave a given question unanswered than to spend five minutes searching through the specifications for clarification and/or trying to interpret the relevant instructions.

Following Question Wording Exactly

Chapter 7, which discusses conceptualization and instrument construction, pointed out the effect of question wording on the responses obtained. Thus, a slight change in the wording of a given question might lead a respondent to answer yes rather than no. You might very carefully phrase your questionnaire items in such a way as to obtain the information you need and to ensure that respondents will interpret items in a manner appropriate to your needs only to have your effort wasted if interviewers rephrase questions in their own words.[4]

Recording Responses Exactly

Whenever the questionnaire contains open-ended questions, that is, questions soliciting the respondent's own answer, it is very important that the interviewer record those answers exactly as given. No attempt should be made to summarize, paraphrase, or correct bad grammar. The response should be written down exactly as given.

Recording exact responses is especially important because the interviewer will not know how the responses are to be coded prior to processing;

---

[4] Years ago, I was asked to talk about a quality-of-life survey I had designed for the elderly in Hawaii. Asked how we would handle the problem of respondents who couldn't speak English, I gave the "textbook answer," explaining that we would have one person translate the questionnaire into the second language and then have someone else translate it back into English. We would then compare the two English versions, make the necessary corrections in the translation, and have it administered by an interviewer who spoke the second language.

Also present in the room was a woman who had been doing a lot of the initial interviews for us, and she volunteered another solution. "If I find the respondent doesn't understand English very well, I just reword it so he can understand." As I sank lower and lower into my seat, she illustrated: "Like this question that says 'In general, how would you characterize your present state of health: Excellent, good, fair, or poor?' I just say 'Eh, you sick?'"

indeed, you might not know this yourself until you have had an opportunity to read a hundred or so responses. For example, the questionnaire might ask respondents how they feel about the traffic situation in their community. One respondent might answer that there are too many cars on the roads and that something should be done to limit their numbers. Another might say that there is a need for more roads. If the interviewer recorded these two responses with the same summary, "congested traffic," you would not be able to take advantage of the important differences in the original responses.

Sometimes a respondent is so inarticulate that the verbal response is too ambiguous to permit interpretation, but the interviewer is able to understand the intent of the response through the respondent's gestures or tone. In such a situation, the exact verbal response should still be recorded, but the interviewer should add marginal comments giving an interpretation and the reasons for arriving at it.

More generally, you will find it useful to have marginal comments to explain aspects of the response not conveyed in the verbal record, such as the respondent's apparent uncertainty in answering, anger, embarrassment, and so forth. In each case, however, the exact verbal response should also be recorded.

Probing for Responses

Respondents will sometimes respond to a question with an inappropriate answer. For example, the question might present an attitudinal statement and ask the respondent to "strongly agree, agree somewhat, disagree somewhat, or strongly disagree." The respondent, however, might reply, "I think that's true." In such situations, you should follow this reply with a statement like "Would you say you *strongly* agree or agree *somewhat*?" If necessary, you can explain that you have instructions to check one of the categories provided. If the respondent adamantly refuses to choose, you should accept that decision politely and write in the exact response given by the respondent.

Probing is more frequently required in eliciting responses to open-ended questions. For example, in response to the previous question about traffic conditions, the respondent might simply reply, "Pretty bad." The interviewer could obtain an elaboration on this response through a variety of probes. Sometimes the best probe is *silence*; if the interviewer sits quietly with pencil poised, the respondent will probably fill the pause with additional comments. (This technique is used effectively by newspaper reporters.) Appropriate verbal probes might be "How is that?" or "In what ways?" Perhaps the most generally useful probe is "Anything else?"

Frequently, the interviewer needs to probe for answers that will be sufficiently informative for analytical purposes. In every case, however, it is imperative that such probes be completely *neutral*. The probe must not in any way affect the nature of the subsequent response. Whenever you anticipate that a given question might require probing in order to elicit appropriate responses, you should present one or more useful probes next to the question in the questionnaire. This practice has two important advantages. First, you will have more time to reflect on and select the best, most neutral probes. Second, all interviewers will use the same probes whenever they are needed. Thus, even if the probe is not perfectly neutral, all respondents will at least be presented with the same stimulus. This logical guideline is the same as that discussed for question wording. Though a question should not be loaded or biased, every respondent must at least be presented with the same question, even a biased one.

Interviewer Training

Let's shift gears slightly now. Most interview surveys require the services of several interviewers, and you may one day find yourself responsible for training and supervising an interviewing staff. Therefore, let's take a little time to consider what is involved in that job.

If you are fortunate enough to organize an interviewing team comprised wholly of intelligent, experienced interviewers, a careful interviewer training course will still be absolutely essential.[5] Every survey and every questionnaire differ from every other one, and interviewers must be retrained for each new survey. The amount of time required for such training will depend on the scope and nature of the survey and the relative experience of the interviewers. A large household survey using a combination of experienced and inexperienced interviewers might normally require about two weeks of training: one week of classroom instruction and practice and another week of practice in the field. Interviewers can sometimes be trained for a short telephone poll in one or two long evening sessions. The following comments touch on the various aspects of interviewer training.

General Instruction

Interviewer training should begin with some general comments about the nature of the survey and its ultimate purpose. You should never regard and treat interviewers as unthinking automatons or as technicians who can be instructed to perform their duties by rote. Generally speaking, if interviewers understand why the survey is being conducted and realize that it is an important undertaking, they will be more diligent and careful in their work.

---

[5] The impact of interviewer training on the quality of data collected has been demonstrated in Jacques Billiet and Geert Loosveldt, "Improvement of the Quality of Responses to Factual Survey Questions by Interviewers," *Public Opinion Quarterly*, vol. 52 (Summer 1988), pp. 190–211.

The general description of the study should include the sponsor of the survey, the primary purpose of the study, and how the sample of respondents was selected. Interviewers should also be given a general picture of the other steps involved in the survey, including how the questionnaire was designed, how the data will be processed, and how data will be analyzed. If interviewers can understand how they fit into the overall process, they will probably feel more a part of the research team and will respond accordingly.

Administrative details should be covered early in the training. How long will the interviewing last? How many hours per week are interviewers expected to work? How often will they be paid? The interviewer who is personally worried about such details will be unable to concentrate on the intricacies of questionnaire items.

Even if all the interviewers are experienced, the training sessions should cover the general guidelines and rules for interviewing. Preparing a set of general guidelines to pass out and discuss with the interviewers is most useful. The most experienced interviewers will often be able to elaborate on some of the rules through examples from their past encounters, and the newer interviewers will get a better idea of what to expect during the actual interviewing.

Studying the Questionnaire
and Specifications

Interviewers and the supervisor and/or project director should go through the entire questionnaire step by step. The purpose of items should be explained, and all possible ambiguities should be discussed. It is essential that this portion of the training be conducted informally, as a discussion rather than a lecture. Interviewers who do not understand a particular point should be encouraged to ask for clarification immediately. Your response should in no way reflect dissatisfaction, or interviewers might sit quietly and not fully understand what is expected of them.

The examination of the questionnaire should involve simultaneous examination of the specifications. You should read the question and the answer categories, explain what is intended by the question, describe some simple situations and the appropriate action by the interviewer, and then describe some more complicated questions, turn to the appropriate section of the specifications, and show how those specifications would resolve each situation.

Much of the discussion during this activity will take the form of "What do I do if the respondent says . . . ?" Where appropriate, you should turn to the specifications and show how the situation should be resolved. Ultimately, the desired handling of the situation presented should be clearly described. The interviewers should know exactly how you would have handled the situation in question.

There is another side to this issue, however. Quite often, a situation will be raised that you cannot easily resolve. The possibility of such a situation might not have occurred to you, and the specifications will not cover it. You should not make snap decisions in such cases; rather, simply promise that the issue will be resolved before the next meeting. You should then carefully work out the appropriate solution, determine how it fits into the overall logic of the question's intent, and be prepared to explain your logic. The resolution should be explained to the interviewers at the next meeting, and the specifications should be updated to take it into account. The worst procedure would be for you to provide off-the-cuff, arbitrary answers without taking care to integrate those answers into the overall logic of the question and the specifications. Such responses will add up until interviewers have too many specific instructions to keep track of; the interviewers will no longer have a general logical understanding to guide them in their own interpretations.

Practice Interviews in Class

After the questionnaire and specifications have been studied in detail, the session should be organized around a series of practice interviews. To begin, two or three interviews should be conducted in front of the whole class. The best beginning would be for you to interview an interviewing supervisor, and vice versa. Because these interviews should serve as models for the interviewers, they should be conducted in precise accord with the general and specific instructions given previously. If you are being interviewed, you should either report your own condition and attitudes or take on a fictional, but consistent, identity. When you are conducting the interview, you should act just as if it were a real interview.

A specific development in the practice interview will sometimes warrant an explanatory comment to the class, but this should not be allowed to break the general mood of a real interview being conducted. At the completion of each interview, the class should be encouraged to discuss the interview, ask questions, and so forth.

Following the series of practice interviews in front of the class, the interviewers should be paired up to practice interviewing each other. These practice interviews need not be done in front of the class but should go on simultaneously. You should walk around the classroom, stopping to listen to the practice interviews, but for the most part it would be better to make notes and then discuss them with the entire class after a round of practice interviews has been completed. The interviewers should be re-paired and more practice interviews conducted. This portion of the training, then, should alternate between practice and discussion.

Practice Interviews in the Field

Upon completion of the classroom instruction, interviewers should be assigned practice interviews in the field, either going to people's homes or calling them on the telephone, depending on the study design. This portion of the training should be exactly like the real thing.

Assignment of respondents and/or addresses should be made just as it will be in the final survey. Normally, the person responsible for selecting the survey sample will provide a list of respondents who were not selected for the main sample.

The interviewers should be given exactly the same materials they will use in the final survey. They will then make their contacts with respondents and conduct their interviews. (Respondents should not be told that they have been selected for a practice interview, because doing this would destroy the realism of the practice.) Interviewers should also practice all administrative procedures at this point; they should fill out time sheets, logs, and so forth. In effect, the practice interviews in the field provide a pilot study of the entire interviewing operation.

Normally, it will be sufficient for interviewers to complete about five practice interviews in the field, although this number can vary in accordance with the nature of the survey. One aspect of the interviewing operation can be modified for practice. In the final survey, interviewers might be asked to make a specified minimum number of calls to a given respondent before judging that person "unavailable." To facilitate practice interviewing, you can omit this aspect from the practice interview. To account for such a modification, however, each interviewer should be given enough assignments to ensure that the specified number of interviews can be conducted.

As interviewers complete their practice interviews, they should review their work with the interviewing supervisor. The supervisor should go through the completed questionnaires with each interviewer to locate and discuss any problems and to answer any new questions the interviewer might have. In the case of a face-to-face interview survey, it might be a good idea to have interviewers report back with their first two or three interviews and then return to the field to complete the rest after review.

Although it is normally a good idea to establish a specific number of practice interviews to be completed, the general rule is that all interviewers must complete enough practice interviews to demonstrate either that they fully understand the task or that they will probably not be able to understand it. Thus, some interviewers might have to conduct more practice interviews than others. This situation is quite appropriate, even if the subsequent practice interviews overlap in time with the beginning of actual interviewing. It is far preferable for some interviewers to begin late than for them to make mistakes in interviewing the final sample. Mistakes might necessitate omitting the incorrectly interviewed respondents, reinterviewing them, or adjusting the sample design—all bad ideas.

The Interviewing Operation

The ongoing interviewing operation must be organized and controlled as carefully as the training session. Although the details of the operation will vary from survey to survey, some general points can be made.

Staff

The number of interviewers required is determined on the basis of (1) the number of interviews to be conducted, (2) the average time required for each interview, (3) the period of time allotted to the entire interviewing operation, and (4) the number of qualified interviewers available.

A good rule of thumb is to determine the number required on the basis of points 1, 2, and 3 and then recruit and train twice that number. During the course of training, many prospective interviewers will drop out voluntarily; others should be asked to drop out. It is generally better to begin interviewing with a few good interviewers than with a lot of poor ones; the latter usually cost more money, take up more supervisory time, and ultimately produce a lower quality of data. After weeding out poor interviewers, you should begin the actual interviewing with more interviewers than you anticipate will be needed throughout. Interviewers will continue to drop out during the study, and it will become apparent that still others should be asked to drop out. Typically, the interviewing staff is terminated in a staggered fashion near the end of the operation; the best interviewers can be asked to stay on longer for the wrap up.

Some surveys will require more than a single supervisor. Although one person should be responsible for supervising the entire operation, that person might be assisted by a staff of supervisors. As a general rule of thumb, one supervisor for every ten interviewers should suffice. In a very large interview survey with several interviewing supervisors, the person in overall charge should not be assigned interviewers to supervise. He or she will have enough work handling the logistics of the operation, coordinating with the project director, possibly recruiting and training additional interviewers, and, of course, supervising the supervisors.

The Actual Interviewing

In a face-to-face household survey, interviewers are given questionnaire booklets and a list of addresses denoting where they are to conduct the interviews. They will be responsible for finding the assigned addresses, knocking on the doors, and introducing themselves. If the interview is a very short one, it might be appropriate to conduct it on the doorstep, but any interview requiring more than a few minutes is more appropriately done in the respondent's home. In addition to making matters easier for the interviewer, conducting interviews indoors also affords the respondent more privacy.

Telephone interviewing, of course, is quite different and varies according to the systems available for the particular survey. At one end of the spectrum, a telephone interview can be very similar to a household survey, with the notable exception that the interviewer calls the respondent instead of driving to the respondent's house. That was pretty much the state of affairs in 1973 when this book was first written.

In recent years, telephone interviewing has evolved enormously, largely due to the possibilities offered by computers. As a consequence, what was generally looked upon as convenient but low in quality has become a widely used and accepted technique. We will take a few minutes, therefore, to consider a **computer-assisted telephone interviewing (CATI)** system.

Imagine yourself wearing a telephone-operator headset and sitting in front of a computer terminal and its video screen. A central computer randomly selects a telephone number and dials it automatically. (Random selection avoids the problem of unlisted telephone numbers.) As the number is being dialed, your introduction appears on the video screen before you ("Hello, my name is . . ."). When the respondent answers the phone, you say hello, introduce the study, and ask the first question displayed on the screen: "Could you tell me how many people live at this address?"

Whenever the respondent answers a question, you merely type that answer into the computer terminal, entering either the verbatim response to an open-ended question or the code category for the appropriate answer to a closed-ended question. The answer is immediately stored in the central computer. The next question appears on the video screen, you ask the question, and the answer is entered into the computer. The interview continues in this manner.

This technology is not science fiction but is being used increasingly by academic, government, and commercial survey researchers. Much of the development work for this survey technique has occurred at the University of California's Survey Research Center in Berkeley, sometimes in collaboration with the U.S. Department of Agriculture and other government agencies.

J. Merrill Shanks and Robert Tortora (1985, p. 4) have described some of the ways in which they have integrated computers into the survey process. In addition to its use in the scenario painted above, the computer is a valuable tool in the different stages involved in the development of questionnaires: drafting, testing, revising, and formatting. In addition, the logistics of the interviewing process (for example, training, scheduling, and supervising interviewers) can also be managed by computer.

Shanks and Tortora go beyond CATI, however, to talk more generally about computer-assisted survey methods. Many of the same techniques discussed with regard to telephone interviews can be applied to face-to-face household interviews. The development of portable microcomputers now permits an interviewer to set up shop in the respondent's home, reading questions off the screen and recording the answers as described above. In addition, it is sometimes appropriate to apply the computer to self-administered questionnaires. In some situations, respondents can be asked to sit at computer terminals and enter their own answers to the questions that appear on the screen.

Supervision and Editing

Procedures should be established concerning the regular reporting of interviewers to you as research director, or to their supervisor in the case of a very large study. In a face-to-face interview survey, the best procedure might be to establish a regular, weekly appointment for each interviewer. At the appointed time, the interviewer would check in with all the completed interviews, go over them with you, and receive a new set of assignments. Reporting is much simpler in the case of a telephone interview with all the interviewing done from a central location, but if interviewers are allowed to work out of their homes the more elaborate procedures just described apply.

You might begin by asking about and discussing any problems that arose in the interviews at hand. You should then read through each completed questionnaire, looking for missing answers, apparent errors, illegible entries, and anything else that makes the questionnaire difficult to use and interpret. You should also look for anything that suggests the interviewer does not understand the task or some part of it. Each error should be pointed out and discussed. The interviewer should end up knowing precisely how the situation should have been handled, so that the same error will not recur.

It is quite reasonable for the supervisor to vary both assignments and review for given interviewers on the basis of their initial performance. An interviewer who is doing a good job might be assigned more interviews at a time, and you might not have to be quite as careful in reading through completed questionnaires in the interviewer's presence. At the same time, you cannot allow a "halo effect" to develop with regard to interviewers. Even if some interviewers do a perfect job in the beginning, you must continue monitoring their later work.

Ultimately, every completed questionnaire must be edited in its entirety. In some cases, this task can be assigned to special editors, but normally the interviewer supervisors should do much, if not all, of the editing.

Reassigning Interviews

Interviewers should be trained in the importance of completing the interviews assigned to them. They should be prepared, in some instances, to persuade uncertain respondents to participate. At the same time, they should be warned against being too forceful in that attempt.

Frequently, an interviewer will call or arrive at a bad time and be turned down by the respondent, or something about the interviewer's appearance or demeanor might lead to an initial refusal. The interviewer should not force the respondent into adamant rejection; it might be possible for the supervisor to telephone the respondent to discuss the survey and gain cooperation.

Some interviewers are simply more successful in gaining cooperation than others. They are able to establish better and faster rapport with respondents. Such people will become identifiable over the course of the survey, and it might be advisable to specifically assign them to difficult interviews. If this is done, their efficiency should not be judged on the same basis as that of other interviewers, since completing difficult interviews typically takes more time. It might also be appropriate to pay them more money on the basis of the difficulty of the task and their special importance to the survey.

Verification of Interviews

All or a portion of the interviews should be verified by the supervisor. This verification can take several forms. As a minimum check, you should call the respondent by telephone, identify yourself, and verify that the interview was, in fact, conducted. In a more rigorous verification procedure, you might reask certain key questions and check the responses against those reported in the questionnaire. Requestioning should not be too extensive, however, because it takes up your time, takes up respondents' time (they have already volunteered time and information), and might even make respondents worry about the confidentiality of the survey.

There are two reasons for verifying interviews. Some interviewers might make honest mistakes in contacting assigned respondents, especially with a complex household sample design; it is also possible to misdial telephone numbers. Other interviewers will simply cheat. You should not assume that all your interviewers will faithfully attempt to contact and interview those people assigned to them. In fact, it is perhaps a rare survey in which no cheating occurs. Some interviewers will conduct only cursory interviews and fill in the questionnaire on that basis. Others will interview more convenient respondents. Still others will complete their questionnaires over a beer in the comfort of their own homes.[6]

Whenever you discover cheating, you should, of course, put an end to it, but at the same time you should try to find out why it occurred in the first place. You might conclude, if you are honest with yourself, that the cheating is attributable in part to your own handling of the interviewing operation. You might have asked too much of the interviewers, provided them with ambiguous instructions, or simply conveyed the impression that you regarded the field-work as below your status. Such discoveries might not save the survey already under way, but they might guide your subsequent surveys.

Firing Interviewers

In practice, it is very difficult to fire interviewers. No one wants to do it, especially after working closely with the interviewer during training and seeing that the interviewer seems to be trying very hard. Nevertheless, some interviewers will inevitably turn out to be unequal to the task. Keeping them at work hurts the financing, timing, and quality of the study.

Whenever a person is not working out as an interviewer, you should bear in mind that interviewing requires very special skills. The lack of these

---

[6] My colleague, Gary Marx, may have set the record in this category. In a national survey of black attitudes during the 1960s, he noticed a series of about twenty questionnaires—all from the same interviewer—that reported the most violently hostile antiwhite attitudes anyone on the research team had ever heard. Before Gary could investigate the matter, we received word that the interviewer and some friends had been arrested in a plot to blow up the Washington Monument and Statue of Liberty.

largely interpersonal skills does not mean that the person is unqualified for other jobs. It might be the case that a person who is a total failure as an interviewer will turn out to be first-rate at some other job in the survey. Both the interviewer and the project might profit from a reassignment rather than a simple termination.

Summary

This chapter has merely touched on some of the important elements involved in an effective interviewing operation. The intention of my comments has not been to train you as an interviewing supervisor but simply to inform you about what is involved. A good interviewing operation requires supervisors and interviewers whose own training and experience provide a far greater elaboration on each point covered so briefly here.

At the same time, the project director must play an active role in the interviewing process. You must know what is involved so that you can adjust your demands to the realities of field interviewing and be in a better position to evaluate the meaning of the data you receive. As was noted frequently in this chapter, your very presence in the interviewing process can have an important effect on morale and on the quality of the work accomplished.

Additional Readings

Norman M. Bradburn, Seymour Sudman, and Associates, *Improving Interview Method and Questionnaire Design* (San Francisco: Jossey-Bass, 1979).

Raymond L. Gorden, *Interviewing: Strategy, Techniques, and Tactics* (Homewood, IL: Dorsey Press, 1969).

Robert M. Groves and Nancy A. Mathiowetz, "Computer Assisted Telephone Interviewing: Effects on Interviewers and Respondents," *Public Opinion Quarterly*, vol. 48 (Spring 1984), pp. 356–369.

Robert L. Kahn and Charles F. Cannell, *The Dynamics of Interviewing* (New York: John Wiley & Sons, 1967).

Paul J. Lavrakas, *Telephone Survey Methods: Sampling, Selection, and Supervision* (Beverly Hills, CA: Sage, 1987).

Lois Oksenberg, Lerita Coleman, and Charles F. Cannell, "Interviewers' Voices and Refusal Rates in Telephone Surveys," *Public Opinion Quarterly*, vol. 50 (Spring 1986), pp. 97–111.

Stephen A. Richardson et al., *Interviewing: Its Forms and Functions* (New York: Basic Books, 1965).

J. Merrill Shanks and Robert D. Tortora, "Beyond CATI: Generalized and Distributed Systems for Computer-Assisted Surveys," prepared for the Bureau of the Census, First Annual Research Conference, Reston, VA, March 20–23, 1985.

Michael F. Weeks, Richard A. Kulka, and Stephanie A. Pierson, "Optimal Call Scheduling for a Telephone Survey," *Public Opinion Quarterly*, vol. 51 (Winter 1987), pp. 540–549.

11

Data Processing

What the microscope was to biology and what the telescope was to astronomy, the computer has been to modern social science. Moreover, you are taking this course at a time when the computer's contribution to survey research is still being discovered. This chapter introduces you to that contribution.

The purpose of this chapter is to describe methods of converting survey research data into a *machine-readable* form, that is, a form that can be read and manipulated by computers and similar machines used in quantitative data analysis. If you were conducting a research project more or less parallel to your reading the chapters of this book, your data, at this point, would be in the form of completed questionnaires, coded transfer sheets, or the like. At the completion of the stage covered in this chapter, your data would have been recorded on microcomputer floppy disks, cassettes, punch cards, or some other device able to be read and analyzed by a machine.

Given the pace of computer development and dissemination today, it is impossible to anticipate the kind of equipment you might have available for your use. This chapter provides you with an overview of several stages in the evolution of computers in survey research, with the intent of saying something about whatever equipment you have to work with. Moreover, the earlier equipment and techniques sometimes reveal the logic of data analysis more clearly than do today's advanced equipment and techniques, in the same sense that you can learn the fundamentals of how an automobile works more easily from a Volkswagen Bug than from the latest, fanciest Maserati.

Computers in Survey Research

For our purposes, the history of computing in social research began in 1801, about the same time the seeds of modern social science itself were germinating. In that year, Frenchman Joseph-Marie Jacquard created a revolution in the textile industry that was to have an impact in the most unlikely corners of life.

To facilitate the weaving of intricate patterns, Jacquard invented an automatic loom that took its instructions from punched cards. As a series of cards passed through the loom's "reader," wooden pegs poked through the holes punched in the cards, and the loom translated that information into weaving patterns. To create new designs, Jacquard needed merely to punch the appropriate holes in new cards, and the loom responded accordingly.

The point to recognize here is that *information* (in this case, a desired weaving pattern) could be *coded* and *stored* in the form of holes punched in a card and subsequently *retrieved* by a machine that read the holes and took action based on the meaning assigned to those holes. This is the fundamental logic we will see repeated in different forms in the following discussion.

The next event in our selective computer history took place in the United States during the 1890 census. As you might know, the U.S. Constitution mandates a complete census of the nation's population every ten years, beginning with the 1790 enumeration of just under 4 million Americans. As the new nation's population grew, however, so did the task of measuring it. The 1880 census enumerated over 62 million people, but it took the Census Bureau *nine years* to finish its tabulations. Clearly, a technological breakthrough was required before the 1890 census. The bureau sought suggestions.

A former Census Bureau employee, Herman Hollerith, had an idea. Hollerith had worked on the 1880 census. As a young engineering instructor at M.I.T., he proposed adapting the Jacquard card to the task of counting the nation's population. As local tallies were compiled, they would be punched into cards. Then a tabulating machine of Hollerith's creation would read the cards and determine the population count for the entire nation.

Hollerith's system was tested in competition with other proposals and found to be the fastest. As a result, the Census Bureau rented $750,000 worth of equipment from Hollerith's new Tabulating Machine Company, and the 1890 population total was reported within six weeks, in contrast to the nine years required for reporting the 1880 census. Hollerith's Tabulating Machine Company, incidentally, continued to develop new equipment, merged with other pioneering firms, and was eventually renamed the International Business Machines Corporation—IBM.

Storing Data on Cards

By the 1950s, punched cards—commonly called *IBM cards*—were being adapted for the storage and retrieval of survey research data; they are still sometimes used for that purpose. Even if you never work with data stored on punch cards, the logic of that method can be useful in helping you understand the ways in which data are more typically stored and retrieved these days. Let's examine how the kinds of data discussed previously in this book might be stored on punched cards.

A punch card like the one shown in Figure 11-1 is divided into eighty vertical columns, usually numbered from left to right. Each vertical column is further divided into ten spaces, numbered 0, 1, 2, 3, 4, 5, 6, 7, 8, 9 from top to bottom. You store data on the cards by punching holes within the columns. Later, computers and other data processing equipment can be used to retrieve data from the card by locating and reading specified columns.

A keypunch machine punches holes in the spaces in columns of punch cards. Using a keyboard similar to that of a typewriter, the keypunch operator can punch specified holes (0, 1, 2, . . .) in specified columns of a given card.

Data are put in machine-readable form by the assignment of one or more specific columns of a data card (a *field*) to a variable and the assignment of punches within that column to the various attributes composing that variable. For example, an experimental subject's sex might be recorded in column 5 of the card. If the subject were male, a 1 might be punched in that column; if female, a 2. Recalling our earlier terminology, notice that columns correspond to *variables* and that the different punches within a column correspond to the *attributes* constituting that variable. A subject's age might be assigned to columns 6 and 7 (a two-column code); if the subject were thirty-five years of age, 3 and 5, respectively, would be punched in those columns. Ages could instead be recorded in categories and stored in a single column, for example, 1-punch for under 20, 2-punch for 20 to 29, and so forth.

A given card, then, represents the data relevant to a given subject—the unit of analysis. If the units of analysis were newspaper editorials being examined in a content analysis, each data card would represent an editorial. The columns of each card would be assigned to specific variables describing that editorial. For example, two columns might be assigned to storing the last two digits of the year in which the editorial appeared.

In survey research, a data card might stand for a questionnaire, with columns assigned to the various items on the questionnaire. Column 34 might be used to store answers to the question "Have you ever smoked marijuana?"

Figure 11-1 *Standard Punch Card for Recording Data*

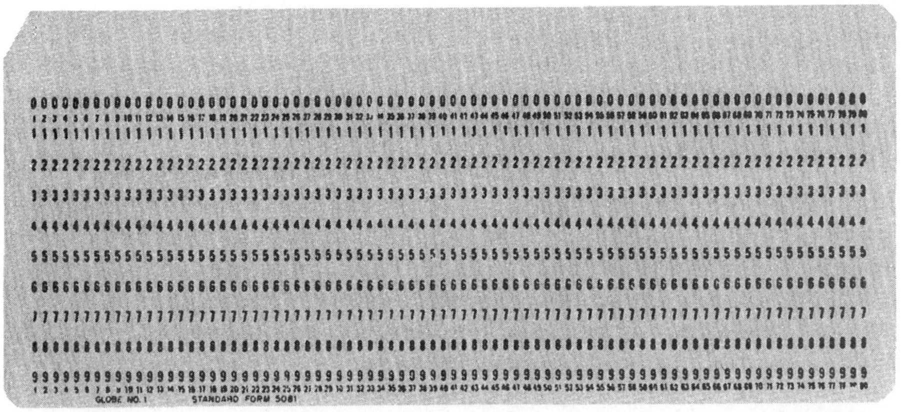

A 1-punch could represent "yes" and a 2-punch "no." The main ideas for you to grasp are that each card represents a single research unit of analysis and that each column (or set of columns) is used for the same variable on each card. Thus, *every* survey respondent's answer to the marijuana question would be stored in column 34.

Retrieving Data from Cards

Several pieces of equipment are capable of reading punch cards. The simplest machine among these is the counter-sorter. Although it is unlikely that you will ever have a chance to use a counter-sorter, you should understand the logic of how the machine works, since it will clarify what more complex machines do. The counter-sorter can be set to read a given column. Then, when the cards are run through it, they are sorted into pockets corresponding to the codes punched in the column specified; a counter indicates the number of cards in each pocket. If sex is recorded in column 5 and the sorter is set for that column, men and women will be sorted into the 1 and 2 pockets, respectively, and the sorter will keep a running total of the cards in each pocket.

The counter-sorter can also be used to examine the relationships between variables. Suppose you wanted to determine whether men or women were more likely to have smoked marijuana. Having separated the respondents by sex, you would set the counter-sorter to read the column that contains responses to the question about smoking marijuana—column 34, for example. All the "men" cards would then be rerun through the counter-sorter to determine whether the respondents had ever smoked marijuana, as indicated by the punches in column 34. The same procedure would be repeated for the women, and the distributions of responses would then be compared.

For years, the counter-sorter was the primary machine used for the analysis of survey research data, but it had three basic limitations. First, it was limited to counting and sorting cards. Although you might use the results for extremely sophisticated analyses, the machine itself couldn't perform sophisticated manipulations of data. Second, the counter-sorter was rather slow in comparison with machines invented later. Third, it was limited to the examination of one card per unit of analysis when looking at the relationships among variables. In effect, you were limited to eighty columns of data per unit of analysis.

Entrance of Computers

Most data analyses today are conducted with computers, ranging from large "mainframe" computers to small, personal microcomputers. You don't need to be told that computers are revolutionizing most areas of modern social life. At the same time, some people worry that computers are "taking over." I hope the following discussion will make it clear that computers are merely *tools*, like typewriters and pocket calculators—but what powerful tools!

The computer, through manipulation programs, can avoid the limitations of the counter-sorter. First, it can go beyond simple counting and sorting to perform intricate computations and provide sophisticated presentations of the results. The computer can be programmed to examine several variables simultaneously and to compute a variety of statistics. Second, data stored on magnetic disks or tapes can be read much faster than was ever possible with cards and the counter-sorter. Computers can also calculate complex statistics a good deal faster and more accurately than humans. Finally, the computer is not limited to the eighty columns of a punch card.

Today, most quantitative social scientific data analyses use computer programs that instruct the computer to simulate the counter-sorter process described earlier—and much more. A number of computer programs are available today for the analysis of social science data. Here are just a few of these programs: ABtab, AIDA, A.STAT, BMDP, CRISP, DAISY, DATA-X, Dynacomp, INTERSTAT, MASS, MicroCase, Microquest, Microstat, MicroSURVEY, Minitab, POINT FIVE, P-STAT, SAM, SAS, SNAP, SPSSX^x, Statgraf, Statpak, StatPro, STATS PLUS, Statview, Survey Mate, SYSTAT, STAT80, STATA, SURVTAB, TECPACS. Whatever program you use, the basic logic of data storage and analysis is the same.

Until the late 1970s, all computer analyses were performed on large, expensive computers—sometimes called mainframe computers—maintained by centralized computer centers; most analyses still are. To make use of such facilities, you need to supply the data for analysis and the instructions for the analysis you want. You might take your data to the computer center in the form of cards or a magnetic tape, or you might simply reference a data set maintained in the center's data library. (Your instructor will tell you how to do this if you are to perform computer analyses in your course.)

Once you arrive at the computer center, you will typically submit your data and analysis instructions to a machine operator who enters the job into the computer, though some centers are set up so that researchers can run their own jobs. After your job has been submitted, your next task is to wait. Sometime later—minutes (rarely), hours, or days, depending on the center's workload— the machine operators will return your completed job, usually in the form of a *printout* of the results of your analysis.

In recent years, two developments have improved this process. Systems place card readers and printers near the researchers, who can then submit jobs and receive results at locations some distance from the computer center. Your departmental office might have such equipment, which allows requests and results to be sent to and from the computer center over coaxial cables.

Time sharing is a further advance in using computers for analysis. In a sense, computers have always operated on a time-sharing basis, with several users sharing the same computer. Initially, however, computer facilities were shared in a serial fashion; the computer would run your job, then mine, then someone else's. Current, more sophisticated computers, however, can perform several different tasks simultaneously. They can read my request, analyze yours, and print out someone else's all at the same time. In fact, the tremendous speed of computers makes it possible to sandwich small operations (that might take only thousandths of a second) in the middle of larger jobs. This capacity has enabled computers to handle requests from hundreds of users simultaneously, often giving the impression that each user has the computer's complete attention. Such interactions with the computer typically take place at *computer terminals*—typewriter devices that either print on paper or display operations on CRT (cathode ray tube) video monitors that look like television screens.

The use of computer terminals on a time-sharing basis supports a variety of uses. Text-editing programs allow users to write, edit, and print out reports. If your computer center supports a time-sharing system, you will probably find a variety of video games stored somewhere in the system.

For our present purposes, however, we will discuss the use of time sharing for social science data analysis. The following is a common pattern. Your data are maintained by the computer center, either on tape or disk. Instead of punching cards requesting a particular analysis, you type instructions on a terminal. After you have prepared the entire set of instructions, you type a final instruction (which will vary from computer to computer) requesting that the job be run. The computer then takes over, eventually producing your results, either to be picked up at the center or printed out on your terminal.

The use of computer time sharing in survey research became even more practical with the development of portable computer terminals no larger than small, portable typewriters. Such terminals communicate with the computer over standard telephone lines. Thus, you can take a portable terminal home or on a trip around the world. To use it, you dial a special number on your regular phone. The computer answers, usually with a shrill tone. You then fit the telephone receiver into two rubber cups, called an *acoustical coupler*, located on the terminal. (Increasingly, portable terminals have communications equipment—called *modems*—built in, so the terminal can be connected directly into a standard telephone jack.) Once connected, anything you type on the terminal is transmitted over telephone lines to the computer. The computer does what you ask and sends the results back to your terminal.

Time-sharing systems can also connect you with nationwide computer *networks*. The possible uses of such networks are mind-boggling, and presently available services only scratch the surface of their potential. You can, for example, dial a special number and connect with a computer that will allow you to search rapidly through past editions of the *New York Times* and locate articles on some topic of interest to you. If you want, you can have the articles printed out on your terminal, or you can leave messages in the network to be picked up by other users.

More relevant to our current discussion is the fact that computer networks make it possible for a researcher sitting at home in California to get a copy of a data set maintained on a computer in Massachusetts and to analyze those data using a program maintained on a computer in Texas. The results of the analysis would be returned to the user in California and could also be stored for examination by a colleague in Michigan. This is not to suggest that the procedure described here is commonplace today, but it is likely to become commonplace early in your research career; before the end of your career, you'll look back on all that as primitive.

Microcomputers

The most useful and exciting development to date has been the *microcomputer*. You're probably familiar with microcomputers and might even have used one. These machines are small, complete computers not much larger than a typewriter (or even smaller). Operations are displayed on a cathode ray tube (CRT), which looks like a television set, or on some other type of video-display screen. (In some systems, you can use a standard television set for the video display.) Early microcomputer systems used standard tape cassettes to store data between uses; these have now been replaced by 5.25-inch "floppy disks" or 3.5-inch "microcassettes"—round sheets of magnetic tape encased in cardboard or plastic. The more advanced systems utilize "hard disks" that store data in the millions of bytes (*megabytes*). The microcomputer on which I typed this text has a built-in hard disk that stores twenty megabytes.

Microcomputers have already been proved effective in a variety of tasks. The same microcomputer used in writing this book is also used for all my correspondence; in addition, it schedules my time and prints daily appointments and to-do lists, maintains my checkbook, handles my personal telephone directory, and performs other tasks too numerous to list. When I am traveling, I can check into a hotel, unplug the regular telephone in my room, plug in my computer, and be in touch with people and computers around the country. When I want a break from writing, I can play chess with the computer or save the world from alien invaders.

In 1986, I wrote: "Microcomputers are just starting to blossom in the arena of social science research. The main drawback has been their small *memory* size, but modern micros are overcoming that problem impressively." That problem has now been resolved, as the following illustration demonstrates.

When I was a graduate student just twenty-five years ago, Berkeley's Survey Research Center had an IBM 1620 computer, which occupied approximately the same space as six or seven refrigerators. Its memory capacity was 24 kilobytes (24K bytes), or approximately 24,000 characters worth of information. Just recently, by contrast, I purchased a battery-operated calculator/computer about the size of a deck of playing cards. Its memory is 32K bytes! The desktop computer I used to write this book has over 5 million bytes (five *megabytes*) of memory (over 200 times the capacity of that old IBM 1620), and its built-in hard disk stores 80 megabytes.

As the memory capacity of microcomputers has mushroomed, data analysis programs have been developed to bring the examination of social science data within the capability of micros. The technological evolution is far from complete. Introducing a special issue of *Sociological Methods & Research* on "Microcomputers and Social Research," David R. Heise says:

> Microcomputers are cheap, reliable, portable, computationally powerful, and easy to use. This profile makes them significantly different from mainframe computers and guarantees wide diffusion. By the end of the decade, microcomputers will have changed the way social scientists do research, the way they teach courses, and the way they work in applied settings. Microcomputers also will create new topics for social analysis as the microcomputer revolution reaches diverse sectors of society.[1]

You now have an overview of the role played by computers in survey research. The remainder of this chapter discusses the steps (and options) involved in converting data into forms amenable to computer analysis. We will discuss the coding process and enumerate the many methods available for transforming data into a machine-readable form.

Coding

For computers to work their magic, they must be able to read the data you have collected in the course of your research. Moreover, computers are at their best with numbers. If a survey respondent tells you that he or she thinks the biggest problem facing Stowe, Vermont, today is "the threat of thermonuclear war," the computer can't understand that response. You must translate through a process called **coding**. The discussion of content analysis in Chapter 2 dealt with the coding process in a manner relevant to our present concern. Recall that the content analyst must develop methods of assigning individual paragraphs, editorials, books, songs, and so forth specific classifications or attributes. In content analysis, the coding process is inherent in data collection or observation. When other research methods are employed, it is often necessary to engage in a coding process after the data have been collected. For example, open-ended questionnaire items result in nonnumerical responses that must be coded before analysis.

As with content analysis, the task here involves reducing a wide variety of idiosyncratic items of information to a more limited set of attributes composing a variable. Suppose, for example, that you have asked respondents "What is your occupation?" The responses to such a question would vary considerably. Although it would be possible to assign a separate numerical code to each separate occupation reported, such a procedure would not facilitate analysis. Analysis typically depends on several subjects having the same attribute.

---

[1] David Heise, ed., *Microcomputers and Social Research*, vol. 9, no. 4 (Beverly Hills, CA: Sage, 1981), p. 395.

The occupation variable has a number of preestablished coding schemes (none of them very good, however). One such scheme distinguishes among professional and managerial occupations, clerical occupations, semiskilled occupations, and so forth. Another scheme distinguishes among different sectors of the economy: manufacturing, health, education, commerce, and so forth. Still others combine both these schemes.

The occupational coding scheme chosen should be appropriate to the theoretical concepts being examined in the study. For some studies, coding all occupations as either white collar or blue collar might be sufficient. For others, self-employed and not self-employed might suffice. A peace researcher might want to know only whether the occupation was dependent on the defense establishment.

Although the coding scheme ought to be tailored to meet particular requirements of the analysis, one general rule of thumb should be noted. If the data are coded in such a way as to maintain a great deal of detail, the code categories can be combined during any analysis that does not require such detail. If the data are coded into relatively few gross categories, however, there is no way to re-create the original detail during analysis. Thus, you would be well advised to code your data in somewhat more detail than you plan to use in the analysis.

Developing Code Categories

There are two basic approaches to the coding process. First, you might begin with a relatively well-developed coding scheme derived from your research purpose. Thus, as suggested above, the peace researcher might code occupations in terms of their relationship to the defense establishment. Or suppose you have been engaging in participant observation of an emerging new religion and have been keeping very careful notes of the reasons new members have given for joining. Perhaps you have developed the impression that new members seem to regard the religion as a family substitute. You might, then, review your notes carefully, coding each new member's comments in terms of whether this aspect of the religion was mentioned. You might also code members' comments in terms of whether or not they have a family.

If you are fortunate enough to have assistance in the coding process, your task would be to refine your definitions of code categories and train your coders so that they will be able to assign given responses to the proper categories. You should explain the meaning of the code categories you have developed and give several examples of each category. To ensure that your coders fully understand what you have in mind, it would be useful for you to code several cases. Your coders should then be asked to code the same cases, without being told how you coded them, and your coders' work should be compared with your own. Any discrepancies will indicate an imperfect communication of your coding scheme to your coders. Even if this exercise shows perfect agreement between you and your coders, however, you should still *check-code* at least a portion of the cases throughout the coding process.

If you are not fortunate enough to have assistance in coding, you should still obtain some verification of your own reliability as a coder. Nobody is perfect, especially not a researcher hot on the trail of a finding. In your study of an emerging religion, suppose you have the impression that people who do not have a regular family will be more likely to regard the new religion as a family substitute. The danger is that whenever you discover a subject who reports no family you will unconsciously try to find some evidence in the subject's comments that the religion is serving as a family substitute. If at all possible, then, you should try to get someone else to code some of your cases to see if that person would make the same assignments you made. (Note how this situation relates to the characteristic of *intersubjectivity* in science.)

The second approach to coding is appropriate whenever you are not sure initially how your data should be coded because you do not know what variables the data represent among your subjects of study. Suppose, for example, that you have asked "What do you think about the idea of a nuclear freeze?" Although you might anticipate coding the responses as positive, negative, or neutral, it is unlikely that you could anticipate the full range of variation in responses. In such a situation, you should prepare a list of perhaps 50 or 100 actual responses to the open-ended question. You could then review that list, noting the different dimensions reflected by the responses. Perhaps you would find that several of the positive responses contained references to the cost of the nuclear arms race or that a number of the negative responses referred to the threat of world communism.

After you have developed a coding scheme based on the list of 50 or 100 responses, you should make sure that each listed response fits into one of the code categories. Then you would be ready to begin coding the remaining responses. If you have coding assistance, the previous comments regarding the training and checking of coders apply here; if you do not, the comments on having your own work checked by someone else apply.

Like the set of attributes composing a variable and the response categories in a closed-ended questionnaire item, code categories should be both exhaustive and mutually exclusive. Every piece of information being coded should fit into *one and only one* category. Problems arise both when a given response appears to fit equally into more than one code category and when it fits into none.

Codebook Construction

The end product of the coding process is the conversion of data items into numerical codes. These codes represent attributes composing variables, which, in turn, are assigned card and column locations within a data file. A **codebook** is a document that describes the locations of variables and lists the code assignments of the attributes composing those variables. A codebook serves two essential functions. First, it is the primary guide used in the coding process. Second, it is your guide to locating variables and interpreting codes in your data file during analysis. If you decide to correlate two variables as a part of your analysis of your data, the codebook tells you where to find the variables and what the codes represent.

Figure 11-2 illustrates portions of the codebook created in connection with a 1987 survey of college faculty members and administrative staff. The format for the codebook reflects the particular program, MicroCase, used to process and analyze the survey data. Other programs use similar formats.

These formats approximate the punch-card columns described earlier and are used to identify the location of data items. We can determine the sex of respondents, for example, by referencing VAR. 5. Also, MicroCase, like many similar programs, allows us to give an abbreviated name to each data item;

Figure 11-2   *Partial Example of a Codebook*

VAR. 1: Curr Appt
What is your current appointment at Chapman?
1. Administration
2. Tenured faculty
3. Untenured, tenure-track faculty
4. Term contract faculty
5. Staff
6. Mixed appointment
7. Other

Var. 2: Acad Rank
What is your academic rank?
1. Assistant Professor
2. Associate Professor
3. Professor
4. Adjunct Professor
5. Other
6. Not applicable

Var. 3: Acad Field
What is your academic field?
1. Business
2. Education
3. Humanities
4. Movement and Exercise Science
5. Natural Sciences
6. Social Sciences
7. Mixed or other
8. Not applicable

VAR. 4: Degree
What is the highest academic degree you hold?
1. Bachelor's
2. Master's
3. Doctorate
4. Other

VAR. 5: Sex
What is your sex?
1. Female
2. Male

Var. 6: No/Directn
The college lacks a sense of direction.
1. Strongly agree
2. Agree
3. Disagree
4. Strongly disagree
5. Don't know

Var. 7: Toward/Res
The college is moving away from teaching toward a greater emphasis on research.
1. Strongly agree
2. Agree
3. Disagree
4. Strongly disagree
5. Don't know

---

thus, ACAD RANK can be used in place of VAR. 2 in accessing the data on respondents' academic ranks. This capability is particularly useful in that you can assign names that you can remember later when you analyze your data.

In addition, every codebook should contain the full wording of the questions asked; as we noted earlier, the wording of questions has a strong impact on the answers returned. Finally, your codebook should indicate the attributes comprising each variable. Thus, in VAR. 7, we see that faculty respondents were asked to "Strongly agree," "Agree," "Disagree," "Strongly disagree," or say they didn't know. Many programs, including MicroCase, allow you to assign abbreviated names to those attributes (for example, SA for "Strongly agree," A for "Agree," and so on) that will be used in displaying the results of your analyses, making tables much easier to read, among other things.

Coding and Data Entry Options

Not long ago, as I have already indicated, nearly all data entry took the form of manual key-punching, with punch cards either analyzed by unit-record equipment or read into computers for more complex analyses. Recent years have brought major advances in data entry. Through the use of computer terminals and microcomputers, data are now typically keyed directly into data files stored on computer disks. As before, however, this data entry is intimately related to coding, and a number of methods can be used for effecting that link. We will look at a few of the possibilities.

Transfer Sheets

The traditional method of data processing involves the coding of data and the transfer of code assignments to a *transfer sheet* or *code sheet*. Such sheets were traditionally ruled off in eighty columns, corresponding to the data card columns, but they can be adapted to other data configurations appropriate to the data entry method. Figure 11-3 provides an illustration of a transfer sheet. Coders write numbers corresponding to the desired code categories in the appropriate columns of the sheets. The code

Figure 11-3 *A Partial Coding Transfer Sheet*

01	02	03	04	05	06	07	08	09	10		71	72	73	74	75	76	77	78	79	80
0	0	1	3	7	8	9	3	1	1		4	5	2	1	1	7	8	7	1	2
0	0	2	4	2	4	2	4	1	2		2	2	5	1	1	2	8	2	2	2
0	0	3	6	6	1	2	3	1	1		1	3	6	2	1	3	9	2	2	3
0	0	4	5	3	4	4	2	2	1		1	4	1	0	3	6	0	4	2	1

sheets are then used for keying data into computer files. This technique is still useful when particularly complex questionnaires or other data source-documents are being processed.

Edge-Coding

Edge-coding does away with the need for code sheets. The outside margin of each page of a questionnaire or other data source-document is left blank or is marked with spaces corresponding to variable names or numbers. Rather than code assignments being transferred to a separate sheet, the codes are written in the appropriate spaces in the margins. The edge-coded source-documents are then used for data entry.

Direct Data Entry

If the questionnaires have been adequately designed, you can often enter data directly into the computer without using separate code sheets or even edge-coding. The precoded questionnaire would contain indications of the columns and the punches to be assigned to questions and responses, and data could be entered directly. This procedure has been greatly enhanced by some of the new computer programs, as illustrated in Figure 11-4. This example is taken from the MicroCase program, but other programs follow a similar procedure. (The boldfaced numbers are those you would enter; the rest is done by the computer.)

Notice in Figure 11-4 that the computer prompts you for data entries by repeating the full question and reminding you of all the possible code numbers that can be entered as answers to that question. As a practical matter, you can enter a data set more rapidly by referring to the information on the screen only from time to time to ensure that you haven't lost your place in the data entry process. When the direct-entry method is to be used, documents must be *edited* before data entry. An editor should read through each document to confirm that every question has been answered (enter a 99 or some other standard code for missing data) and to clarify any unclear responses.

If most of the document is amenable to direct entry (for example, closed-ended questions presented in a clear format), you could also code a few open-ended items and still enter data directly. In such a situation, you should enter the code for a given question in a specified location near the question in order to ease the data entry job.

The layout of the document is extremely important for effective direct data entry. If most of the question and response categories are presented on the right side of the page, but one set is presented on the left side, persons entering the data frequently miss the deviant set. (*Note:* Many respondents will make the same mistake, so a questionnaire carefully designed for data entry will be more effective for data collection, as well.)

For content analysis and similar situations in which coding takes place during data collection, it makes sense to record the data in a form amenable to

Figure 11-4 *Computerized Data Entry Illustrated*

Case 1

What is your current appointment at Chapman?
- 1. Admin
- 2. Tenured
- 3. Ten'r trk
- 4. Term fac
- 5. Staff
- 6. Mixed
- 7. Other

Var. 1. Curr Appt   **2**

What is your academic rank?
- 1. Asst Prof
- 2. Assoc Prof
- 3. Professor
- 4. Adj Prof
- 5. Other
- 6. Not appl

Var. 2. Acad Rank   **3**

---

direct entry. A precoded form might be appropriate, or, in some cases, the data might be recorded directly on transfer sheets.

Data Entry by Interviewers

The easiest data entry method in survey research—computer-assisted telephone interviewing, or CATI—has already been discussed in Chapter 10. As you will recall, interviewers with telephone headsets sit at computer terminals, which display the questions to be asked and wait for the interviewer to type in the respondents' answers. In this fashion, data are entered directly into data files as soon as they are generated. Closed-ended data are ready for immediate analysis. Open-ended data can also be entered, but they require an extra step.

Suppose the questionnaire asks respondents "What would you say is the greatest problem facing America today?" The computer terminal would prompt the interviewer to ask that question. Then, instead of expecting a simple numerical code as input, the computer would allow the interviewer to type in whatever the respondent said, for example, "crime in the streets, especially the crimes committed by . . . ." Subsequently, coders sitting at computer terminals would retrieve the open-ended responses one-by-one and assign numerical codes as discussed earlier in this chapter.

Coding to Optical Scan Sheets

Sometimes data entry can be achieved effectively through the use of an *optical scanner,* a machine that reads black pencil marks on a special code sheet and creates data files to correspond to those marks. (These sheets are frequently called *op-sense* or *mark-sense* sheets.) Coders can transfer coded data to such special sheets by blacking in the appropriate spaces. The sheets are then fed into an optical scanner, and data files are created automatically.

Although an optical scanner gives greater accuracy and speed to manual data entry, it has disadvantages as well. First, some coders find it very difficult to transfer data to the special sheets. Locating the appropriate column can be difficult, and once the appropriate column is found, the coder must search for the appropriate space to blacken.

Second, the optical scanner has relatively rigid tolerances. Unless the black marks are sufficiently black, the scanner might make mistakes. (You will have no way of knowing when this has happened until you begin your analysis.) Moreover, if the op-sense sheets are folded or mutilated, the scanner might refuse to read them at all.

Direct Use of Optical Scan Sheets

It is sometimes possible to use optical scan sheets a little differently and possibly avoid the difficulties they present coders. Persons asked to complete questionnaires can be instructed to record their responses directly on such sheets. Either standard sheets can be provided with instructions on their use, or special sheets can be prepared for the particular study. Questions can be presented with the different answer categories, and the respondents can be asked to blacken the spaces provided beside the answers they choose. If such sheets are properly laid out, the optical scanner can read and enter the answers directly. This method might be even more feasible in recording experimental observations or in compiling data in a content analysis.

Precoding for Data Entry

In laying out the format for a questionnaire, giving special attention to the method of data processing that will be used is essential, especially if you plan to enter data directly from the questionnaire. The following suggestions will make direct data entry easier.

The questionnaire should be precoded to facilitate data entry (or coding, for that matter). Items of data should be assigned to named or numbered data fields in advance, and notations in the questionnaire should indicate those assignments.

In addition to making data field notations, you might want to indicate the numeric assignments for the various responses within a given field, for example, 1 = Strongly agree, 2 = Agree, and so on. Whenever a given field contains three or fewer response categories, precoding can normally be omitted. Whenever the field contains more than three response categories, it is normally safest to enter the punch assignments. Figure 11-5 presents examples of precoding appropriate for direct punching.

Several things should be noted in Figure 11-5. First, note that the field assignments are presented in parentheses; Var14 to Var17 are assigned to approval/disapproval of the political figures, and Var18 is assigned to political party identification. The response assignments for question 9 are presented above the response category columns. In question 10, which has only one set of responses, the assignments are presented beside the spaces for checking answers.

Finally, note that all responses to the two questions are located on the same side of the page. This is important to facilitate data entry. You will be able to enter the data from both questions without moving your eyes back and forth across the page. If the location of responses moves around on the page, you will work more slowly and be more likely to miss items.

Whenever precoding is used on a self-administered questionnaire, it is normally a good idea to mention the precoding in an introductory note to the questionnaire. ("The numbers shown in parentheses and beside answer categories should be ignored; they are included only to assist the processing of your answers.") In any event, precoding should be kept as inconspicuous as possible

Figure 11-5   *Illustration of Precoding a Questionnaire*

9. Beside each of the political figures listed below, please indicate whether you Strongly Approve (SA), Approve (A), Disapprove (D), Strongly Disapprove (SD), of that person's political philosophy, or check (DK) if you Don't Know.

(Var14-Var17)	SA	A	D	SD	DK
	1	2	3	4	5
a. George Bush	[ ]	[ ]	[ ]	[ ]	[ ]
b. Michael Dukakis	[ ]	[ ]	[ ]	[ ]	[ ]
c. Jesse Jackson	[ ]	[ ]	[ ]	[ ]	[ ]
d. George Wallace	[ ]	[ ]	[ ]	[ ]	[ ]

10. What is your political party affiliation, if any?

(Var18)

Democratic Party	1 [ ]
Republican Party	2 [ ]
Peace and Freedom Party	3 [ ]
Green Party	4 [ ]
American Independent Party	5 [ ]
Other	6 [ ]
None	7 [ ]

to avoid confusing the respondent. If the questionnaire is set in type, the precoding should be entered in a smaller typeface.

Even if the questionnaire is not to be punched directly, the format should take intended processing methods into account. Most of the suggestions made here will facilitate hand coding. If the questionnaire is to be read by an optical-sensing machine, then you must check your format against the requirements of the machine.

Data Cleaning

Whichever data-processing method you have used, you will now have a set of machine-readable data that purports to represent the information collected in your study. The next important step is the elimination of errors—"cleaning" the data. No matter how the data have been entered, or how carefully, some errors are inevitable. Depending on the data-processing method, errors can result from incorrect coding, incorrect reading of written codes, incorrect sensing of blackened marks, and so forth. Two types of cleaning should be done: *possible-code cleaning* and *contingency cleaning*.

Possible-Code Cleaning

Any given variable has a specified set of legitimate attributes, translated into a set of possible codes. In the variable *sex,* there might be three possible codes: 1 for male, 2 for female, and 0 for no answer. If a case has been coded 7, say, in the column assigned to sex, clearly an error has been made.

Possible-code cleaning can be accomplished in two different ways. First, as we have seen, the modern computer programs available for data entry can check for errors as the data are being entered. If you tried to enter a 7 for sex in MicroCase, for example, the computer would beep and refuse the erroneous code. Other computer programs are designed to test for illegitimate codes in data files that weren't checked during data entry.

Second, if you do not have access to these kinds of computer programs, you can achieve a possible-code cleaning by examining the distribution of responses to each item in your data set. Thus, if you find that your data set contains 350 people coded 1 on sex (for female), 400 people coded 2 (for male), and one person coded 7, you will suspect that 7 is an error.

Whenever you discover errors, the next step is to locate the appropriate questionnaire, determine what code should have been entered, and make the necessary correction.

Contingency Cleaning

Contingency cleaning is more complicated. The logical structure of the data might place special limits on the responses of certain respondents. For example, a questionnaire might ask for the number of children women have had. All female respondents, then, should have a response coded (or a special code for failure to answer), and no male respondent should have an answer recorded (or should have a special code indicating that the question is inappropriate). If a given male respondent is coded as having borne three children, either an error has been made and should be corrected, or your study is about to become more famous than you ever dreamed.

Contingency cleaning is typically accomplished through computer programs that require a rather complicated set of if-then statements. In some cases, the programs that support direct data entry can be instructed to check automatically for proper contingencies, just as they check for legitimate codes.

Although data cleaning is an essential step in data processing, it can be safely avoided in certain cases. Perhaps you will feel you can safely exclude the very few errors that appear in a given item if the exclusion of those cases will not significantly affect your results, or perhaps some inappropriate contingency responses can be safely ignored. If some men have been given motherhood status, you can limit your analysis of this variable to women. However, you should not use these comments as rationalizations for sloppy research. "Dirty" data will almost always produce misleading research findings.

Summary

The preceding discussion of data-processing options and requirements should emphasize the absolute necessity for proper survey planning. You cannot reasonably first collect your data and then begin considerations of how they will be processed. The data-processing method should be selected before the questionnaire is designed.

There is a common tendency among researchers to breathe a sigh of relief once the questionnaires have been completed and are safely stored in the research office. The danger of not getting the data has been avoided; now all that is needed is to put them in order for analysis. In fact, this sense of well-being could be unfounded. Even if the data have been collected in an appropriate format, it might still be impossible to put them in a meaningful order. Only if you have carefully thought out your data processing in advance and designed a data-collection technique appropriate to those plans can you safely feel relieved.

Additional Readings

Paul C. Cozby, *Using Computers in the Behavioral Sciences* (Palo Alto, CA: Mayfield, 1984).

Perry Edwards and Bruce Broadwell, *Data Processing,* 2d ed. (Belmont, CA: Wadsworth, 1982).

David Heise, ed., *Microcomputers in Social Research* (Beverly Hills, CA: Sage, 1981).

12

Pretests and

Pilot Studies

By now, probably every research textbook admonishes researchers to conduct some form of testing of the research design prior to the major research effort. The arguments for pretesting are compelling. No one wants to invest large sums of money and considerable effort in a hefty research design only to fail to achieve the research objectives due to some unforeseen error.

In this chapter, we will distinguish two types of testing: pretests and pilot studies. **Pretests** will refer to initial testing of one or more aspects of the study design, such as the questionnaire, the sample design, a computer program for analysis, and so forth. **Pilot studies** will refer to miniaturized walk-throughs of the entire study design.

Conducting Pretests

As defined above, pretests represent initial tests of one or more aspects of the research design. Most commonly, this has meant the administration of a draft questionnaire to a group of subjects, but the concept of pretesting is more broadly applicable.

Pretesting the Sample Design

Often, a sample design might seem reasonable on paper only to prove unmanageable in practice. If the design calls for the creation of a sampling frame, a portion of that frame might be created in a pretest. For example, in an area cluster sample design, you might attempt to update block-size estimates within one or more census tracts before setting your staff to work on updating all block sizes. If the design calls for the listing of households, you might send listers into the field to list selected blocks in order to uncover any unexpected difficulties.

If you will be selecting your sample from an existing list, you should carefully scrutinize the list for problems. If stratification is intended, you should make a trial run of the stratification procedure. Can stratification be done manually? Try it. Will a computer program be required? Test it.

The same steps are in order with regard to sample selection. If a computer program is to be used, it should be tested with partial or even hypothetical data. If manual selection is intended, a pretest on partial or even hypothetical data will point to problems and possibly even suggest that the task is impossible or extremely difficult; computer selection might be more feasible.

The pretest of a sample design, then, can indicate whether that design is possible, provide an assessment of its difficulty, and give a rough estimate of the time and cost that will be involved. Pretests serve another function, also. Inevitably, you will fail to recognize in advance all the decisions you must make in the course of your survey. Although you might have been explicit in defining your survey population, the art of sampling will uncover hidden problems of definition. In an area sample, for example, will you interview transients (such as tourists) as well as residents? Will foster children be considered family members? In conducting a university study, you will soon discover that "students" and/or "faculty" are extremely difficult creatures to define. Hopefully, a pretest will turn up such problems (some of them at least), and you will be able to make your decisions as to definition at the outset. With a pretest, your decisions can be more carefully considered, and you will be able to ensure that they are acted upon consistently throughout the final study.

Pretesting the Research Instrument

The purpose of pretesting the research instrument should be obvious, but the methodological options might not be as clear. The following points are offered by way of clarification.

1. Either the entire instrument or a portion thereof might be pretested. Perhaps you will be primarily concerned about the applicability of a particular set of questions. If so, you might devote most of your attention to several pretests of that portion of the questionnaire and modifications of it. While this is a legitimate endeavor, you should be aware that the context in which the questions appear will affect their reception. Thus, while initial testing of portions of the questionnaire is undoubtedly worthwhile, it would be better to wrap up the activity with one or more pretests of the whole instrument.

2. Preferably, the instrument should be pretested in the manner intended for the real study; self-administered questionnaires and

interview schedules should be pretested in the appropriate manners. Nevertheless, this remark in no way prohibits the initial testing of the instrument through a different method. It is especially worthwhile to pretest an early draft of what ultimately will be a self-administered questionaire by interviewing. This procedure will permit a better determination of problems in that an interviewer can detect confusion and probe into the nature of that confusion on the spot. Ultimately, however, the instrument should be pretested in the form that will be used.

3. An open-ended format can be used profitably in the pretest to determine appropriate response categories for what will ultimately become a closed-ended question. Respondents might be asked to give their own answers to a question; those answers would then be added by the researcher, and standardized response categories would be created. However, it is important that the eventual closed-ended questions also be tested. Thus, a new pretest should be conducted to uncover any bugs in the standardized categories.

4. The selection of subjects for instrument pretests can profitably be kept flexible and varied. For the most part, controlled sampling should be ignored altogether at this point. (Quick studies are quite a different matter.) I make this point because researchers often cut short their pretesting due to the difficulties of sampling. The single guideline I would recommend for selection of subjects is that subjects should be reasonably appropriate respondents for the questions under consideration. Ideally, every individual item should initially be pretested among the members of the research staff, although such a test is not sufficient. If the study is aimed at a particular population, then any members of that population or any persons similar to that population can serve as pretest subjects. However, it should be understood that this point is aimed not only at making pretesting easier but also at making it more extensive.

In more rigorous pretesting of the research instrument, little attention should be given to strict representativeness; rather, an attempt should be made to achieve the broadest range of respondent types—including those who might represent a small minority of the population. In a study of political attitudes, for example, you should pretest the instrument on subjects drawn from the entire political spectrum of your population. This means that the extremes—right and left—will be overrepresented in the test. This is done to ensure that the instrument will make sense to, and be useful in understanding, all types of respondents in the population. (Of course, the instrument should be tested among "moderates" as well.) The goal of pretesting is to improve the research instrument rather than to provide descriptions of the population.

5. Pretesting lends itself to comparative testing of different methods for obtaining desired data. Different drafts of the questionnaire can be tested simultaneously, and successive revisions and retesting over time should have the same effect.

6. It might be useful to utilize the same subjects more than once in pretesting the instrument. This is of special value in that the overall subject profile thus derived offers a general basis for evaluating responses to specific items. On the other hand, the final draft should be pretested among new subjects to take into account the inevitable learning process that will take place among the earlier subjects.

These, then, are some of the options and guidelines relevant to the execution of pretests. I would emphasize that pretesting should be a multistage, cumulative process; you should not think in terms of *the* pretest. A later section considers the evaluation of questionnaire pretests.

Pretesting Data Collection

Much has been written about techniques of data collection, and experienced researchers have probably regularized their own basic methods. Nevertheless, every study is different in the particulars of data collection.

If a mail-out questionnaire is to be used, you should physically try out the procedures for assembling and mailing the questionnaires. Only in this manner can the many steps involved be properly organized. Tasks such as rubber-stamping, label sticking, letter folding, and envelope stuffing can be more difficult in practice than in theory.

An experienced researcher can probably organize such work, lay out the work space, and assign tasks to the staff in a mental run-through. Nevertheless, only a physical test can demonstrate that questionnaires do not really fit into envelopes, that return envelopes do not fit into outgoing envelopes, and so forth. If a questionnaire identification number must be matched with the mailing label on the envelope, only a physical practice will point to the dangers (and, one hopes, preventions) of mixing up these numbers.

Because the nature of the problems to be uncovered varies so much with the nature of the study design, continuing the listing of examples is perhaps pointless. Suffice it to say that the researcher who does not pretest the execution of a questionnaire mailing runs a considerable risk.

The necessity for pretesting data-collection methods is even greater in an interview survey. Since this falls in the area of interviewer training, testing, and supervision, the discussion of such pretesting has been presented in Chapter 10.

Pretesting Data Processing

Researchers have a tendency to be very tense about the data-collection phase of a survey (Will the data come in?) and relatively more sanguine about the later stages. There is a feeling that once the question-naire or interview schedules are in hand, the pressure is off, and problems can be coped with in a more leisurely manner. In large part this is true, but such an attitude can be seriously dysfunctional if it prevents you from pretesting your data-processing procedures.

The nature of the data-processing pretest depends, of course, on the nature of the data processing itself. Following one track, you will want to pretest your coding and data-entry operations. The results of such a pretest have important implications for the format of the research instrument. Often a few minor modifications to the layout of the questionnaire will vastly improve the efficiency of coding and data entry. Pretesting will point to the specific needs for precoding, calling for more or indicating that less would suffice. The only way to make such determinations is to have coders practice coding, either with hypothetical data or with completed pretest questionnaires. If practice coding is delayed until the final questionnaires are coming in, however, the whole point of the exercise is lost.

If the questionnaires are designed for direct data entry, the above com-ments apply. In addition, a pretest will point to the needs related to editing of questionnaires prior to entry. If data entry is to be done from coded transfer sheets, the test will point to any special coding requirements.

I cannot overemphasize the need for a pretest if the data are to be entered through the use of optical scanning equipment. A pretest should point to machine tolerances with regard to mutilated score sheets and lightly marked spaces. One major survey transferred questionnaire data to op-sense sheets and did not learn until the completion of coding that the coders used pencils with the wrong degree of softness. The entire project staff spent two weeks of full-time drudgery "blacking over" the initial markings.

If open-ended questions are asked in the questionnaire, some early attention should be given to the organization of their coding. A pretest will suggest the most efficient order for coding the open-ended and closed-ended questions.

Post-data-entry processing, such as data transfer to tape and data clean-ing, should also be pretested. General-purpose computer programs may be available for accomplishing these steps, but only an actual test will ensure that the particular survey data will be appropriately organized for the utilization of such programs. If the study calls for the construction of multilevel data files (for example, household file, family file, and person file), rather complex manipula-tion of data might be required; this manipulation must be pretested.

Finally, pretests of the data-processing stage(s) should show the way to an efficient interrelation of the separate elements. The logging of completed questionnaires, initial editing, coding, data entry, cleaning, and whatever other elements are involved in the processing of a given study must be coordinated. In one sense, this means that the outputs of one stage must be appropriate inputs to the next. In another, more mundane, sense it includes such decisions as who will carry the questionnaires from one room to another, where the questionnaires will be stored at any given time, and so on.

Pretesting the Analysis

You might consider it bizarre to talk about pretesting the analysis, especially considering this book's insistence that analysis is an open-ended and often serendipitous activity. Nevertheless, I am quite serious in suggesting this step. Although you inevitably do things with your data that you did not initially intend to do, you must ensure that you will be able to do those things that you *did* intend to do.

Though I will devote more space to analysis with regard to pilot studies, two kinds of analytical pretesting should be mentioned here. First, you should walk through all the steps from raw data to finished product with regard to table construction, index building, scale construction, regressions, factor analysis, and any other analytical modes you have in mind. You should go through each step from the data records to the written presentation. The purpose of this walk-through is to ensure that you *can* get there from here. Often, the form in which information is to be collected does not lend itself to the analysis intended. Some examples follow.

A population survey might collect data on each member of the household. The data would be standardized, with a separate record for each household member. In such a format, however, you might not be able to determine easily the *number* of persons in each household. Also, while it might be possible to count the number of person-records for each household, it might not be possible to utilize that information as an analytical variable. (*Solution:* Have coders count persons and code the person count in a separate household file.) By the same token, you might have difficulty relating household data (for example, age of household head) to the individual person files unless steps are taken in advance to permit this.

As another example, suppose that in pretesting your analysis you discover that you want to present age as a *mean* to permit comparisons with other data; if you have asked for age in gross categories rather than asking for exact age, you could be in trouble.

The second type of analytical pretesting concerns the use of data manipulation hardware and software—testing the computer and programs you intend to use in your analysis. You might find that the questionnaire format has to be modified to support your analytical intentions; learning that during pretesting is better than finding it out after you have collected all your data in the actual survey.

Conducting Pilot Studies

Conducting pretests of the various individual aspects of the study design and analysis is extremely important, as I have tried to indicate. Ideally, extensive pretests of each aspect are in order. Moreover, you should be continually on the alert for the implications of the pretesting of one aspect for other aspects and work to interrelate all aspects involved. Nevertheless, the best method of ensuring valid interrelationships is to conduct a pilot study—a miniaturized walk-through of the entire study from sampling to reporting. The pilot study should differ from the final survey only in scale, with fewer cases studied (and less time used). The following comments are elaborations on this basic principle.

Pilot-Study Sampling

Unlike a pretest, a pilot study should be directed at a representative sample of the target population. The pilot-study sample, then, should be selected in exactly the same fashion as is intended for the final survey. One exception might be made in this regard, however.

Because you might want to avoid studying the same respondents in both the pilot study and the final survey, you could select both samples at the same time as insurance against that possibility. This can be accomplished in one of two ways. First, you might select your final sample—or at least your primary sampling units (such as census blocks)—and then select a pilot-study sample from the remainder of the population. Second, you might select an initial sample that contains enough elements for both samples and subsample from that list for your pilot study.

If overlap is not considered a problem, then the pilot-study sample should be selected in exactly the same manner as you intend for the final survey. In either event, you will have tested your sample design (or actually drawn your final sample).

Pilot-Study Research Instrument

The pilot study should involve the administration of a research instrument as nearly identical as possible to the one intended for the final survey. If a rather elaborate mailing piece is to be used, however, you should recognize that it might not be feasible to produce the number required for the pilot study and then produce a revised batch for the final survey.

The pilot-study questionnaire should contain all the intended questions in the wording, format, and sequence that pretesting has indicated are best for the final survey. The pilot study should not be a vehicle for trying out new items that have not been pretested.

Having said this, we must note one exception. It is quite reasonable for the pilot-study instrument to contain more questions than are intended to be included in the final survey. Often, uncontrolled pretesting is not sufficient for determining which of several methods of obtaining certain data will be most useful for the study. In some instances, then, you might include more than one method in the same questionnaire (provided they do not present the respondent with a gross repetition), with the intention of evaluating them through the pilot study.

A better, though more complex, way of handling this situation would be to use different versions of the questionnaire among subsamples of the pilot study. In this fashion, pilot-study respondents will be reacting to what might become the final survey instrument, providing a better test.

If a single pilot-survey instrument is used to evaluate alternative methods of obtaining given data, you must be careful not to create a significantly larger instrument. The lessons you learn from the administration of a long, repetitive pilot-study instrument might be inappropriate to the final, shorter instrument.

Pilot-Study Data Collection and Data Processing

Like other phases of the pilot study, data collection and data processing should be a miniaturized walk-through of the final survey design. To the extent that the research instruments are comparable, they should be administered exactly as you intend for the final survey. Where they differ, you must be especially conscious of the implications of those differences, attempting to learn from inference rather than experience. The completed questionnaires should be coded, and the data should be entered, transferred, cleaned, and analyzed exactly as planned for the final research.

Pilot-Study Analysis

Bizarre as it may seem, *organized* research should begin with a drafting of the written report. The draft report should contain the logical arguments of the study, blank or hypothetical tables, and all the verbal glue that holds them together. The organized researcher thus ensures familiarity with what information is needed, from whom it is needed, and the form in which it is needed. The pilot-study analysis, like the final analysis, should consist of filling in the empirical blanks, noting unexpected developments, and elaborating on these developments.

The pilot-study analysis should be carried out with all the vigor and imagination intended for the final analysis. Because the pilot-study sample is drawn with the intention of representing the target population, the results of the pilot-study analysis should be essentially the same as those of the final survey. The latter should be a more substantial replication of the former.

In reality, however, study designs are never perfect, nor is the logical reasoning they are based on. As a result, the pilot-study analysis never quite turns out as expected. More often, it points to errors in reasoning and/or design. The methods for determining these errors are discussed in the following section. At this point, I would emphasize the importance of pushing the pilot survey as far as possible in order to uncover as many errors as possible before committing major resources to the final survey.

If needed data are missing in the pilot study, they should be approximated, if possible, to permit the greatest possible elaboration on the intended analysis. If the analysis suggests that income levels should be taken into account, for example, and you have not collected income data, you should use education or occupation as a rough approximation. If you have no approximate data, then construct hypothetical tables, presenting all the possible outcomes true to income. Determine how each possible outcome would be explained and what other data should be considered in the explanation. If furtheranalyses are suggested by this approach, follow up on them through the pilot-study data; approximate, if necessary, or create new hypothetical tables and try to explain them.

It is altogether too easy to note a problem and its immediate implications and resolve to correct the problem in the final survey. This procedure is dangerous. To profit from doing a pilot study, you must consider the second-order, third-order, and fourth-order implications of the problem and its possible solutions.

Evaluating Pretests and Pilot Studies

This section offers some guidelines and criteria for the evaluation of pretests and pilot studies. Although I will offer no laws for resolving this issue, I will attempt to link research objectives to evaluation criteria. In doing this, moreover, the discussion will be limited to the uses of analysis in evaluating the research instrument.

The guts of survey research are the collection, manipulation, and understanding of data. Above all else, pretests and pilot studies are aimed at ensuring the creation of useful data. We will now consider ways of recognizing useful data.

Question Clarity

In order to be useful, questions should make sense to respondents, even though the most important implications of the questions might not be evident to them. By the same token, the answer categories (if provided) should make sense, both in themselves and in their relation to the question and to each other. Some danger signs that might be apparent in pretest or pilot-study results follow.

Failure to Answer.   Typically, every respondent skips some questions, and every question is skipped by someone; however, when a given question produces a number of "no answers," it is a clue to problems in survey design.

Multiple Answers.   Even when respondents are asked to select only one answer from a list of alternatives, some will persist in selecting more than one. Again, if one question produces a number of multiple answers, you should suspect that either your answer categories are not mutually exclusive or the question is being misunderstood. The solution to this problem varies with the type of multiple responses. If the same two categories are frequently chosen together, perhaps they can be better distinguished or, alternatively, combined. If several combinations of answers are being selected, then something more basic is wrong, and the whole question should be examined.

"Other" Answers.   It is often appropriate in closed-ended questions to offer the respondent the alternative of volunteering his or her own answer. Whether or not this option is offered, having a large number of responses written in (or volunteered to an interviewer) indicates that the categories provided are not sufficiently exhaustive. If the "other" answers fit conceptually into one or more logical categories that do not overlap the existing ones, perhaps they should be added to the list of alternatives.

You should not add answers to the list, of course, if doing so conflicts with the objectives of the study or of the particular question. For example, communications researchers in the United States sometimes ask respondents to indicate their primary source of political information, offering choices from a list that includes newspapers, friends, radio, magazines, and political circulars. Television is purposely omitted from the list on the assumption that most respondents would choose it if it were included. If respondents volunteer television as a response, you still would not want to include it in your list. If a sufficiently large number of respondents give television as a response, however, you might decide to specifically exclude it in the basic question (for example, "Aside from television, which of the following . . . ?").

Qualified Answers.   Respondents often qualify their answers, both in selecting a provided category and in answering an open-ended question. Asked to report their "annual income," respondents might note that they are reporting last year's income or their estimate of this year's. Asked to report their father's age, they might note that they are reporting their *step*father's age or their father's age at death. Asked for their political party affiliation, they might note that they have checked their party registration but that they generally vote independently of that registration; on the other hand, they might note just the opposite: that they are not registered in a party but generally vote for the party indicated.

Qualified answers such as these point to a lack of clarity in the questions and/or the answers provided. One such response might be sufficient to warrant a revision of the question if the lack of clarity could result in a different understanding of the question. Several such responses demand revision.

Direct Comments.   Respondents often point directly to problems in question wording or format (for example, "This is a lousy question"). Though you must be relatively thick-skinned in reviewing such comments, you should be on the alert for particular questions that generate more than their share.

Questionnaire Format

All the danger signs discussed earlier can point to errors in format as well as errors in wording. Missing answers, for example, might indicate that the "flow" of the questionnaire is not proper. This is especially true for contingency questions, which often are (and *should be*) set off to the side of the questionnaire page. Better instruction, boxes, arrows, and similar devices should resolve this problem.

Sometimes respondents will answer a series of yes/no questions by simply checking some yeses. You might ask what kinds of organizations respondents belong to and provide a list of different types with "yes/no" beside each. If respondents only check yes in some cases, the problem consists of not knowing whether the organizations not checked are ones they do not belong to or whether they skipped over them inadvertently in going through the list. More explicit instruction (for example, "Please answer for *each*") might alleviate this problem. If you are especially concerned for the accuracy of responses in such a situation, asking about membership in each organization in separate questions, rather than using a list, should solve the problem; it will, however, lengthen the questionnaire.

Variance in Responses

A primary concern in evaluating tests of a research instrument should be the distribution of answers evoked by each question. Are the responses about evenly distributed among the several answer categories provided, or did most respondents select the same answers? The consideration of this aspect of pretest and pilot-study results, though essential, depends on the objectives of the study and of the particular questions.

If the question is designed to measure factual information that permits an independent evaluation (for example, asking students to report their grade point averages when you know the distribution of the population), then the pilot-test responses provide a clue as to the clarity of the question and/or the reliability of the respondents (assuming, of course, that the test data are taken from a representative sample).

If the particular question, as worded, is intrinsically interesting, you might be satisfied with whatever distribution of responses you obtain. It might be interesting and valuable to know, for example, that everyone in the population indicates approval of a proposed law before the legislature, and you might want to replicate this finding in your final survey.

In social research, however, you are more often interested in the *relationships among variables*. You want to know why some people favor a particular product and others do not. You want to know why some people are conservative and others are not. If no variation exists in the responses elicited by a given question, you cannot explain the answers. You cannot explain differences that do not appear in your data collection.

In seeking relationships among variables, the most interesting part of social research, variance in responses must be obtained. You would do well to recall earlier discussions in this book that declared that social data are created rather than neutrally collected, as well as discussions that declared that social concepts and variables do not exist in any absolute sense. It is in this sense that you must *create* variance in the responses provided by your research instrument. If your questions tapping religiosity suggest that everyone is religious, you must tighten your criteria for religiosity, rewording the questions in such a way as to push more respondents into apostasy. This point is made bluntly here because it is sometimes difficult to accept. Unfortunately, the traditional literature related to the scientific method has created an unfounded belief that researchers must conceptualize variables and positions on those variables that closely reflect *reality*. Thus, many new researchers spend considerable time attempting to precisely define concepts such as "religiosity" and then define what constitutes "the religious person."

The epistemological base of this book is the conviction that concepts such as "religiosity" and "religious" cannot be defined in any ultimate sense. The best we can hope for in the present era is the generation of perhaps several useful conceptualizations of religiosity, thus permitting us to speak roughly of *more* and *less* religious people. (The same is true of concepts such as social status, liberalism, alienation, knowledgeability, and so forth.) Conceptualizations are "useful" to the extent that they help you understand your empirical data and facilitate the construction of coherent theories.

In order to explain why some people are more religious than others, you must define and measure religiosity in such a way as to permit you to score some respondents as more religious and others as less religious, even if this scoring method does not completely jibe with your personal belief about what the "religious person" is ultimately.

If you are interested in the long run in explaining variance, then you should try to maximize the variance among your respondents. In a question offering dichotomous answers, you should seek to obtain a fifty-fifty distribution of responses. In a question with more answer possibilities, you should aim for an even distribution. Doing so will provide you with a greater potential for later analysis. In evaluating your pretest or pilot-study results, then, you might want to manipulate the variance produced by given questions. The following are some guidelines for accomplishing the desired variance.

1. Variance can be manipulated by changing the emphasis of the question. Suppose, for example, that we want to examine students' commitments to the vocational training aspect of a liberal arts college education. We might ask them to agree or disagree with the statement "Learning a vocational skill is an important part of my college education." If nearly all students in a pilot study agreed with the statement, it would not be possible for us to analyze the responses, due to a lack of variance. The variance might be manipu-

lated by "toughening" the statement: "Learning a vocational skill is the most important part of my college education." We should expect that fewer students would agree with the second statement than with the first.

By the same token, if the modified statement had been asked in a pilot study and very few students agreed with it, we might consider changing it to the easier form.

2. Changing the emphasis of answer categories can have the same effect. Suppose respondents in a public health survey were asked "How often do you wash your hands before eating?" and were given the following answers: "always, sometimes, never." If few or no respondents in a pilot study said "always," this answer might profitably be changed to "almost always."

3. Expanding the list of answer categories almost inevitably increases variance in responses. Increasing the answer possibilities from "agree/disagree" to "strongly agree, agree, disagree, strongly disagree" will almost always spread out the responses. (You should realize, however, that you may end up explaining the difference between "strongly agree" on the one hand and "agree" to "strongly disagree" on the other.)

Similarly, respondents are sometimes reluctant to select an extreme answer, especially in attitudinal or orientational items. Asked to characterize their political orientations as either "very liberal," "moderately liberal," "moderately conservative," or "very conservative," relatively few respondents are likely to select either of the two extreme characterizations. If the terms "radical left" and "radical right" were added to the extremes of the list, we would expect more people to select the "very liberal" and "very conservative" categories; at the very least, more respondents should be found selecting either "radical left" or "very liberal" than would have selected "very liberal" in the first list.

The examples listed involve the expansion of the answer categories toward both extremes of the variable. It is sometimes useful to skew the answers in only one direction, that is, to expand only one extreme of the variable. Thus, in the example given, campus researchers might add only "New Left" or "socialist" to the initial list of answer categories.

Variance in responses is a very important criterion for evaluating the worth of questions. The manner in which variance is evaluated, however, still depends on the purpose of the study or of the question. If your purpose is description, you might consider little variance interesting and important when the form of the given question is clearly appropriate, or you might take the lack of variance as an indication that the question is poorly worded and that the description produced is misleading. If your purpose is explanation, however, you must have variance. Your respondents must appear different if you are to explain their differences.

Internal Validation of Items

Thus far, we have discussed only the independent evaluation of individual items. Items can also be evaluated through examination of their relationships with other items. Suppose you have included in your pilot-study questionnaire several items that you believe measure "alienation." One function of the pilot-study evaluation would be to determine whether each item serves that purpose. If all items do indeed measure the same disposition among respondents, then the answers to one item should be correlated with the answers to others.

Typically, all or most of the items designed to measure a single variable will be empirically related to one another, but the strengths of the relationships will vary among the pairs of items. If one item is very weakly related to the others, you might conclude that it does not really measure the variable and drop it from the questionnaire. (The possibility that it is the only item measuring the variable is one of those problems that researchers must learn to live with.) You should not resolve to drop all such items automatically, however. Quite possibly, you will decide on substantive grounds that the item in question represents a somewhat different dimension of the variable, empirically unrelated but conceptually important, or you might decide that the lack of a relationship is descriptively interesting and want to replicate the findings in your final survey.

At the opposite extreme, a very high correlation between two items might suggest that including both in the final questionnaire is unnecessary. Deleting one or more redundant items would save space that could be used for items that were overlooked in the pilot-study questionnaire.

Evaluating the Analysis and Reporting

The evaluation of a pilot study ultimately rests on the same criteria as the eventual evaluation of a finished research project—its ability to tell an interesting and useful story. Though the preceding comments point to some specific evaluative criteria, each specific point is contained within the context of achieving the success of the pilot-study analysis and reporting. As we noted earlier, you should analyze and write up your pilot-study data as if they represented the final survey. Basically, those sections of the pilot-study design that permit the construction of an interesting and useful story should be retained; those sections that prevent such construction should be modified.

Problems of question wording, format, variance, validity, and relationships among items are ultimately evaluated in terms of the dysfunctions they bring to the story line. Only a full analysis and reporting can determine this.

You should go through all the intended steps of the analysis, constructing tables, indexes, scales, factors, correlations, regressions, and so forth and presenting explanations for the findings. I repeat: This is the only adequate method for evaluating pilot-study results.

By seriously analyzing the pilot-study data, you will be in a position to recognize that certain data have been overlooked at a time when modifying the design is still possible. Equally important, you will become aware of the analytical surprises that are in store for you in the final survey analysis. If your primary research hypothesis is simply not supported by the pilot-study data, you should analyze those data to the fullest extent possible in an attempt to discover why the hypothesis was not supported. Has the measurement of variables been adequate? Do other factors interfere with the expected relationship among variables? Was the hypothesis simply unfounded? Although you might not be able to answer all these questions satisfactorily, the pilot-study analysis should provide you with some clues as to what else ought to be examined in the final survey. At the very least, the pilot-study findings should give you time to tone down your declarations of the hypothesis around lunch at the faculty club. (You might also withdraw your offer of a paper for the annual professional meetings.)

Summary

It is appropriate to end this chapter on a realistic note. Though the pretesting and pilot studies described here would surely lead to more professional and more valuable survey research, few, if any, past research examples could be held up as exemplifying the rules and guidelines discussed. You should be aware of the reasons why past researchers typically have not made adequate use of pretests and pilot studies and why future researchers probably will not do so.

Good survey research is almost always time-consuming and expensive; pretesting and pilot studies—especially good pilot studies—seem to add to both time and cost. Since survey research (especially conducting interview surveys) is labor-intensive, the survey researcher must always cope with the problem of personnel management. It might be impossible for you to conduct a good pilot study and keep your interviewers, coders, data entry people, and so on waiting several months or years while you carry out an extensive pilot-study analysis. More likely, the overall study budget and time period will not permit a great deal of pretesting and a good pilot study. Also, issues, attitudes, sampling frames, and so forth might change significantly between the pilot study and a delayed final survey. These are but a few of the reasons why pretesting and pilot studies are not presently used properly and probably will not be used to their fullest potential in the future.

At the same time, the above constraints still allow for better use of these methods than is typically the case, and the following suggestions should be considered. First, pretesting should be a multistage, continuing process throughout the study design, as noted earlier. You should not think in terms of conducting *the* pretest as a research ritual but should instead take every opportunity to pretest each aspect of the study design under whatever testing conditions might be available.

Second, if there is insufficient time for a detailed analysis of the pilot-study results prior to the revision and printing of the final questionnaire, the following steps should be considered. As soon as pilot-study questionnaires begin arriving, they should be coded and entered. Do not wait until all are returned. Beginning immediately might point out some processing difficulties that can be modified promptly. Once a substantial batch of questionnaires, say a hundred or so, has been processed, an initial analysis should begin. The marginal distributions should be run immediately to permit an initial evaluation of the individual items. These tentative conclusions can be verified more substantially as new batches of pilot-study questionnaires become available.

As soon as sufficient numbers of questionnaires have been processed, a more detailed explanatory analysis should begin; again, this should be done even on a partial data file. If it is not possible to conduct an exhaustive analysis prior to the deadlines for revising and printing the final survey questionnaire, the pilot-study analysis should not end when the revised questionnaire is sent to the printer. If the sample design and data-collection method have been sufficiently organized in advance, printing the final questionnaire typically provides you with one of the few "dead spots" in the project schedule. This period could very profitably be spent in a more detailed analysis of the pilot study, even though you have committed yourself to the final questionnaire form. Sometimes, you can correct missing or faulty data through modifications in the data-collection procedures without changing the questionnaire itself. Moreover, this detailed analysis will give you a head start on your final analysis.

Having noted that the ideal procedures outlined in this chapter cannot be followed religiously, we should stress the value of adhering to them as much as possible. This chapter has attempted to suggest some guidelines and procedures for conducting and evaluating pretests and pilot studies in survey research. Because this topic has received little or no attention in previous research literature, the present offering must be viewed as fragmentary and tentative. Until such time as future writers round out the picture with reports of their own experiences, this initial effort will hopefully be of use.

PART FOUR

Survey Research

Analysis

Part Four of this book is devoted to a number of topics involved in the analysis of survey data. In the chapters to follow, we will pick up the study at the point at which you find yourself the proud possessor of computerized data files and will discuss some of the steps along the way to publication of your findings.

Chapter 13, which discusses the logic of measurement and association, is a continuation of the discussion in Chapters 7 and 8 on measurement. We will again find ourselves concerned with the meaning of responses to survey questionnaire items with regard to how these responses can be taken as indications of variables such as religiosity, prejudice, alienation, and so forth. In Chapter 13, however, we will examine this issue from an empirical standpoint, whereas our earlier discussion was limited to conceptual manipulations.

Chapter 14 addresses the logic of contingency tables. Some researchers regard this analytical format as rather rudimentary, but we will give it special attention as a basis for understanding the logic of all survey analysis. The discussion will begin at the very beginning, with univariate analysis, and then proceed to bivariate analysis and on to more elaborate multivariate analyses. In the process, we will move from the logic of description to the logic of explanation. You should emerge from reading this chapter with two important abilities: You should be able to construct contingency tables from your own survey data, and you should be able to understand tables constructed and published by other researchers.

The key to Part Four is Chapter 15, which deals with the *elaboration model*. Here we will examine the basic logic of scientific explanation in the survey context. Basically, we will seek an understanding of the empirical relationship between two variables through the controlled introduction of additional variables. Having noted that lower-class women are more religious than upper-class women, for example, we will see how the manipulation of additional variables can shed some light on why this is the case.

For the most part, this book deals with the *logic* of research rather than the statistics involved. I believe that if you fully understand the logic of research, you will be in an excellent position to marshal whatever statistical manipulations are relevant to your needs. From this perspective, Chapter 16 provides an overview of some common statistics. The chapter does not give a full explanation of the computations of such statistics (several excellent statistics texts serve this function) but attempts to place them within the logic of analysis as discussed throughout this book. In this regard, special attention is given to tests of statistical significance, those seemingly convenient shortcuts to understanding.

As noted above, this text primarily examines survey analysis through the medium of contingency tables, because I believe that the logic of analysis is more clearly seen in this way. At the same time, other, more complex, modes of analysis provide the analyst with greater explanatory potential, and such modes are more appropriate in many situations. Several of these additional methods of analysis are discussed in overview in Chapter 17. I have not attempted to provide a cookbook in this instance but have hoped rather to place such methods within the same logic that informs the discussion of contingency tables. Hopefully, you will come away from the presentation with a general understanding of the constraints and potentials of such methods and will take advantage of other relevant texts in actually employing these methods where appropriate.

Chapter 18 is devoted to a topic that has received relatively little attention—the reporting of survey findings. It discusses some general guidelines in the communication of research and considers the different types of presentations that can be made.

13

The Logic of

Measurement

and Association

The heart of survey analysis lies in the twin goals of description and explanation. The survey analyst makes measurements of variables and then examines the associations among them. As we noted in Chapter 1, however, considerable confusion exists as to the nature of the activities involved in this process.

Now that you have had some exposure to the various aspects of survey design, especially conceptualization and instrument design, it will be useful to return to the traditional image of the scientific method. Earlier, we reviewed that image utilizing a survey example; here we will examine the traditional image schematically and then examine an alternative, more appropriate, image of science in practice. With regard to the latter, we will consider Paul Lazarsfeld's notion of "the interchangeability of indexes."

The Traditional Image

The traditional perspective on the scientific method is based on a set of serial steps that scientists are thought to follow in their work. These steps are summarized in the following list.

1. Theory construction
2. Derivation of theoretical hypotheses
3. Operationalization of concepts
4. Collection of empirical data
5. Empirical testing of hypotheses

Before refuting this traditional perspective, we should examine it in somewhat more detail.

Elements of the Traditional Model

Theory Construction.   Faced with an interesting aspect of the natural or social world, the scientist presumably creates an abstract, deductive theory to describe it. This exercise is supposedly purely logical. Assume for the moment that you are interested in deviant behavior. You presumably construct, on the basis of existing sociological theory, a theory of deviant behavior. Among other things, this theory includes a variety of concepts relevant to the causes of deviant behavior.

Derivation of Theoretical Hypotheses.   On the basis of your overall theory of deviant behavior, you presumably derive hypotheses relating the various concepts comprising your theory. This, too, is a purely logical procedure. Expanding on the above example, suppose that you logically derive the hypothesis that juvenile delinquency is a function of supervision, specifically, as supervision increases, juvenile delinquency decreases.

Operationalization of Concepts.   The next step in the traditional view of the scientific method involves the specification of empirical indicators to represent your theoretical concepts. Whereas theoretical concepts must be somewhat abstract and perhaps vague, your empirical indicators must be precise and specific. Thus, in our example, you might operationalize the concept "juvenile" as anyone under eighteen years of age, "delinquency" as being arrested for a criminal act, and "supervision" as the presence of a nonworking adult in the home.

The effect of operationalization is to convert the theoretical hypothesis into an empirical one. In the present case, the empirical hypothesis would be as follows: Among persons under eighteen years of age, those living in homes with a nonworking adult will be less likely to be arrested for a criminal act than will those without a nonworking adult in the home.

Collection of Empirical Data.   Based on the operationalization of theoretical concepts, you then presumably collect data relating to the empirical indicators. In the present example, you might conduct a survey of persons under eighteen years of age. Among other things, the survey questionnaire would ask each subject whether the person lived in a home with a nonworking adult and whether the person had ever been arrested for a criminal act.

Empirical Testing of Hypotheses.   After the data have been collected, the final step involves the statistical testing of the hypothesis. You determine, empirically, whether those juveniles with nonworking adults in the home are less likely to have been arrested for criminal acts than those lacking nonworking adults. The confirmation or disconfirmation of the empirical hypothesis is then used for purposes of accepting or rejecting the theoretical hypothesis.

A Schematic Presentation

The traditional image of science is depicted in schematic form in Figure 13-1. Note that you begin with a particular interest about the world, create a general theory about it, and use that deductive theory to generate a hypothesis regarding the association between two variables. This hypothesis is represented in the form $Y = f(X)$. This expression is read "Y is a function of X," meaning that values of Y are determined or caused by values of X. In our example, delinquency (Y) is a function of supervision (X). Next you operationalize the two variables by specifying empirical measurements to represent them in the real world. You collect data relevant to such measurements and, finally, test the expected relationship empirically.

Figure 13-1   *The Traditional Image of Science*

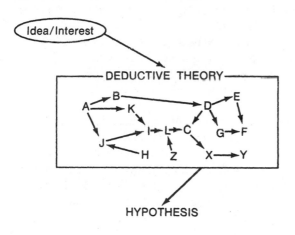

Two Basic Problems

The preceding description of the traditional view of the scientific method might persuade you that scientific research is a relatively routine activity. All you must do is move faithfully through steps 1 to 5, and you will have discovered truth.

Unhappily, scientific research is not that neat, nor is the empirical world that we study that neat. The traditional view of the scientific method simply does not represent what really goes on in scientific research—social or otherwise. Two basic problems prevent the dream from coming true.

First, theoretical concepts seldom, if ever, permit unambiguous operationalization. Whereas concepts are abstract and general, every specification of empirical indicators must represent an approximation. In our previous example, it is unlikely that the variable "supervision" is adequately represented by the presence of a nonworking adult in the home. The mere presence of such an adult does not ensure supervision of the juvenile, and in some homes lacking such an adult, other arrangements might be made for the juvenile's supervision.

Likewise, being arrested for a criminal act cannot be equated with the abstract concept "delinquency." Some juveniles might engage in delinquent behavior without being arrested. Others might be arrested falsely. Moreover, the specification of "juvenile" as a person under eighteen years of age is arbitrary. Other specifications might have been made, and probably none would be unambiguously correct.

Furthermore, it is not sufficient to argue that we should have specified "better" indicators of our concepts. The key point is that theoretical concepts almost never have perfect indicators. Every empirical indicator has some defects; all could be improved upon, and the search for better indicators is endless.

The second problem is that the empirical associations between variables are almost never perfect. In the previous example, if all juveniles with nonworking adults in the home had never been arrested and all those without such adults present had been arrested, we might conclude that the hypothesis was confirmed. If both groups had exactly the same arrest records, we might reject the hypothesis. Neither eventuality is likely in practice, however. Nearly all variables are empirically related to one another "to some extent." Specifying the "extent" that represents acceptance of the hypothesis and the "extent" that represents rejection, however, is an arbitrary act. (See Chapter 16 for a discussion of tests of statistical significance.)

In reality, then, we use approximate indicators of theoretical concepts to discover partial associations. These problems conspire with one another against you. Suppose that you specify an extent of association that will constitute acceptance of the hypothesis and the empirical analysis falls short. You will quite naturally ask yourself whether different indicators of the concepts might have resulted in the specific extent of association.

The purpose of the preceding comments has been to indicate that the traditional view of the scientific method is inappropriate to research in practice. Research does not simply happen via a dogged traverse through steps 1 to 5. This realization should not be a source of dismay, however, but should serve as a challenge to you. It should not be taken as a denial of the possibility of scientific research but should instead lay the basis for enlightened, truly scientific, research.

Measurement and association are interrelated concepts that must be handled simultaneously and logically. Rather than moving in order through a fixed set of steps, you should move back and forth through them endlessly. Your theoretical constructions are often built around previously observed associations between empirical indicators, partial theoretical constructions can suggest new empirical data to be examined, and so forth. It is hoped that after each activity you understand your subject matter a little better. The "critical experiment" that ultimately determines the fate of an entire theory is rare indeed.

Scientific research, then, is a never-ending enterprise aimed at the understanding of some phenomenon. To that end, you continually measure and examine associations, maintaining constant awareness of the nature of their interrelations. The following sections should clarify the nature of the interrelations.

The Interchangeability of Indexes

Paul Lazarsfeld, in his discussions of the "interchangeability of indexes," has provided an important conceptual tool for our understanding of the relationship between measurement and association as well as a partial resolution to the two problems discussed in the previous section.[1] His comments grow out of the recognition that any given concept has several possible indicators.

Let us return for the moment to the notion of a theoretical hypothesis $Y = f(X)$. Lazarsfeld recognizes that there are several possible indicators of supervision; we might write these indicators as x_1, x_2, x_3, and so forth. While reasons might exist for believing that some possible indicators are better than others, they are essentially interchangeable. Thus, you face the dilemma of which indicators to use in the testing of the hypothesis $Y = f(X)$.

The solution to the dilemma lies in the use of all the indicators. Thus, you would test the following empirical hypotheses: $y = f(x_1), y = f(x_2), y = f(x_3)$, and so forth. Rather than having one test of the hypothesis, you have several, as indicated schematically in Figure 13-2.

---

[1] Paul Lazarsfeld, "Problems in Methodology," in Robert K. Merton, ed., *Sociology Today* (New York: Basic Books, 1959), p. 375.

Figure 13-2 The Interchangeability of Indexes

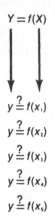

$$Y = f(X)$$

$$y \overset{?}{=} f(x_1)$$
$$y \overset{?}{=} f(x_2)$$
$$y \overset{?}{=} f(x_3)$$
$$y \overset{?}{=} f(x_4)$$
$$y \overset{?}{=} f(x_5)$$

You might already have anticipated a new dilemma. If the scientist following the traditional view of the scientific method faced the problem that the single empirical association might not be perfect, the scientist using the present method will be faced with several empirical associations, none of which will be perfect and some of which might conflict with one another. Thus, even if you have specified a particular extent of association as sufficient to confirm the hypothesis, you might discover that the tests involving x_1, x_3, and x_5 meet that specified criterion but the tests involving x_2 and x_4 do not. Your dilemma is seemingly compounded. In fact, however, the real situation might be clarified.

In terms of the notion of "the interchangeability of indexes," the theoretical hypothesis is accepted as a *general* proposition if it is confirmed by all the specific empirical tests. If, for example, juvenile delinquency is a function of supervision in a broadly generalized sense, then juvenile delinquency should be empirically related to every empirical indicator of supervision.

If, however, the scientist discovers that only certain indicators of supervision are related to juvenile delinquency, then the kinds of supervision for which the proposition holds have been specified. In practice, this can help you reconceptualize "supervision" in more general terms. Perhaps, for example, juvenile delinquency is a function of structural constraints, and some kinds of supervision are indicators of constraints while others are not.

It is very important to realize what you will have accomplished through the process described here. Rather than routinely testing a fixed hypothesis relating supervision to delinquency, you will have gained a more comprehensive understanding of the nature of that association. This accomplishment will be meaningful, however, only if we view the goal of science as increasing understanding rather than simply constructing theories and testing hypotheses.

Implications

The primary implication of the preceding comments is that measurement and association are importantly intertwined. The measurement of a variable makes little sense outside the empirical and theoretical contexts of the associations to be tested. Asked "How should I measure social class?" the experienced scientist will reply, "What is your *purpose* for measuring it?" The proper way of measuring a given variable depends heavily on the variables to be associated with it. One more example should further clarify this point.

A controversy has recently raged in the sociology of religion concerning the relationship between religiosity and prejudice. A book by Charles Y. Glock and Rodney Stark titled *Christian Beliefs and Anti-Semitism*[2] reported empirical data indicating that Christian church members holding orthodox beliefs were more likely to be anti-Semitic than were less orthodox members. The book's findings stirred considerable discussion within the churches and resulted in follow-up research on the same topic by other researchers.

One subsequent research project arrived at the opposite conclusion from that of Glock and Stark, reporting that as orthodoxy increased, prejudice decreased. Upon closer examination, however, it was noted that the measures of orthodoxy were based on acceptance of questionnaire statements reflecting the traditional Christian doctrines of "All men are brothers" and "Love thy neighbor." Not surprisingly, survey respondents who accepted the statements based on these doctrines appeared less prejudiced than respondents who rejected them. Normally, these research findings would be (and were) challenged on the grounds of "contamination." The two variables being examined (religious orthodoxy and tolerance) actually measured the same or similar qualities. Calling one set of indicators "orthodoxy" and the other "tolerance" does not show that prejudice decreases with increasing orthodoxy in any general sense.

The discussions of this chapter suggest a somewhat different reaction to the research findings. Asking *how* orthodoxy and prejudice are associated with each other rather than asking whether they are associated, we would conclude that orthodoxy measured in terms of the Glock and Stark indicators (belief in God, Jesus, miracles, and the like) is positively associated with prejudice, while orthodoxy measured as commitment to the norms of brotherly love and equality is negatively associated with prejudice. Both conclusions are empirically correct, though neither conclusion answers the more general question of whether religion and prejudice are related.

The final remaining step, of course, is to evaluate the relative utility of the conclusions. The finding that orthodoxy and prejudice are negatively associated would probably be disregarded as either tautological or trivial. (Of

---

[2] New York: Harper & Row, 1967.

course, the measurement of orthodoxy in terms of brotherly love and equality might be extremely useful in some other context.)

Summary

The goal of this rather short chapter has been to provide a healthy perspective on the twin goals of measurement and association in science. In this regard, I have sought to give you a better understanding of the scientific enterprise per se. The chapter began with a review of the traditional textbook discourse on the scientific method and indicated that this traditional view does not accurately reflect scientific research in practice. In place of the traditional perspective, I have tried to offer an alternative model, which I believe will be more useful in actual research activities.

My motivation in dwelling on this issue stems from the hardships of previous inexperienced researchers who have accepted the traditional perspective as a true picture of how "real" scientists proceed only to be severely disappointed in their own research. I do not mean to imply that scientific research is sloppy or "unscientific," only that it is far from routine. The scientist is quite different from the technician.

Additional Readings

Earl Babbie, *Observing Ourselves: Essays in Social Research* (Belmont, CA: Wadsworth, 1986).

Paul Lazarsfeld, "Problems in Methodology," in Robert K. Merton, ed., *Sociology Today* (New York: Basic Books, 1959).

Paul F. Lazarsfeld, Ann K. Pasanella, and Morris Rosenberg, eds., *Continuities in the Language of Social Research* (New York: Free Press, 1972), sec. I.

14

Constructing and
Understanding Tables

Most survey analysis falls within the general rubric of multivariate analysis; the bulk of Part Four is devoted to the varieties of multivariate analysis. The term simply refers to the simultaneous examination of several variables. The analysis of the simultaneous associations among age, education, and prejudice is an example of multivariate analysis.

You should recognize that multivariate analysis is not a specific form of analysis; specific techniques for conducting a multivariate analysis include factor analysis, smallest-space analysis, multiple correlation, multiple regression, and path analysis, among others. The basic logic of multivariate analysis can best be seen through simple tables, called contingency tables or cross-tabulations. Thus, the present chapter is devoted to the construction and understanding of such tables. Furthermore, multivariate analysis cannot be fully understood without a firm understanding of two even more fundamental analytic modes, univariate analysis and bivariate analysis. The bulk of this chapter, therefore, discusses these analytic modes.

Univariate Analysis

Univariate analysis is the examination of the distribution of cases on only one variable at a time. We will begin our discussion with the logic and formats for the analysis of univariate data.

Distributions

The most basic format for presenting univariate data is to report all individual cases, that is, to list the attribute for each case under study in terms of the variable in question. Suppose you are interested in the ages of corporate executives surveyed in a study of business practices. The most direct manner of reporting the ages of corporate executives would be to list them: 63, 57, 49, 62, 80, 72, 55, and so forth. Although such a report would provide your reader with the fullest details of the data, it would be too cumbersome for most purposes. You could arrange your data in a somewhat more manageable form without losing any detail by reporting that five executives were 38 years old, seven were 39, eighteen were 40, and so forth. Such a format would avoid duplicating data on this variable.

An even more manageable format, with a certain loss of detail, would be to report executives' ages as *marginals,* which are **frequency distributions** of grouped data: 246 executives under 45 years of age, 517 between 45 and 50 years of age, and so on. In this case, your reader would have fewer data to examine and interpret but would not be able to fully reproduce the original ages of all the executives. For example, the reader would have no way of knowing how many executives were 41 years of age.

The above example presented marginals in the form of raw numbers. An alternative form would be the use of *percentages.* Thus, for example, you could report that x percent of your corporate executives were under 45, y percent were between 45 and 50, and so forth (see Table 14-1). In computing percentages, you must frequently make a decision about the *base* from which to compute the percentage, that is, the number that represents 100 percent. In the most straightforward examples, the base is the total number of cases under study. A problem arises, however, whenever some cases have missing data. Assume, for example, that you have conducted a survey in which respondents were asked to report their ages. If some respondents failed to answer that question, you have two alternatives. First, you might still base your percentages on the total number of respondents, reporting those who failed to give their ages as a percentage of the total. Second, you could use the number of persons giving an answer to the question as the base from which to compute the percentages. You should still report the number who did not answer, but these respondents would not figure in the percentages.

The choice of a base depends wholly on the purposes of the analysis. If

Table 14-1 *An Illustration of a Univariate Analysis*

Ages of Corporate Executives (hypothetical)

Under 35	9%
36–45	21
46–55	45
56–65	19
66 and older	6
	100% = (433)
No data =	(18)

you want to compare the age distribution of your survey sample with comparable data describing the population from which the sample was drawn, you will probably want to omit the "no answers" from the computation. Your best estimate of the age distribution of all respondents is the distribution for those answering the question. Since "no answer" is not a meaningful age category, its presence among the base categories would confuse the comparison of sample and population figures. (See Table 14-1 for an example.)

Central Tendency

Beyond simply reporting marginals, you might choose to present your data in the form of summary **averages** or measures of *central tendency*. Your options in this regard are the **mode** (the most frequent attribute, either grouped or ungrouped), the arithmetic **mean**, or the **median** (the middle attribute in the ranked distribution of observed attributes). How the three averages would be calculated from a set of data is described here.

Suppose you are conducting a small pilot study that involves teenagers as respondents. Ages in your sample range from 13 to 19, as indicated below.

Age	Number
13	3
14	4
15	6
16	8
17	4
18	3
19	3

Having listed the actual ages of the thirty-one respondents, how old would you say they are in general, or "on the average"? We will look at three different ways you might answer that question.

The easiest average to calculate is the *mode*, the most frequent value. As the list shows, there were more 16-year-olds (eight of them) than any other age, so the modal age is 16, as indicated in Figure 14-1.

Figure 14-1 also demonstrates the calculation of the mean, which involves three steps: (1) Multiply each age by the number of respondents who have that age, (2) total the results of those multiplications, and (3) divide that total by the number of respondents. As shown in Figure 14-1, the mean age in this example is 15.87.

The *median* represents the "middle" value; half are above it, half below. If we had the *precise* ages of each respondent (for example, 17 years and 124 days), we would be able to arrange all thirty-one respondents in order by age,

Figure 14-1 *Three "Averages"*

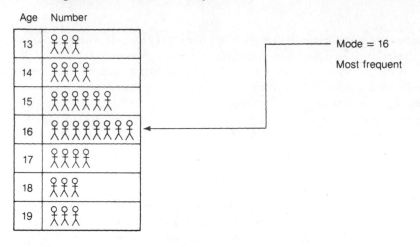

Age Number

Mode = 16

Most frequent

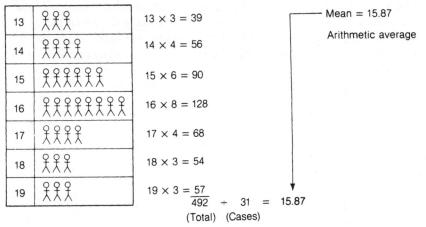

Age Number

13 × 3 = 39

14 × 4 = 56

15 × 6 = 90

16 × 8 = 128

17 × 4 = 68

18 × 3 = 54

19 × 3 = 57

492 ÷ 31 = 15.87

(Total) (Cases)

Mean = 15.87

Arithmetic average

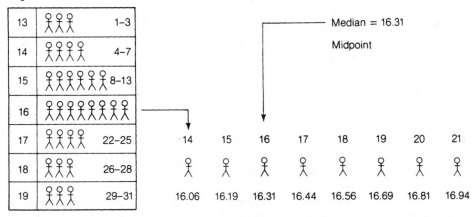

Age Number

Median = 16.31

Midpoint

and the median for the whole group would be the age of the middle respondent. As you can see, however, we do not know precise ages; our data constitute "grouped data" in the sense that three people who are not precisely the same age have been grouped in the category "13 years old," for example.

Figure 14-1 illustrates the logic of calculating a median for grouped data. Since there are thirty-one respondents altogether, the "middle" respondent would be number 16 if respondents were arranged by age; fifteen would be younger and fifteen would be older. In the bottom portion of Figure 14-1, you'll see that the middle person is one of the eight 16-year-olds. In the enlarged view of that group, we see that number 16 is the third from the left.

We do not know the precise ages of the respondents in this group; the statistical convention in such cases is to assume that they are evenly spread along the width of the group. In this instance, the *possible* ages of the respondents go from 16 years and no days to 16 years and 364 days. Strictly speaking, the range is 364/365 years. As a practical matter, it is sufficient to call it one year.

If the eight respondents in this group were evenly spread from one limit to the other, they would be one-eighth of a year apart—a 0.125-year interval. The illustration shows that if we place the first respondent half the interval from the lower limit and add a full interval to the age of each successive respondent, the final respondent is half an interval from the upper limit.

What we have done, therefore, is calculate hypothetically the precise ages of the eight respondents, assuming that their ages are spread out evenly. Having done that, we merely note the age of the middle respondent (16.31) and use that as the median age for the group.

Whenever the total number of respondents is an even number, of course, there is no middle case. In that situation, you merely calculate the mean of the two values it falls between. Suppose there were one more 19-year-old, for example. The midpoint would then fall between number 16 and number 17. The mean, in that case, would be calculated as $(16.31 + 16.44)/2 = 16.38$.

In the research literature, you will find both means and medians presented. Whenever means are presented, you should be aware that they are susceptible to the effect of extreme values—a few very large or very small numbers. For example, the 1988 mean per capita gross national product (GNP) for the United States was $17,500, contrasted with $400 for Sri Lanka, $5,080 for Ireland, and $11,910 for Australia. The tiny oil sheikdom of the United Arab Emirates, however, had a mean per capita GNP of $14,410, yet most residents of that country are impoverished, as can be inferred from the fact that their infant mortality rate was almost four times that of the United States in 1988. The high mean per capita GNP in the United Arab Emirates reflects the enormous petro-wealth of a few.[1]

---

[1] Data are taken from Carl Haub and Mary Mederios Kent, "1988 World Population Data Sheet" (a poster), Washington, DC: Population Reference Bureau.

Dispersion

Averages offer the special advantage to the reader of reducing the raw data to the most manageable form. A single number (or attribute) can represent all the detailed data collected in regard to the variable. This advantage comes at a cost, of course, since the reader cannot reconstruct the original data from an average. This disadvantage can be somewhat alleviated by the reporting of summaries of the **dispersion** of responses. The simplest measure of dispersion is the *range*—the distance separating the highest value from the lowest value. Thus, besides reporting that our respondents have a mean age of 15.87, we might also indicate that their ages ranged from 13 to 19. A somewhat more sophisticated measure of dispersion is the *standard deviation*. The logic of this measure was discussed in Chapter 5 in describing the standard error of a sampling distribution.

There are many other measures of dispersion, such as the *quartile deviation* (or *semi-interquartile range*). A quartile is one-fourth of the cases under study. If we were studying IQ, for example, the first quartile would be comprised of those subjects with the highest IQs, the fourth quartile those subjects with the lowest IQs. The quartile diviation is calculated as half the distance between the bottom of the first quartile and the top of the fourth quartile. So, if the highest quartile had IQs ranging from 120 to 150, and the lowest quartile had scores ranging from 60 to 90, the quartile deviation would be (120 − 90)/2 = 15.

Notice, incidentally, that since the four quartiles are of equal size, the first and second quartiles contain that half of the respondents with the highest scores and the third and fourth quartiles contain that half of the respondents with the lowest scores. This means that the mid-point between the second and third quartiles is the *median* score.

Continuous and Discrete Variables

The calculations described above are not appropriate for all variables. To understand why, we must examine two types of variables: *continuous* and *discrete*. Age is a continuous, ratio variable; it increases steadily in tiny fractions instead of jumping from category to category as does a discrete variable such as sex or military rank. If discrete variables are being analyzed—a nominal or ordinal variable, for example—then some of the techniques discussed above are not applicable. Strictly speaking, medians and means should be calculated only for interval and ratio data, respectively. If the variable in question were sex, for example, raw number or percentage marginals would be appropriate and useful analyses. Calculating the mode would be a legitimate, though not very revealing, analysis, but reports of mean, median, or dispersion summaries would be inappropriate. Researchers can sometimes learn something of value by violating rules like these, but you should do so with caution.

Subgroup Descriptions

Univariate analyses serve the purpose of *describing* the survey sample and, by extension, the population from which the sample was selected. Bivariate and multivariate analyses are aimed primarily at explanatory issues. Before turning to explanation, however, we should consider the intervening case of subgroup description.

Often, you want to describe subsets of your survey sample. In a straightforward univariate analysis, you might want to present the distribution of responses to a question relating to equal rights for women and men. In your detailed exploration of the answers, it would make sense to examine the responses of men and women in the sample separately. In examining attitudes toward the Ku Klux Klan, it would make sense to describe black and white respondents, as well as respondents from different regions of the country, separately.

In computing and presenting stratified descriptions, you follow the same steps as outlined in the section on univariate analysis; the difference is that the steps are followed independently for each relevant subgroup. For example, all men in the sample would be treated as a total sample representing 100 percent, and the distributions of responses or summary averages would be computed for the men. The same would be done for the women. You could then report, for example, that 75 percent of the women approved of sexual equality and that 63 percent of the men approved. Each group would have been subjected to a simple, univariate analysis. Frequency distributions for subgroups are often referred to as *stratified* marginals.

In some situations, you present stratified marginals or other subgroup analyses for purely descriptive purposes. The reporting of census data often has this purpose. The average value of dwelling units on different census blocks might be presented for descriptive purposes. You might then note the average house value for any given block.

More often, however, the purpose of subgroup descriptions is comparative. In the case of sexual equality, you would clearly be interested in determining whether the women were *more likely* than the men to approve of the proposition. Moreover, such comparisons are not motivated by idle curiosity in most cases. Typically, a comparison is based on an expectation that the stratification variable will have some form of causal effect on the description variable. Whether the respondent is a man or a woman should affect the reported attitude toward equality of the sexes. Similarly, whether respondents are black or white should affect their attitudes toward the Ku Klux Klan. Whenever the analysis is motivated by such expectations, we move into the realm of explanation rather than description.

Before moving on to the logic of bivariate, causal analysis, we should consider another example of subgroup comparisons that will allow us to address some table-formatting issues.

"Collapsing" Response Categories

"Textbook examples" of tables are often simpler than those you will typically find in published research reports or in your own analyses of data; this section and the section following address two common problems and suggest solutions.

We will begin by turning to Table 14-2, which reports data collected in a multinational poll conducted by the *New York Times,* CBS News, and the *Herald Tribune* in 1985 concerning attitudes about the United Nations. The question reported in Table 14-2 dealt with general attitudes about the way the United Nations was handling its job.

The question we want to address is how people in the five nations reported on in Table 14-2 compare in their support for the kind of job the United Nations is doing. As you review the table, you might find that there are simply too many numbers to allow you to see any meaningful pattern.

Part of the problem with Table 14-2 is the relatively small percentages of respondents selecting the two extreme response categories, *very* good job and *very* poor job. While it might be tempting to merely read the second line of the table, those saying "good job," doing so would be improper. Looking only at the second row, we would conclude that West Germany and the United States were the most positive (46 percent) about the United Nations' performance, followed closely by France (45 percent), with Britain (39 percent) less positive than those three, and Japan (11 percent) the least positive of all.

This procedure is inappropriate in that it ignores those respondents who gave the most positive answer of all—"very good job." In a situation like this, you should combine, or "collapse," the two ends of the range of variation. In this instance, you would combine "very good" with "good" and "very poor" with "poor." If you were to do this in analysis of your own data, it would be wise to add the raw frequencies together and recompute percentages for the combined categories; in analyzing a published table such as Table 14-2, you can simply add the percentages, as illustrated in Table 14-3.

Table 14-2   *Attitudes Toward the United Nations: "How is the U.N. doing in solving the problems it has had to face?"*

	West Germany	Britain	France	Japan	United States
Very good job	2%	7%	2%	1%	5%
Good job	46	39	45	11	46
Poor job	21	28	22	43	27
Very poor job	6	9	3	5	13
Don't know	26	17	28	41	10

SOURCE: "5-Nation Survey Finds Hope for U.N.," *New York Times,* 26 June 1985, p. 6.

Table 14-3 Collapsing Extreme Categories

	West Germany	Britain	France	Japan	United States
Good job or better	48%	46%	47%	12%	51%
Poor job or worse	27	37	25	48	40
Don't know	26	17	28	41	10

The collapsed categories illustrated in Table 14-3 allow us to rather easily read across the several national percentages indicating those who say the United Nations is doing at least a good job. The United States now appears the most positive; Germany, Britain, and France are only slightly less positive and are nearly indistinguishable from one another; Japan stands alone in its quite low assessment of the United Nations' performance. Though the conclusions to be drawn now do not differ radically from those we might have drawn from simply reading the second line of Table 14-2, we should note that Britain now appears relatively more supportive.

I would like to spare you a common mistake. Suppose you had hastily read the second row of Table 14-2 and noted that the British had a somewhat lower assessment of the job the United Nations was doing than did the Americans, West Germans, and French. You might feel obliged to think up an explanation for why this was so, possibly creating an ingenious psychohistorical theory about the painful decline of the once powerful and dignified British Empire. Then, once you had touted your "theory," someone else might point out that a proper reading of the data would show the British to be no less positive in their assessment than the other European nations. This is *not* a hypothetical risk. It happens frequently, but you can avoid it by collapsing answer categories where appropriate.

Handling "Don't Knows"

Tables 14-2 and 14-3 illustrate another common problem in the analysis of survey data. It is usually a good idea to give people the option of saying "don't know" or "no opinion" when you ask for their opinions on issues, but what do you do with those answers when you analyze the data?

Notice the wide range of variation in the national percentages saying "don't know" in Table 14-3—from only 10 percent in the United States to 41 percent in Japan. The presence of substantial percentages responding "don't know" can confuse the results of such a table. For example, are the Japanese so much less likely to say the United Nations is doing a good job simply because such a large number of Japanese didn't express any opinion?

There is an easy way to recalculate percentages with the "don't knows" excluded. Look at the first column of percentages in Table 14-3—West Germany's answers to the question. Notice that 26 percent of the respondents said they didn't know. This means that respondents who said "good job" or "bad job," taken together, represent only 74 percent (100 minus 26) of the whole. If we divide the 48 percent saying "good job or better" by 0.74 (the proportion expressing any opinion), we can say that 63 percent "of those with an opinion" said the United Nations was doing a good or very good job (48%/0.74 = 63%). Table 14-4 presents the whole table with the "don't knows" excluded.

Notice that these new data offer a somewhat different interpretation than do the previous tables. Specifically, it now appears that France and West Germany are the most positive in their assessments of the United Nations, with the United States and Britain a bit lower. Though Japan still stands out as lowest in this regard, it has moved from 12 percent positive up to 20 percent.

At this point, having seen three versions of the data, you might be asking yourself which is the *right* one. The answer depends on your purpose in analyzing and interpreting the data. For example, if it is not essential for you to distinguish "very good" from "good," combining them makes sense, because it will then be easier to read the table.

Whether to include or exclude the "don't knows" is harder to decide in the abstract. The fact that such a large percentage of the Japanese had no opinion might be a very important finding if you wanted to find out whether people were familiar with the work of the United Nations. On the other hand, if you wanted to know how people might vote on an issue, it might be more appropriate to exclude the "don't knows" on the assumption that they wouldn't vote or would divide their votes between the two sides of the issue.

In any event, the *truth* contained within your data is that a certain percentage said they didn't know and the remainder divided their opinions to the extent reported. Often, it is appropriate to report your data in both forms—with and without the "don't knows"—so your readers can draw their own conclusions.

We will now slightly adjust the logic of our examination of subgroup comparisons and turn to a discussion of bivariate analysis.

Table 14-4   *Omitting the "Don't Knows"*

	West Germany	Britain	France	Japan	United States
Good job or better	63%	55%	65%	20%	57%
Poor job or worse	36	45	35	81	44

Bivariate Analysis

Explanatory, bivariate analysis is basically the same as subgroup description, with certain special constraints. In subgroup descriptions, you are completely free to pick whatever stratification variable you desire and describe each subgroup in terms of any other variable. In the example of sexual equality, you might describe men and women separately in terms of the percentages approving or disapproving (see Table 14-5), or you might describe those approving and those disapproving separately in terms of the percentages of men and women (see Table 14-6).

Either table would be a legitimate presentation of subgroup descriptions. The data presented in the two tables would be read differently, however. From Table 14-5, we would note that 63 percent of the men in the sample approved of sexual equality, compared with 75 percent of the women. From Table 14-6, we would note that of those approving 46 percent are men, whereas of those disapproving 60 percent are men. Or we might note that 37

Table 14-5 *"Do you approve or disapprove of the proposition that men and women should be treated equally in all regards?"*

	Men	Women
Approve	63%	75%
Disapprove	37	25
	100%	100%
	(400)*	(400)

*The figures shown in parentheses represent the base for percentaging. In this instance, there are 400 men altogether, 63 percent (252 of the men) of whom "approve." Thirty-seven percent (148 of the men) "disapprove."

Table 14-6 *Reversed Direction of Percentaging*

	Approve	Disapprove
Men	46%	60%
Women	54	40
	100%	100%
100% = (552)	(248)

percent of the 400 men disapprove, compared with 25 percent of the 400 women (Table 14-5), and that of those approving 54 percent are women, whereas of those disapproving 40 percent are women (Table 14-6).

In an explanatory, bivariate analysis, however, only Table 14-5 would make sense. The reasoning behind this assertion can best be presented as a series of propositions.

1. Women generally are awarded an inferior status in American society; thus, they should be more supportive of the proposed equality of the sexes.

2. A respondent's sex should therefore affect his or her response to the questionnaire item; women should be more likely to approve than men.

3. If the male and female respondents in the survey are described separately in terms of their responses, a higher percentage of the women than the men should approve.

Following this logic, Table 14-5 divides the sample into two groups, men and women, and then describes the attitudes of the two groups separately. The percentages expressing approval in the two groups are then compared, and we see that women are indeed more likely than men to approve.

If Table 14-6 were presented as an explanatory, bivariate analysis, the logic of the table would be as follows. Attitudes on sexual equality affect the sex of the person holding that attitude. Approving of sexual equality will tend to make the person a woman more than it will make the person a man. This reasoning is, of course, absurd. Respondents' sexes are predetermined long before attitudes regarding sexual equality are formed. Different attitudes about sexual equality can have no effect on whether the person holding a given attitude will be a man or a woman.

Note, however, that Table 14-6 would be legitimate from the standpoint of subgroup description and even for purposes of *prediction*. If for some reason we knew a given respondent's attitude toward sexual equality and wanted to predict whether that person was a man or a woman, Table 14-6 would be an appropriate source upon which to base such a prediction. If we knew that the respondent approved of sexual equality, we would predict that the respondent was a woman. (*Note:* If we made several independent predictions of this sort, we would be wrong 46 percent of the time.) If we knew the respondent disapproved of sexual equality, we would guess that the respondent was a man (and be wrong 40 percent of the time in repeated tests).

For purposes of explanation, however, only Table 14-5 is legitimate. In explanation, you must understand the logic of *independent* and *dependent* variables. Basically, you attempt to explain values on the dependent variable on the basis of values on the independent variable. In this sense, you reason that the independent variable causes the dependent variable (typically in a probabilistic sense). In the above example, attitudes toward sexual equality comprise the dependent variable, and respondents' sexes comprise the independent variable. Thus, sex causes attitudes toward sexual equality.

The determination of which of two variables is dependent and which is independent is sometimes difficult and even arbitrary, but some guidelines are in order. To begin with, whenever there is a clear time order relating to the two variables, the variable whose values are determined earlier in time is always the independent variable; the one whose values are determined later in time is always the dependent variable. The notion of causation backward in time is illogical. Because respondents' sexes are determined prior to their attitudes regarding sexual equality, sex must be the independent variable.

One implication of this guideline is that two variables occurring simultaneously in time cannot be linked causally. A person's sex and race cannot be analyzed per se in an explanatory fashion. Of course, the parents' race and the child's sex could logically be analyzed in this manner if you believed, for example, that black parents were more likely to have male children than were white parents. In such a case, a time order of variables could be determined. (There appears to be no logical or empirical evidence to support this particular expectation, however.)

In many instances, no clear time order relates the two variables. For example, if you wanted to examine the causal relationship between education and prejudice, the time order of the two variables would be somewhat more ambiguous than in the example of sex and attitudes. You might argue that increasing education will make a person less prejudiced; education would then be the independent variable and prejudice the dependent variable. On the other hand, it might be the case that prejudice affects the amount of education a person will seek or receive. You could argue that education would be anathema to the deeply prejudiced person and/or that a prejudiced person would be more likely to flunk out of school. Thus, a case could be made for prejudice being the independent variable and education the dependent variable.

In situations in which the time order of variables is not clear, the designation of independent and dependent variables must be made and presented on a logical basis. Often, the case for doing this cannot be made sufficiently strong to satisfy all readers. In other instances, you might believe that the two variables affect each other in a cyclical manner. For example, you might say that for some respondents education affects prejudice, while for others the opposite is true; for still others an even more complex dynamic might be at work: An unprejudiced person could be led to get more education and that education could further reduce that person's prejudices.

No matter what the situation is regarding the inherent time order of the variables or a logical positing of time order, every explanatory, bivariate table implicitly designates an independent and a dependent variable. If you believe that no time order connects the variables, you must arbitrarily designate a quasi–time order in constructing your table. The following discussion assumes that one variable has been designated as dependent, disregarding the basis for such designation.

Constructing Tables

Constructing explanatory, bivariate tables consists of the following steps:

1. The sample is divided into values or categories of the independent variable.
2. Each subgroup is then described in terms of the values or categories of the dependent variables.
3. Finally, the table is read by comparing the independent variable subgroups in terms of a given value of the dependent variable.

Recall our earlier analysis of sex and attitudes toward sexual equality. Following the above steps, with sex designated as the independent variable and attitudes toward sexual equality the dependent variable, we proceed as follows:

1. The sample is divided into men and women.
2. Each sex subgroup is described in terms of approval or disapproval of sexual equality.
3. Men and women are compared in terms of the percentages approving of sexual equality.

Figure 14-2 gives a graphic presentation of the steps involved in the creation of percentage tables. The number of cases has been reduced for the purpose of simplicity.

One problem that often confuses inexperienced researchers should be commented on. Should a table be percentaged "down" or "across"? Should a column of percentages total 100 percent, or should a row of percentages? The answer to these questions is altogether arbitrary. In this book, I have tended to standardize the procedure by percentaging down so that columns of percentage figures equal 100 percent, but this convention is only a matter of personal taste and habit.

A useful guideline follows from this general issue, however. If a table is percentaged down, it should be read across. If a table is percentaged across, it should be read down. Using Table 14-5 as an example, we find that it has been percentaged down in the sense that the percentages in each column total 100 percent. The table is interpreted by reading across: Sixty-three percent of the men approve compared with 75 percent of the women.

Dogged adherence to this general rule for table construction and interpretation will help you avoid a common error. Many inexperienced researchers would read Table 14-5 as follows: "Sixty-three percent of the men approve of sexual equality compared with 37 percent who disapprove. Therefore, men are more likely to approve." This interpretation is misleading. Though it is true that men are more likely to approve of sexual equality *than to disapprove,* this finding has no significance outside of a simple description of men's attitudes. The more important observation is that men are less likely to approve *than are women.* Because the table is percentaged down, it should be read across.

Figure 14-2 *Percentaging a Table*

A. Some men and women who either favor (=) sexual equality or don't (≠) favor it.

B. Separate the men and the women (the independent variable).

Women   Men

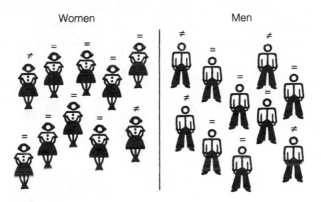

C. Within each gender group, separate those who favor equality from those who do not (the dependent variable).

Women   Men

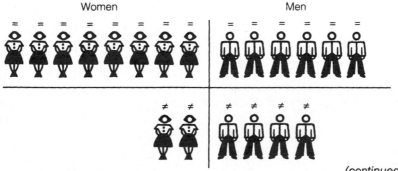

(continued)

Figure 14–2 *Percentaging a Table (Continued)*

D. Count the numbers in each cell of the table.

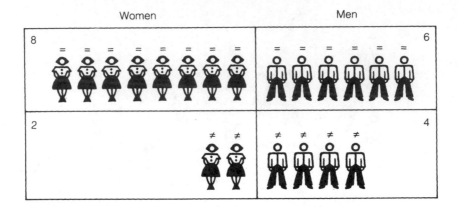

E. What percentage of the women favor equality?

F. What percentage of the men favor equality?

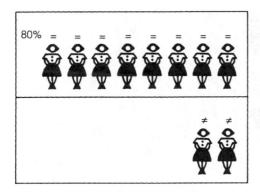

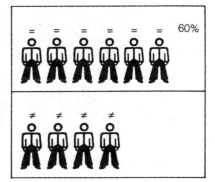

G. Conclusions

While a majority of both men and women favored sexual equality, women were more likely than men to do so.

Thus, gender appears to be one of the causes of attitudes toward sexual equality

	Women	Men
Favor equality	80%	60%
Don't favor equality	20%	40%
Total	100%	100%

Bivariate Table Formats

There is no standardized format for presenting the kind of percentaged tables we've been examining (often called contingency tables or cross-tabulations). As a result, a variety of formats is found in research literature. As long as a table can be easily read and interpreted, there is probably no reason to strive for standardization. At the same time, however, a number of guidelines should be followed in the presentation of most tabular data. (See Table 14-7 for an example of a good table.)

1. Tables should have headings or titles that sufficiently describe what is contained in the table.

2. The original content of the variables should be clearly presented—in the table itself, if at all possible, or in the text with a paraphrase in the table. This is especially critical when a variable is described from responses to an attitudinal question, because the meaning of the responses will depend largely on the wording of the questionnaire item.

3. The values or categories of each variable should be clearly indicated. Complex response categories will have to be abbreviated, but the meaning should be clear in the table, and, of course, the full response should be reported in the text.

4. When percentages are presented in the table, the base upon which they are computed should be indicated. Note that it is redundant to present all the raw numbers for each percentaged category, since they can be reconstructed from the percentages and the bases. Moreover, the presentation of both numbers and percentages often makes a table confusing and more difficult to read.

5. If any respondents are omitted from the table due to missing data ("no answer," for example), their numbers should be indicated in the table.

Table 14-7 *"Do you approve or disapprove of the general proposition that men and women should be treated equally in all regards?"*

	Men	Women
Approve	63%	75%
Disapprove	37	25
	100%	100%
	(400)	(400)
No answer =	(12)	(5)

Multivariate Analysis

The logic of multivariate analysis is the topic of later chapters of this book, especially Chapter 15. At this point, however, it will be useful to discuss briefly the construction of multivariate tables, that is, tables constructed from several variables.

Multivariate tables can be constructed on the basis of a more complicated subgroup description, following essentially the same steps outlined for bivariate tables. Instead of one independent variable and one dependent variable, however, we will have more than one independent variable. Instead of explaining the dependent variable on the basis of a single independent variable, we will seek an explanation based on more than one independent variable.

Returning to the example of attitudes toward sexual equality, suppose you believed that age would also affect such attitudes in that young people would be more likely to approve of sexual equality than would older people. As the first step in table construction, we would divide the total sample into subgroups based on the various values of both independent variables simultaneously: young men, old men, young women, and old women. Then the several subgroups would be described in terms of the dependent variable, and comparisons would be made. Table 14-8 is an example of a hypothetical multivariate table that might result.

Following the convention of this textbook, Table 14-8 has been percentaged down and should, therefore, be read across. The interpretation of this table warrants several conclusions.

1. Among both men and women, younger people are more supportive of sexual equality than are older people. Among women, 90 percent of those under 30 and 60 percent of those 30 and over approve.

2. Within each age group, women are more supportive than men. Among respondents under 30 years of age, 90 percent of the women approve, compared with 78 percent of the men. Among those 30 and over, 60 percent of the women and 48 percent of the men approve.

Table 14-8 *"Do you approve or disapprove of the proposition that men and women should be treated equally in all regards?"*

	Women		Men	
	Under 30	30 and Over	Under 30	30 and Over
Approve	90%	60%	78%	48%
Disapprove	10	40	22	52
	100%	100%	100%	100%
	(200)	(200)	(200)	(200)
No answer =	(2)	(3)	(10)	(2)

3. As measured in the table, age would appear to have a stronger effect on attitudes than sex would. For both men and women, the effect of age can be summarized as a 30-percentage-point difference. Within each age group, the percentage-point difference between men and women is 12.

4. Both age and sex have independent effects on attitudes. Within a given value of one independent variable, different values of the second will affect attitudes.

5. Similarly, the two independent variables have a cumulative effect on attitudes. Young women are the most supportive, and older men are the least supportive.

Chapter 15 on the elaboration model examines the logic of multivariate analysis in much greater detail. Before we conclude this section, however, it would be useful to note an alternative format for presenting such data.

Each table presented in this chapter is somewhat inefficient. Because the dependent variable—attitude toward sexual equality—is dichotomous (has two values), knowing one value permits you to easily reconstruct the other. Thus, if we know that 90 percent of the women under 30 years of age approve of sexual equality, then we know automatically that 10 percent disapprove (assuming we have removed "don't know" and other such responses). Reporting the percentage of those who disapprove, then, is unnecessary. On the basis of this recognition, Table 14-8 could be presented in the alternative format shown in Table 14-9.

In Table 14-9, the percentages approving of sexual equality are reported in the cells representing the intersections of the two independent variables. The numbers in parentheses below each percentage represent the number of cases upon which the percentages are based. Thus, for example, you know that there were 200 women under 30 years of age in the sample and that 90 percent of those women approved of sexual equality. This tells you in addition that 180 of the 200 women approved and that the other 20 (or 10 percent) disapproved. Table 14-9 is easier to read than Table 14-8, and it does not sacrifice any detail.

Table 14-9 *"Do you approve or disapprove of the proposition that men and women should be treated equally in all regards?"*

Percent Who "Approve"	Women	Men
Under 30	90	78
	(200)	(200)
30 and over	60	48
	(200)	(200)

Summary

This chapter has introduced the logic and mechanics of table construction. At first glance, tables seem too simple to warrant extended discussion. In fact, they are rather complex. As a result, tables are frequently misconstructed and misinterpreted.

I have sought to present the logic of table construction and interpretation, beginning with univariate analysis, moving to subgroup description, and then discussing explanatory, bivariate analysis and multivariate analysis. The following chapters of this book depend heavily on an understanding of contingency tables, so it is essential that you feel comfortable with them.

At this point, it is worth repeating the essential steps involved in the construction and interpretation of explanatory tables.

1. Divide the sample into subgroups based on the values of the independent variable(s).

2. Describe each subgroup on the basis of values of the dependent variable.

3. Compare the independent variable subgroups in terms of a given value of the dependent variable.

Finally, you should commit to memory the following rule of thumb: Percentage down and read across, or percentage across and read down.

Additional Readings

Stephen Cole, *The Sociological Method* (Chicago: Markham, 1972).

James Davis, *Elementary Survey Analysis* (Englewood Cliffs, NJ: Prentice-Hall, 1971).

Sanford Labovitz and Robert Hagedorn, *Introduction to Social Research* (New York: McGraw-Hill, 1971).

Hans Zeisel, *Say It with Figures* (New York: Harper & Row, 1957).

15

The Elaboration Model

This chapter is devoted to a perspective on survey analysis that is variously referred to as "the elaboration model," "the interpretation method," "the Columbia school," or "the Lazarsfeld method." This varied nomenclature derives from the fact that the method discussed aims at *elaboration* on an empirical relationship among variables in order to interpret that relationship in the manner developed by Paul Lazarsfeld at Columbia University.

The elaboration model is used to make the relationship between two variables understandable through the simultaneous introduction of additional variables. It was developed primarily through the medium of contingency tables, but in later chapters we will see how it can be used with other statistical techniques.

It is my firm belief that the elaboration model offers you a clearer picture of the logic of survey analysis than any other available method. Especially when used with contingency tables, this method portrays the logical processes of scientific analysis. Moreover, if you are able to fully comprehend the use of the elaboration model using contingency tables, you should be better able to use and understand more sophisticated statistical techniques.

History of the Elaboration Model

Reviewing the historical origins of the elaboration model is especially instructive for a realistic appreciation of scientific research in practice. During World War II, Samuel Stouffer organized and headed a special social research branch within the United States Army. Throughout the war, this group conducted a large number and variety of surveys among American servicemen. Although the objectives of these studies varied somewhat, they generally focused on the factors affecting soldiers' effectiveness in combat.

Several of the studies examined the issue of morale in the military. Because morale was believed to affect combat effectiveness, the improvement of morale would presumably increase the effectiveness of the war effort. Stouffer and his research staff, then, sought to uncover some of the variables that affected morale. In part, the group sought to confirm empirically some commonly accepted propositions, including the following:

1. Promotions surely affected soldiers' morale, and soldiers serving in units with low promotion rates would have relatively low morale.

2. Given racial segregation and discrimination in the South, black soldiers being trained in Northern training camps would have higher morale than those being trained in the South.

3. Soldiers with more education would be more likely to resent being drafted into the army as enlisted men than would soldiers with less education.

Each of these propositions made sense logically, and common wisdom held each to be empirically true. Stouffer decided to test each proposition empirically; to his surprise, not one was confirmed.

First, soldiers serving in the Military Police, where promotions were the slowest in the Army, had fewer complaints about the promotion system than did soldiers serving in the Army Air Corps, where promotions were the fastest in the Army. This finding was derived from responses to a question asking whether the soldier believed the promotion system to be generally fair. Second, the general morale of black soldiers serving in Northern training camps and of those serving in Southern training camps seemed to differ little, if at all. Third, less educated soldiers were more likely to resent being drafted into the Army than were soldiers with more education.

Faced with data such as these, many researchers would no doubt have tried to hide the findings as reflecting poorly on their scientific abilities. Others would have run tests of statistical significance and then tried to publish the results. Stouffer, instead, asked why he had obtained such data.

He found the answer to this question within the concepts of "reference group" and "relative deprivation." Simply put, Stouffer suggested that soldiers evaluated their positions in life not in accord with absolute, objective standards but on the basis of their relative position vis-à-vis others around them. The people they compared themselves with were their reference group, and they felt relative deprivation if they did not compare favorably with others in that group.

Within these concepts of reference group and relative deprivation, Stouffer found an answer to each anomaly in his empirical data. Regarding promotion, he suggested that soldiers judged the fairness of the promotion system on the basis of their own experiences relative to others around them. In the Military Police, where promotions were few and slow, few soldiers knew of a less qualified buddy who had been promoted faster than they had. In the

Army Air Corps, however, the rapid promotion rate meant that many soldiers knew of less qualified buddies who had been promoted faster than seemed appropriate. Thus, ironically, the MP's said the promotion system was generally fair, and the Air Corpsmen said it was not.

A similar explanation seemed appropriate in the case of the black soldiers. Rather than comparing conditions in the North with those in the South, they compared their own status, as black soldiers, with the status of the black civilians around them. In the South, where discrimination was at its worst, being a soldier somewhat insulated blacks from adverse cultural norms in the surrounding community. Whereas Southern black civilians were grossly discriminated against and denied self-esteem, good jobs, and so forth, black soldiers had a slightly better status. In the North, however, many of the black civilians the soldiers encountered were holding down well-paying defense jobs; with discrimination less severe, being a soldier did not help a black's status in the community.

Finally, reference group and relative deprivation seemed to explain the anomaly of highly educated draftees accepting their induction more willingly than those with less education. Stouffer reasoned as follows:[1]

1. A soldier's friends will, on the whole, have about the same educational status as his own.
2. Soldiers with less education will be more likely to engage in semi-skilled, production-line occupations and farming than those with more education.
3. During wartime, many production-line industries and farming were declared vital to the national interest; production-line workers in those industries and farmers would be exempted from the draft.
4. A soldier with little education was more likely to have friends who were in draft-exempt occupations than a soldier with more education.
5. The draftee of little education would be more likely to feel discriminated against than would the draftee with more education by virtue of each comparing himself with his friends.

These were the explanations that Stouffer suggested to unlock the mystery of the three anomalous findings. Because they were not part of a preplanned study design, however, he lacked empirical data for testing them. Nevertheless, Stouffer's logical exposition provided the basis for the later development of the elaboration model: understanding the relationship between two variables through the controlled introduction of other variables.

---

[1] Samuel A. Stouffer et al., *The American Soldier* (Princeton, NJ: Princeton University Press, 1949), vol. 1, pp. 122 ff., esp. p. 127.

Table 15-1 *Summary of Stouffer's Data on Education and Acceptance of Induction*

	Hi Ed	Lo Ed
Should *not* have been deferred	88%	70%
Should have been deferred	12	30
	100%	100%
	(1761)	(1876)

Tables 15-1, 15-2, 15-3, and 15-4 are modified with permission of the Macmillan Company from *Continuities in Social Research: Studies in the Scope and Method of "The American Soldier"* by Robert K. Merton and Paul F. Lazarsfeld. Copyright 1950 by The Free Press, a Corporation.

Table 15-2 *Hypothetical Relationship Between Education and Deferment of Friends*

		Hi Ed	Lo Ed
	Yes	19%	79%
Friends deferred?			
	No	81	21
		100%	100%
		(1761)	(1876)

The formal development of the elaboration model was the work of Paul Lazarsfeld and his associates at Columbia University. In a methodological review of Stouffer's army studies, Lazarsfeld and Patricia Kendall presented hypothetical tables that would have proved Stouffer's contention regarding the relation between education and acceptance of induction had the empirical data been available.[2]

Kendall and Lazarsfeld began with Stouffer's data showing the negative association between education and acceptance of induction (see Table 15-1). Following Stouffer's explanation, Kendall and Lazarsfeld created a hypothetical table, compatible with the empirical data, to show that education was related to whether a soldier had friends who were deferred. In Table 15-2, note

---

[2] Patricia L. Kendall and Paul F. Lazarsfeld, "Problems of Survey Analysis," in Robert K. Merton and Paul F. Lazarsfeld, eds., *Continuities in Social Research: Studies in the Scope and Method of "The American Soldier"* (New York: Free Press, 1950), pp. 133–196.

that 19 percent of those with high education reported having friends who were deferred, compared with 79 percent of those with less education.

Stouffer's explanation next assumed that soldiers with friends who had been deferred would be more likely to resent their own induction than would those who had no deferred friends. Table 15-3 presents the hypothetical data from Kendall and Lazarsfeld that would have supported that assumption.

The hypothetical data presented in Tables 15-2 and 15-3 confirm the linkages specified by Stouffer in his explanation. First, soldiers with low education were more likely to have friends who had been deferred than were those with more education. Second, having friends who were deferred made a soldier more likely to think that he should have been deferred. Stouffer had suggested that these two relationships would clarify the original relationship between education and acceptance of induction. Kendall and Lazarsfeld created the hypothetical table that would confirm that ultimate explanation (see Table 15-4).

Recall that the original finding was that draftees with high education were more likely to accept their induction into the army as fair than were those with less education. In Table 15-4, however, we note that level of education has

Table 15-3   *Hypothetical Relationship Between Deferment of Friends and Acceptance of One's Own Induction*

	Friends Deferred?	
	Yes	No
Should not have been deferred	63%	94%
Should have been deferred	37	6
	100%	100%
	(1819)	(1818)

Table 15-4   *Hypothetical Data Relating Education to Acceptance of Induction Through the Factor of Having Friends Who Were Deferred*

	Friends Deferred		No Friends Deferred	
	Hi Ed	Lo Ed	Hi Ed	Lo Ed
Should not have been deferred	63%	63%	94%	95%
Should have been deferred	37	37	6	5
	100%	100%	100%	100%
	(335)	(1484)	(1426)	(392)

no effect on the acceptance of induction among those who report having friends deferred. Among both educational groups, 63 percent say they should not have been deferred. Similarly, educational level has no significant effect on acceptance of induction among those who reported having no friends deferred; 94 and 95 percent of the two education groups say they should not have been deferred.

On the other hand, among those with high education the acceptance of induction is strongly related to whether or not their friends were deferred— 63 percent versus 94 percent. The same is true among those with less education. The hypothetical data in Table 15-4, then, support Stouffer's contention that education affected acceptance of induction only through the medium of having friends deferred. Highly educated draftees were less likely to have friends who had been deferred and, by virtue of that fact, were more likely to accept their own induction as fair. Draftees with less education were more likely to have friends who had been deferred and, by virtue of that fact, were less likely to accept their own induction.

It is important to recognize that neither Stouffer's explanation nor the hypothetical data denied the reality of the original relationship. As educational level increased, acceptance of one's own induction also increased. The nature of this empirical relationship, however, was interpreted through the introduction of a third variable. That variable, deferment of friends, did not deny the original relationship; it merely clarified the mechanism through which the original relationship occurred. This, then, is the heart of the elaboration model and of multivariate analysis.

Having observed an empirical relationship between two variables, you seek to understand the nature of that relationship through the effects produced by introducing other variables. Mechanically, you accomplish this by first dividing your sample into subsets on the basis of the control or test variable. For example, having friends deferred or not is the control variable in our present example; the sample is divided into those who have deferred friends and those who do not. The relationship between the original two variables is then recomputed separately for each subsample. The tables produced in this manner are called the partial tables, and the relationships found in the partial tables are called the partial relationships. The partial relationships are then compared with the initial relationship discovered in the total sample.

The Elaboration Paradigm

This section presents guidelines for you to follow in understanding an elaboration analysis. To begin, you must know whether the test variable is antecedent (prior in time) to the other two variables or whether it is intervening between them; these differences suggest different logical relationships in the multivariate model. If the test variable is intervening, as in the case of education, deferment of friends, and acceptance of induction, then the relationships of Figure 15-1 are posited. The logic of this multi-

Figure 15-1

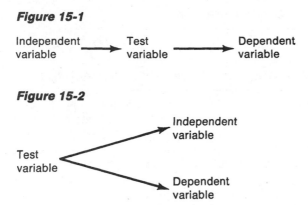

Figure 15-2

variate relationship is as follows: The independent variable (educational level) affects the intervening test variable (having friends deferred or not), which in turn affects the dependent variable (accepting induction).

If the test variable is antecedent to both the independent variable and the dependent variable, a very different multivariate relationship is posited (see Figure 15-2). In this second situation, the test variable affects both the independent variable and the dependent variable.[3] Because of their individual relationships to the test variable, the independent and dependent variables are empirically related to each other, but no causal link exists between them. Their empirical relationship is merely a product of their coincidental relationships to the test variable. (Subsequent examples will further clarify this point.)

Table 15-5 is a guide to understanding an elaboration analysis. The two columns in the table indicate whether the test variable is antecedent or intervening in the sense described above. On the left side of the table is indicated the nature of the partial relationships compared with the original relationship between the independent and dependent variables. The technical notations assigned to each case are given in the body of the table.

Replication

Whenever the partial relationships are essentially the same as the original relationship, the term **replication** is assigned to the result, regardless of whether the test variable is antecedent or intervening. The meaning here is essentially the same as common sense would dictate. The original relationship has been replicated under test conditions. If, in our previous example, education still affected acceptance of induction among both those

---

[3] Realize, of course, that the terms "independent variable" and "dependent variable" are, strictly speaking, used incorrectly in the diagram. In fact, we have one independent variable (the test variable) and two dependent variables. The incorrect terminology has been used only to provide continuity with the preceding example.

Table 15-5 *The Elaboration Paradigm*

Partial Relationships Compared with Original	Test Variable	
	Antecedent	Intervening
Same relationship	Replication	
Less or none	Explanation	Interpretation
Split*	Specification	

*One partial is the same or greater, while the other is less or none.

who had friends deferred and those who did not, then we would say that the original relationship had been replicated. Note, however, that this finding would not confirm Stouffer's explanation of the original relationship. Having friends deferred or not would not be the mechanism through which education affected the acceptance of induction.

Researchers frequently use the elaboration model rather routinely in the hope of replicating their findings among subsets of the sample. If you discovered a relationship between education and prejudice, for example, you might introduce test variables such as age, region of the country, race, religion, and so forth to test the stability of that original relationship. If the relationship were replicated among young and old, among persons from different parts of the country, and so forth, you might thereby conclude that the original relationship was genuine and general.

Explanation

Explanation is the term used to describe a *spurious relationship,* that is, an original relationship that is "explained away" through the introduction of a test variable. Two conditions are required for this situation to exist. The test variable must be antecedent to both the independent and the dependent variables, and the partial relationships must be zero or significantly less than was found in the original. Three examples should make this clear.

First, there is an empirical relationship between the number of storks in different areas and the birthrates for those areas. The more storks in an area, the higher the birthrate. This empirical relationship might lead one to assume that the number of storks affects the birthrate. An antecedent test variable explains away this relationship, however. Rural areas have both more storks and higher birthrates than do urban areas. Within rural areas, no relationship exists between the number of storks and the birthrate, nor is there such a relationship within urban areas.

Second, there is a positive relationship between the number of fire trucks responding to a fire and the amount of damage done. If more trucks respond, more damage is done. One might assume from this that the fire trucks themselves cause the damage. An antecedent test variable, however—the size of the fire—explains away the original relationship. Large fires do more damage than small ones, and more fire trucks respond to large fires than to small ones. Looking only at large fires, the original relationship would vanish (or perhaps reverse itself); the same would be true looking only at small fires.

Finally, there is an empirical relationship between the region of the country in which medical school faculty members attended medical school and their attitudes toward Medicare.[4] To simplify matters, we will examine only the East and the South. Of faculty members attending Eastern medical schools, 78 percent said they approved of Medicare, compared with 59 percent of those attending Southern medical schools. This finding makes sense in view of the fact that the South seems generally more resistant to such programs than the East, and medical school training should presumably affect a doctor's medical attitudes. This relationship, however, is explained away through the introduction of an antecedent test variable—the region of the country in which a faculty member was raised.

Of faculty members raised in the East, 89 percent attended medical school in the East, and 11 percent attended medical school in the South. Of those raised in the South, 53 percent attended medical school in the East and 47 percent in the South. Moreover, the area in which faculty members were raised is related to attitudes toward Medicare. Of those raised in the East, 84 percent approved of Medicare, compared with 49 percent of those raised in the South.

Table 15-6 presents the three-variable relationship among region in which faculty members were raised, region of medical school training, and attitudes toward Medicare. Faculty members raised in the East are quite likely

Table 15-6   *Region of Origin, Region of Schooling, and Attitudes Toward Medicare*

Percent Who Approve of Medicare		Region in Which Raised	
		East	South
Region of Medical School Training	East	84	50
	South	80	47

SOURCE: Babbie, *Science and Morality in Medicine*, p 181.

---

[4] Earl R. Babbie, *Science and Morality in Medicine* (Berkeley: University of California Press, 1970), see esp. p. 181.

to approve of Medicare, regardless of where they attended medical school. By the same token, those raised in the South are relatively less likely to approve of Medicare, but again the region of their medical school training has little or no effect. These data indicate that the original relationship between region of medical training and attitudes toward Medicare was spurious; it was due only to the coincidental effect of region of origin on both region of medical training and attitudes toward Medicare. When region of origin is *held constant,* as in Table 15-6, the original relationship disappears in the partial relationships.

Interpretation

Interpretation is similar to explanation, except for the time placement of the test variable and the implications that follow from that difference. The earlier example of education, friends deferred, and acceptance of induction is an excellent illustration of interpretation. In the terms of the elaboration model, the effect of education on acceptance of induction is not explained away; it is still a genuine relationship. In a real sense, educational differences *cause* differential acceptance of induction. The intervening variable, deferment of friends, merely helps us interpret the mechanism through which the relationship occurs.

Note an important point here. You might have begun your analysis with the observation that having friends deferred made draftees less willing to accept their own induction as fair. In the attempt to better understand this original finding, you might have introduced education as an antecedent test variable. Had you done this, however, you would have found that the relationship between friends being deferred and acceptance of induction was *replicated* among the highly educated and among the less educated soldiers (see Table 15-4). You would also have noted that highly educated soldiers were less likely to have friends deferred, but the original relationship would not have been explained away.

As a final example of interpretation, researchers have observed in the past that children from homes with working mothers are more likely to become delinquent than children whose mothers do not work. This relationship can be interpreted, however, through the introduction of "supervision" as a test variable. Among children who are supervised, delinquency rates are not affected by whether or not their mothers work. The same is true among children who are not supervised. It is the relationship between working mothers and lack of supervision that produced the original relationship.

Specification

Sometimes the elaboration model produces partial relationships that differ significantly from each other. For example, one partial relationship might look very much like the original two-variable relationship, while the second partial relationship is near zero. This situation is referred to as

Table 15-7   Social Class and Mean Church Involvement among
Episcopal Women

	Social Class Levels				
	Low 0	1	2	3	High 4
Mean Involvement	.63	.58	.49	.48	.45

SOURCE: Glock et al., To Comfort and to Challenge, p. 85.

Table 15-8   Social Class and the Holding of Office in Secular
Organizations

	Social Class Levels				
	Low 0	1	2	3	High 4
Percent who have held office in a secular organization	46	47	54	60	83

SOURCE: Glock et al., To Comfort and to Challenge, p. 92.

specification in the elaboration paradigm. You have specified the conditions
under which the original relationship occurs.

In a study of the sources of religious involvement, Glock and his associates discovered that among Episcopal church members involvement decreased
as social class increased.[5] Table 15-7 examines mean levels of church involvement found among women parishioners at different social class levels. Glock
interpreted this finding in the context of others in the analysis and concluded
that church involvement provides an alternative form of gratification for people who are denied gratification in the secular society. This interpretation
explained why women were more religious than men, why old people were
more religious than young people, and so forth. Glock reasoned that people of
lower social class (measured by income and education) had fewer chances to
gain self-esteem from the secular society than did people of higher social
class. To illustrate this point, he noted that social class was strongly related
to whether a woman had ever held an office in a secular organization (see
Table 15-8).

Glock then reasoned that if social class was related to church involvement only by virtue of the fact that lower-class women were denied opportuni-

---

[5] Charles Y. Glock, Benjamin B. Ringer, and Earl R. Babbie, To Comfort and to Challenge
(Berkeley: University of California Press, 1967), p. 92.

Table 15-9   *Church Involvement by Social Class and Holding Secular Office*

	Social Class Levels				
	Low				High
Mean Church Involvement	0	1	2	3	4
Have held office	.46	.53	.46	.46	.46
Have not held office	.62	.55	.47	.46	.40

SOURCE: Glock et al., *To Comfort and to Challenge*, p. 92.

ties for gratification in the secular society, the original relationship should not hold among women who were getting social gratification. As a rough indicator of the receipt of gratification from the secular society, he used the variable of the holding of secular office. In terms of this test, social class should be unrelated to church involvement among women who had held such office (see Table 15-9).

Table 15-9 presents an example of specification. Among women who have held office in secular organizations, there is essentially no relationship between social class and church involvement. In effect, the table specifies the conditions under which the original relationship holds as a lack of gratification in the secular society.

The term *specification* is used in the elaboration paradigm regardless of whether the test variable is antecedent or intervening. The meaning is the same in either case: You have specified the particular conditions under which the original relationship holds.

Refinements to the Paradigm

The preceding sections present the primary logic of the elaboration model as developed by Lazarsfeld and his colleagues. Morris Rosenberg has offered an excellent presentation of the paradigm described above and goes beyond it to suggest additional variations.[6] Rather than reviewing the comments made by Rosenberg, it might be useful at this point to consider the logically possible variations. Some of these comments can be found in Rosenberg's book; others were suggested by it.

First, the basic paradigm assumes an initial relationship between two variables. A more comprehensive model, however, might usefully differentiate between positive and negative relationships. Moreover, Rosenberg suggests the application of the elaboration model to an original relationship of *zero*, with

---

[6] Morris Rosenberg, *The Logic of Survey Analysis* (New York: Basic Books, 1968).

the possibility that relationships will appear in the partials. Rosenberg cites as an example of this a study of union membership and attitudes toward having Jews on the union staff.[7] The initial analysis indicated that length of union membership did not relate to the attitude. Those who had belonged to the union less than four years were as willing to accept Jews on the staff as were those who had belonged to the union longer than four years. The *age* of union members, however, was found to *suppress* the relationship between length of union membership and attitudes toward Jews. Overall, younger members were more favorable to Jews than were older members. At the same time, of course, younger members were less likely to have been in the union as long as older members. Within specific age groups, however, those in the union longest were the most supportive of having Jews on the staff. Age, in this case, was a **suppressor variable,** concealing the relationship between length of membership and attitudes toward Jews.

Second, the basic paradigm focuses on partials as being the same strength or weaker than the original relationship but does not provide guidelines for specifying what constitutes a significant difference between the original relationship and the partials. Every researcher using the elaboration model will frequently be in the position of making an arbitrary decision as to whether a given partial is significantly weaker than the original. This dilemma, then, suggests another dimension to the paradigm.

Third, the limitation of the basic paradigm to partials that are the same as or weaker than the original relationship neglects two other possibilities. A partial relationship might be *stronger* than the original, or a partial relationship might be the reverse of the original—negative where the original was positive.

Rosenberg provides a hypothetical example of this situation by first suggesting that a researcher might find working-class respondents more supportive of the Civil Rights movement than middle-class respondents.[8] He further suggests that *race* might be a **distorter variable** in this instance, distorting the true relationship between class and attitudes. Presumably, black respondents would be more supportive of the movement than whites, but blacks would also be overrepresented among working-class respondents and underrepresented among the middle class. Middle-class black respondents might be more supportive of the movement than working-class blacks, however, and the same relationship might be found among whites. Holding race constant, then, we would conclude that support for the Civil Rights movement was greater among the middle class than among the working class.

These new dimensions all further complicate the notion of specification. If one partial is the same as the original while the other partial is even stronger than the original, how should you react? You have specified one condition

---

[7] Ibid., pp. 88–89.
[8] Ibid., pp. 94–95.

under which the original relationship holds up, but you have also specified another condition under which the relationship holds up even better.

Finally, the basic paradigm focuses primarily on dichotomous test variables. In fact, the elaboration model is not so limited, either in theory or in use, but the basic paradigm becomes more complicated when the test variable divides the sample into three or more subsamples. The paradigm becomes more complicated yet when more than one test variable is used at the same time.

These comments are not made with the intention of faulting the basic elaboration paradigm. To the contrary, the intention is to impress upon you the fact that the elaboration model is not a simple algorithm, or set of procedures through which analysis is accomplished. The elaboration model is a logical device used to assist you in understanding your data. A firm understanding of the elaboration model will facilitate a sophisticated survey analysis. It does not suggest which variables should be introduced as controls, however, nor does it suggest definitive conclusions as to the nature of the elaboration results. For these things, you must look to your own ingenuity. Such ingenuity, moreover, will come only through extensive experience. By pointing out the oversimplifications in the basic elaboration paradigm, I have sought to bring home the point that the model provides only a logical framework. Sophisticated analysis will be far more complicated than the examples used to illustrate the basic paradigm.

At the same time, the elaboration paradigm is a very *powerful* logical framework. If you fully understand the basic model, you will be in a far better position to understand other techniques such as correlations, regressions, factor analyses, and so forth. The next chapter attempts to place techniques such as partial correlations and partial regressions in the context of the elaboration model.

Elaboration and Ex Post Facto Hypothesizing

Before we leave the topic of the elaboration model, we should make one further note regarding its power in connection with an unfortunate sacred cow in the traditional norms of scientific research. The reader of methodological literature will find countless references to the fallacy of *ex post facto hypothesizing*. The intentions of such injunctions are sound, but the inexperienced researcher is sometimes led astray.

Observing an empirical relationship between two variables and then simply suggesting a reason for that relationship is sometimes called ex post facto hypothesizing—generating a hypothesis linking two variables after their relationship is already known. Recall from an early discussion in this book that all hypotheses must be subject to disconfirmation. Unless you can specify empirical findings that would disprove your hypothesis, it is essentially useless. It is reasoned, therefore, that once you have *observed* a relationship between two variables any hypothesis regarding that relationship cannot be disproved.

This is a fair assessment in situations in which you do nothing more than dress up your empirical observations with deceptive hypotheses after the fact. Having observed that women are more religious than men, for example, you should not simply assert that women are more religious than men because of some general dynamic of social behavior and then rest your case on the initial observation.

The unfortunate spin-off of the injunction against ex post facto hypothesizing is its inhibition of good, honest hypothesizing after the fact. Inexperienced researchers are often led to believe that they must make all their hypotheses before examining their data, even if this means making a lot of poorly reasoned ones. Furthermore, they are led to ignore any empirically observed relationships that do not confirm some prior hypothesis.

Surely, few researchers would wish that Sam Stouffer had hushed up his anomalous findings regarding morale among soldiers in the army. Stouffer noted peculiar empirical observations and set about hypothesizing the reasons for his findings. His reasoning has proved invaluable to subsequent researchers.

A further, more sophisticated, point should be made here, however. Whereas anyone can generate hypotheses to explain observed empirical relationships in a body of data, the elaboration model provides the logical tools for testing those hypotheses within the same body of data. A good example can be found in our earlier discussion of social class and church involvement. Glock explained the original relationship in terms of social deprivation theory. If he had stopped at that point, his comments would have been interesting but hardly persuasive. He went beyond that point, however, and noted that if the hypothesis were correct then the relationship between social class and church involvement should disappear among women receiving gratification from the secular society, specifically, those who had held office in a secular organization. This relationship was then subjected to an empirical test. Had the new hypothesis not been confirmed by the data, Glock would have been forced to reconsider.

These additional comments should further illustrate the point that data analysis is a continuing process that demands all the ingenuity and perseverance you can muster. The scenario of a researcher carefully laying out hypotheses and then testing them in a ritualistic fashion results only in ritualistic research.

In case you are concerned that the strength of ex post facto proofs is less than that of traditional kinds, let me repeat the earlier assertion that "scientific proof" is a contradiction in terms. Nothing is ever proved *scientifically*. Hypotheses, explanations, theories, or hunches can all escape a stream of attempts at disproof, but none can be proved in any absolute sense. The acceptance of a hypothesis, then, is really a function of the extent to which it has been tested and not disconfirmed. No hypothesis should be considered sound on the basis of one test, whether the hypothesis was generated before or after the observation of empirical data. With this in mind, you should not deny yourself some of the most fruitful avenues available to you in data analysis. You should always try to reach an honest understanding of your data, develop meaningful theories for more general understanding, and not worry about the manner in which you have reached that understanding.

Summary

In this chapter, we have examined the fundamental logic of survey analysis—the elaboration model. Though we have limited our analyses to simple percentage tables, we will see in the following chapter that this basic logic applies equally in the case of more complex statistical analyses.

The logic of elaboration is as follows:

1. We begin with the observation of an empirical relationship between two variables, say X and Y.

2. We seek to understand the nature of that relationship through the process of holding other variables constant. This allows us to compare the original relationship between X and Y and the *partial relationships* found among subsets based on the control variable.

3. If the partial relationships are essentially the same as the original one, we say that the original relationship has been *replicated,* and we conclude that the relationship between X and Y is genuine and generalizable.

4. If only one of the partial relationships is essentially the same as (or stronger than) the original, while the other partial relationship is essentially zero, we call the result *specification*; we have specified the conditions under which X causes Y.

5. If the original relationship disappears in the partials and if the control variable is antecedent (prior) to X and Y, that outcome is called an *explanation*, meaning that we have explained away a spurious (ingenuine) relationship.

6. If the original relationship disappears in the partials and if the control variable intervenes chronologically between X and Y, that outcome is called an *interpretation*, meaning that we have discovered the means by which X causes Y.

Additional Readings

Charles Y. Glock, ed., *Survey Research in the Social Sciences* (New York: Russell Sage Foundation, 1967), ch. 1.

Travis Hirschi and Hanan Selvin, *Principles of Survey Analysis* (New York: Free Press, 1973).

Herbert Hyman, *Survey Design and Analysis* (New York: Free Press, 1955).

Paul F. Lazarsfeld, Ann K. Pasanella, and Morris Rosenberg, eds., *Continuities in "The Language of Social Research"* (New York: Free Press, 1972), sec. II.

Morris Rosenberg, *The Logic of Survey Analysis* (New York: Basic Books, 1968).

Samuel A. Stouffer, *Social Research to Test Ideas* (New York: Free Press, 1962).

16

Social Statistics

Many people are intimidated by empirical research because they are uncomfortable with mathematics and statistics. Indeed, many research reports are filled with unspecified computations. The role of statistics in survey research is very important, but it is equally important for that role to be seen in its proper perspective. Empirical research is first and foremost a logical, rather than a mathematical, operation. Mathematics is merely a convenient and efficient language for describing the logical operations inherent in good data analysis. Statistics, an applied branch of mathematics, is especially appropriate to a variety of research analyses.

In this chapter, we will look at two types of statistics: *descriptive* and *inferential*. Descriptive statistics is a medium for describing data in manageable forms. Inferential statistics, on the other hand, assists you in drawing conclusions from your observations; typically, it involves drawing conclusions about a population from the study of a sample drawn from that population.

Descriptive Statistics

As I have already stated, descriptive statistics is a method for presenting quantitative descriptions in a manageable form. Sometimes you want to describe single variables, and sometimes you want to describe the associations that connect one variable with another. We will look at some of the ways in which those descriptions are carried out.

Data Reduction

Scientific research often involves the collection of large masses of data. Suppose we had surveyed 2,000 people, asking each person 100 questions—not an unusually large study. We would now have a staggering 200,000 answers! No one could possibly read all 200,000 answers and reach any meaningful conclusion about them. Therefore, much scientific

Table 16-1 Partial Raw Data Matrix

	Sex	Age	Education	Income	Occupation	Political Affiliation	Political Orientation	Religious Affiliation	Importance of Religion
Person 1	1	3	2	4	1	2	3	0	4
Person 2	1	4	2	4	4	1	1	1	2
Person 3	2	2	5	5	2	2	4	2	3
Person 4	1	5	4	4	3	2	2	2	4
Person 5	2	3	7	8	6	1	1	5	1
Person 6	2	1	3	3	5	3	5	1	1

analysis involves the *reduction* of data from unmanageable details to manageable summaries.

To begin our discussion, we will look briefly at the raw data matrix created by a quantitative research project. Table 16-1 presents a partial data matrix. Notice that each row in the matrix represents a person (or other unit of analysis), each column represents a variable, and each cell in the matrix represents the coded attribute or value a given person has on a given variable. The first column in Table 16-1 represents a person's sex. Suppose "1" represents male and "2" represents female. That means that persons 1 and 2 are male, person 3 is female, and so forth.

In the case of age, person 1's "3" might mean 30–39 years old, and person 2's "4" might mean 40–49. However age had been coded (see Chapter 11), the code numbers shown in column 2 of Table 16-1 would describe each person represented there.

Notice that the data have already been reduced somewhat by the time a data matrix like the one in Table 16-1 has been created. If age has been coded as suggested above, the specific answer "33 years old" has already been reduced to the category "30–39." The people responding to the survey might have given sixty or seventy different ages, but the ages have now been reduced to six or seven categories.

Chapter 14 has already discussed some ways of further summarizing univariate data: averages such as the mode, the median, and the mean, and measures of dispersion such as the range and the standard deviation. It is also possible to summarize the association between variables.

Measures of Association

The association between any two variables can be represented by a data matrix produced by the joint frequency distributions of the two variables. Table 16-2 presents such a matrix. It provides all the information needed to determine the nature and extent of the relationship between education and prejudice.

Notice, for example, that twenty-three people (a) have no education and (b) scored high on prejudice; seventy-seven people (a) had graduate degrees and

Table 16-2 *Hypothetical Raw Data on Education and Prejudice*

	Education Level				
Prejudice	*None*	*Grade School*	*High School*	*College*	*Graduate Degree*
High	23	34	156	67	16
Medium	11	21	123	102	23
Low	6	12	95	164	77

(b) scored low on prejudice. Like the raw-data matrix in Table 16-1, this matrix gives you more information than you can easily comprehend. If you study the table carefully, however, you will note that as education increases from "None" to "Graduate degree" there is a general tendency for prejudice to decrease, but no more than a general impression is possible. A variety of descriptive statistics permit the summarization of this data matrix. Selecting the appropriate measure depends initially on the nature of the two variables.

We will turn now to some of those options available for summarizing the association between two variables. If you want more information on any of these topics, you might want to refer to a social statistics textbook.[1] Each measure of association we will discuss is based on the model of **proportionate reduction of error** (PRE). To see how this model works, assume that I asked you to guess respondents' attributes on a given variable, for example, whether they answered yes or no to a given questionnaire item.

First assume that you know the overall distribution of responses in the total sample, say, 60 percent yes and 40 percent no. You would make the fewest errors if you always guessed the *modal* (most frequent) response—yes.

Second, assume that you also know the empirical relationship between the first variable and some other variable, say, sex. Now each time I ask you to guess whether a given respondent said yes or no, I'll tell you whether the respondent is a man or a woman. If the two variables are related, you should make fewer errors the second time. It is possible, therefore, to compute the PRE by knowing the relationship between the two variables; the greater the relationship, the greater the reduction of error.

This basic PRE model is modified slightly to take into account different levels of measurement—nominal, ordinal, or interval. The following sections consider each level of measurement and present one measure of association appropriate to each. You should realize that the measures discussed are only three of many appropriate measures.

Nominal Variables.   If the two variables consist of nominal data (for example, sex, religious affiliation, race), **lambda** (λ) would be one appropriate measure. Lambda is based on your ability to guess values on one of the variables, that is, the PRE achieved through knowledge of values on the other variable. A simple hypothetical example will illustrate the logic and method of lambda. Table 16-3 presents hypothetical data relating sex to employment status. Overall, we note that 1,100 people are employed and 900 are unemployed. If you were to predict whether people were employed knowing only the overall distribution on that variable, always predicting "employed" would

---

[1] Two social statistics textbooks that you might find useful: Margaret Platt Jendrek, *Through the Maze: Statistics with Computer Applications* (Belmont, CA: Wadsworth, 1985); Hubert Blalock, *Social Statistics* (New York: McGraw-Hill, 1979).

Table 16-3   Hypothetical Data Relating Sex
to Employment Status

	Men	Women	Total
Employed	900	200	1,100
Unemployed	100	800	900
Total	1,000	1,000	2,000

result in fewer errors than always predicting "unemployed." Nevertheless, this strategy would result in 900 errors out of 2,000 predictions.

Suppose you had access to the data in Table 16-3 and were told each person's sex before you made your prediction of employment status. Your strategy would then change. For every man you would predict "employed," and for every woman you would predict "unemployed." In this instance, you would make 300 errors—the 100 unemployed men and the 200 employed women—or 600 fewer errors than you would make without knowing the person's sex.

Lambda, then, represents the reduction in errors as a proportion of the errors that would have been made on the basis of the overall distribution. In this hypothetical example, lambda would equal .67, that is, 600 fewer errors divided by the 900 total errors based on employment status alone. Lambda thus measures the statistical association between sex and employment status in this example.

If sex and employment status were statistically independent, we would find the same distribution of employment status for men and women. In this case, knowing the sex of the respondents would not affect the number of errors made in predicting employment status, and the resulting lambda would be zero. If, on the other hand, all men were employed and all women were unemployed, by knowing each person's sex you would avoid errors in predicting employment status. Specifically, you would make 900 fewer errors (out of 900), so lambda would be 1.0, representing a perfect statistical association.

Lambda is only one of several measures of association appropriate to the analysis of two nominal variables.

Ordinal Variables.   If the variables being related are ordinal (for example, social class, religiosity, alienation), **gamma** (γ) is one appropriate measure of association. Like lambda, gamma is based on your ability to guess values on one variable knowing values on another. Instead of exact values, however, gamma is based on the ordinal arrangement of values. For any given *pair* of cases, you guess that their ordinal ranking on one variable will correspond, positively or negatively, to their ordinal ranking on the other. For example, if you suspect that religiosity is positively related to political conservatism and if

Table 16-4 *Hypothetical Data Relating Social Class to Prejudice*

Prejudice	Lower Class	Middle Class	Upper Class
Low	200	400	700
Medium	500	900	400
High	800	300	100

Person A is more religious than Person B, you guess that A is also more conservative than B. Gamma is based on the number of paired comparisons that fit this pattern versus those that contradict it.

Table 16-4 presents hypothetical data relating social class to prejudice. The general nature of the relationship between these two variables is that as social class increases, prejudice decreases. There is a negative association between social class and prejudice.

Gamma is computed from two quantities: (1) the number of pairs having the same ranking on the two variables and (2) the number of pairs having the opposite ranking on the two variables. The number of pairs having the same ranking is computed by multiplying the frequency of each cell in the table by the sum of all cells appearing below and to the right of it and then summing all these products. In Table 16-4, the number of pairs with the same ranking would be computed as follows: 200(900 + 300 + 400 + 100) + 500(300 + 100) + 400(400 + 100) + 900(100), or 340,000 + 200,000 + 200,000 + 90,000 = 830,000.

The number of pairs having the opposite ranking on the two variables is computed by multiplying the frequency of each cell in the table by the sum of all cells appearing below and to the left of it and then summing all these products. In Table 16-4, the numbers of pairs with opposite rankings would be computed as follows: 700(500 + 800 + 900 + 300) + 400(800 + 300) + 400(500 + 800) + 900(800), or 1,750,000 + 440,000 + 520,000 + 720,000 = 3,430,000.

Gamma is computed from the numbers of same-ranked pairs and opposite-ranked pairs as follows:

$$\text{gamma} = \frac{\text{same} - \text{opposite}}{\text{same} + \text{opposite}}$$

In our example, gamma equals (830,000−3,430,000) divided by (830,000 + 3,430,000), or −.61. The negative sign in this answer indicates the negative association suggested by the initial inspection of the table. Social class and prejudice, in this hypothetical example, are negatively associated with one another. The numerical figure for gamma indicates that 61 percent more of the pairs examined had the opposite ranking than had the same ranking.

Note that while values of lambda vary from 0 to 1, values of gamma vary from −1 to +1, representing the *direction* as well as the magnitude of the association. Because nominal variables have no ordinal structure, it makes no sense to speak of the direction of the relationship. (A negative lambda would indicate that you made more errors in predicting values on one variable knowing values on the second than you made in ignorance of the second; that is not logically possible.)

Consider the following example of the use of gamma in contemporary survey research. To study the extent to which widows sanctified their deceased husbands, Helena Znaniecki Lopata administered a questionnaire to a probability sample of 301 Chicago-area widows.[2] In part, the questionnaire asked the respondents to characterize their deceased husbands in terms of the *semantic differentiation scale* shown in Table 16-5. Respondents were asked to describe their deceased spouses by circling a number for each pair of opposing characteristics. Note that the series of numbers connecting each pair of characteristics is an ordinal measure.

Next, Lopata wanted to discover the extent to which the various measures were related to each other. Appropriately, she chose gamma as the measure of association. Table 16-6 shows how she presented the results of her investigation. The format shown is called a **correlation matrix.** For each pair of measures, Lopata has calculated the gamma. "Good" and "useful," for example, are related to each other by a gamma equal to .79. The matrix is a

Table 16-5 *Semantic Differentiation Scale*

			Characteristic				
Positive Extreme							*Negative Extreme*
Good	1	2	3	4	5	6	7	Bad
Useful	1	2	3	4	5	6	7	Useless
Honest	1	2	3	4	5	6	7	Dishonest
Superior	1	2	3	4	5	6	7	Inferior
Kind	1	2	3	4	5	6	7	Cruel
Friendly	1	2	3	4	5	6	7	Unfriendly
Warm	1	2	3	4	5	6	7	Cold

[2] Helena Znaniecki Lopata, "Widowhood and Husband Sanctification," *Journal of Marriage and the Family,* May 1981, pp. 439–450.

Table 16-6   *Gamma Associations among the Semantic Differentiation Items of the Sanctification Scale*

	Useful	Honest	Superior	Kind	Friendly	Warm
Good	.79	.88	.80	.90	.79	.83
Useful	—	.84	.71	.77	.68	.72
Honest		—	.83	.89	.79	.82
Superior			—	.78	.60	.73
Kind				—	.88	.90
Friendly					—	.90

SOURCE: Helena Znaniecki Lopata, "Widowhood and Husband Sanctification," *Journal of Marriage and the Family*, May 1981, pp. 439–450.

convenient way of presenting the intercorrelations among several variables, and it is found frequently in the research literature. In this case, we see that all the variables are quite strongly related to each other, though some pairs are more strongly related than others.

Gamma is only one of several measures of association appropriate to ordinal variables.

Interval or Ratio Variables.   If interval or ratio variables (for example, age, income, grade point average) are being associated, one appropriate measure of association is Pearson's **product-moment correlation** (r). The derivation and computation of this measure of association is too complex for the scope of this book, so we will make only a few general comments.

Like both gamma and lambda, r is based on guessing the value of one variable by knowing the other. For continuous interval or ratio variables, however, it is unlikely that you would be able to predict the *precise* value of the variable. On the other hand, predicting only the ordinal arrangement of values on the two variables would not take advantage of the greater amount of information conveyed by an interval or ratio variable. In a sense, r reflects *how closely* you can guess the value of one variable based on your knowledge of the value of the other.

To understand the logic of r, consider the manner in which you might hypothetically guess values that particular cases have on a given variable. With nominal variables, we have seen that you might always guess the modal value. For interval or ratio data, you would minimize your errors by always guessing the mean value of the variable. Although this practice produces few, if any, perfect guesses, the extent of your errors will be minimized.

In computing lambda, we noted the number of errors produced by always guessing the modal value. In the case of r, errors are measured in terms of the sum of the squared differences between the actual value and the mean. This sum is called the *total variation*. To understand this concept, we must expand the scope of our examination. We will return to this discussion when we examine *regression analysis* in Chapter 17. At this point, however, we turn our attention from descriptive statistics to inferential statistics.

Inferential Statistics

Many, if not most, survey research projects involve the examination of data collected from a sample drawn from a larger population. A sample of people might be interviewed in a survey; a sample of divorce records might be coded and analyzed; a sample of newspapers might be examined through content analysis. Researchers seldom, if ever, study samples just to describe the samples per se; in most instances, their ultimate purpose is to make assertions about the larger population from which the sample has been selected. Frequently, then, you will want to interpret your univariate and multivariate sample findings as forming the basis for *inferences* about some population.

This section examines the statistical measures used in making such inferences and their logical bases. We will begin with univariate data and move to multivariate data in discussing tests of statistical significance.

Univariate Inferences

The opening sections of Chapter 14 dealt with methods of presenting univariate data. Each summary measure was intended to offer a method for describing the sample studied. We will now use those measures to make broader assertions about the population.

If 50 percent of a sample of people say they have had colds during the past year, 50 percent is also our best estimate of the proportion of colds in the total population from which the sample was drawn. (This estimate assumes a simple random sample, of course.) It is rather unlikely, however, that *precisely* 50 percent of the population have had colds during the year. If a rigorous sampling design for random selection has been followed, we will be able to estimate the expected range of error when we apply the sample finding to the population.

Chapter 5 on sampling theory covered the procedures for making such estimates, so we will only review them here. In the case of a percentage, the quantity

$$\sqrt{\frac{p \times q}{n}}$$

where p is a percentage, q equals $1 - p$, and n is the sample size, is called the *standard error*. As we noted in Chapter 5, this quantity is very important in the estimation of sampling error. We can be 68 percent confident that the population figure falls within plus or minus one standard error of the sample figure, 95 percent confident that it falls within plus or minus two standard errors, and 99.9 percent confident that it falls within plus or minus three standard errors.

Any statement of sampling error, then, must contain two essential components: the *confidence level* (for example, 95 percent) and the *confidence interval* (for example, ±2.5 percent). If 50 percent of a sample of 1,600 people say that they have had colds during the year, we might say that we are 95 percent confident that the population figure is between 47.5 percent and 52.5 percent. (Appendix B provides a table you can use for making these estimates.)

Recognize that we have moved in this example beyond simply describing the sample to making estimates (inferences) about the larger population. In doing so, we must be wary of the implications of several assumptions.

First, the sample must be drawn from the population about which inferences are being made. A sample taken from a telephone directory cannot legitimately be the basis for statistical inferences about the population of a city because some people do not have telephones and others have unlisted numbers.

Second, the inferential statistics assume the use of simple random sampling, which is virtually never the case in sample surveys. The statistics assume sampling with replacement, which is almost never done, but that is probably not a serious problem. Systematic sampling is used more frequently than random sampling; that, too, probably presents no serious problem if done correctly. Stratified sampling, since it improves representativeness, clearly presents no problem. Cluster sampling does present a problem, however, because the estimates of sampling error might be too small. Quite clearly, street-corner sampling does not warrant the use of inferential statistics. The standard error calculation also assumes a 100 percent completion rate; this problem increases in seriousness as the completion rate decreases.

Third, inferential statistics address sampling error only; they do not take into account **nonsampling** errors. Thus, although it might be correct to state that between 47.5 and 52.5 percent of the population (95 percent confidence) would report having had colds during the previous year, we could not so confidently guess the percentage who had actually had them. Some will report colds when they really had something else; others will forget the colds they did have. Because nonsampling errors are probably larger than sampling errors in a respectable sample design, you need to be especially cautious in generalizing from your sample findings to the population.

Tests of Statistical Significance

There is no scientific answer to the question of whether a given association between two variables is significant, strong, important, interesting, or worth reporting. Perhaps the ultimate test of significance rests with your ability to persuade your audience (present and future) of the association's significance. At the same time, a body of inferential statistics, called *tests of significance,* is available to assist you in this regard. As the name suggests, parametric statistics are statistics that make certain assumptions about the parameters describing the population from which the sample is selected.

Although **tests of statistical significance** are widely reported in social scientific literature, the logic underlying them is rather subtle and is often misunderstood. Tests of significance are based on the same sampling logic discussed elsewhere in this book. To help you understand that logic, we will return for a moment to the concept of sampling error in regard to univariate data.

Recall that a sample statistic normally provides the best single estimate of the corresponding population parameter but the statistic and the parameter seldom correspond precisely. Thus, we report the probability that the parameter falls within a certain range (confidence interval). The degree of uncertainty within that range is due to normal sampling error. The corollary of such a statement of probability is, of course, that it is *improbable* that the parameter would fall outside the specified range purely as a result of sampling error. Thus, if we estimate that a parameter (99.9 percent confidence) lies between 45 percent and 55 percent, we say by implication that it is *extremely improbable* that the parameter is actually, say, 90 percent assuming that our only error of estimation is due to normal sampling. The preceding is the basic logic behind tests of significance.

A Graphic Illustration.   We can probably illustrate this logic of statistical significance best in a series of diagrams representing the selection of samples from a population. The elements in the logic to be illustrated are as follows:

1. Assumptions regarding the *independence* of two variables in the population study
2. Assumptions regarding the *representativeness* of samples selected through conventional probability sampling procedures
3. The observed *joint distribution* of sample elements in terms of the two variables

Figure 16-1 represents a hypothetical population of 256 people, half women and half men. The diagram indicates how each person feels about women enjoying equality with men. In the diagram, those favoring equality are represented by open circles, and those opposing it are represented by circles shaded in.

The question we will investigate is whether any relationship exists between gender and feelings about equality for men and women. More specifically, we will examine whether women are more likely to favor equality than men, since they would presumably benefit more from equality. Take a moment to look at Figure 16-1 and try to discern the answer to that question.

Figure 16-1 *A Hypothetical Population of Men and Women Who Either Favor or Oppose Sexual Equality*

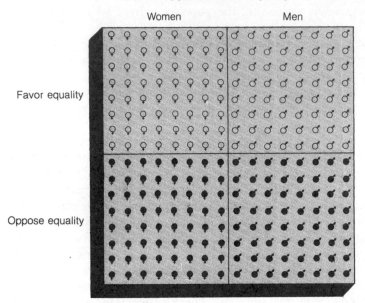

	Women	Men
Favor equality	50%	50%
Oppose equality	50%	50%
	100%	100%

Legend	
♀	Woman who favors equality
♂	Man who favors equality
●	Woman who opposes equality
♂	Man who opposes equality

The illustration in the figure indicates that there is no relationship between gender and attitudes about equality. Exactly half of the members of each group favor equality and half oppose it. Recall the earlier discussion of proportionate reduction of error. In this instance, knowing a person's gender would not reduce the "errors" we would make in guessing that person's attitude toward equality. The table at the bottom of Figure 16-1 provides a tabular view of what you can observe in the graphic diagram.

Figure 16-2 represents the selection of a one-fourth sample from the hypothetical population. In terms of the graphic illustration, a "square" selected from the center of the population provides a representative sample. Notice that our sample contains sixteen of each type of person. Half are men and half are women, and half of each gender group favor equality while the other half oppose it.

Figure 16-2   *A Representative Sample*

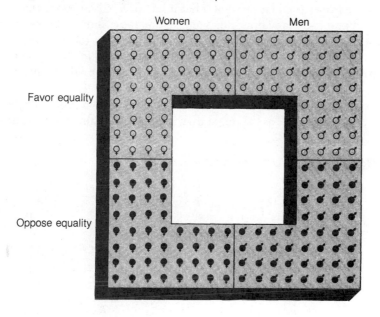

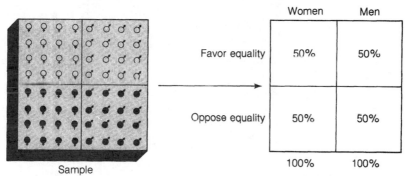

The sample selected in Figure 16-2 would allow us to draw accurate conclusions about the relationship between gender and equality in the larger population. Following the sampling logic you learned in Chapter 5, we would note that there was no relationship between gender and equality in the sample; thus, we would conclude that there was no relationship in the larger population (we have presumably selected a sample in accord with the conventional rules of sampling).

Of course, real-life samples are seldom such perfect reflections of the populations from which they are drawn. It would not be unusual for us to have selected, say, one or two extra men who opposed equality and a couple of extra women who favored it—even if there was no relationship between the two variables in the population. Such minor variations are part and parcel of probability sampling, as you learned in Chapter 5.

Figure 16-3 represents a sample that falls far short of the mark in reflecting the larger population. Note that it has selected far too many supportive women and too many opposing men. As the table shows, three-fourths of the women in the sample support equality, while only one-fourth of the men do so. If we had selected this sample from a population in which the two variables were unrelated to each other, we would be sorely misled by the analysis of our sample.

Recall that it is highly unlikely that a properly drawn probability sample will ever be as inaccurate as the one shown in Figure 16-3. In fact, if we actually

Figure 16-3 An Unrepresentative Sample

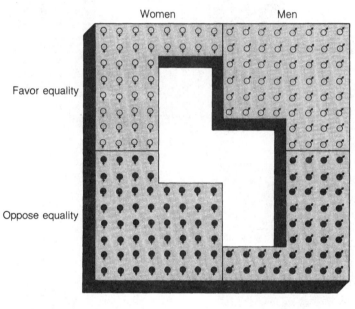

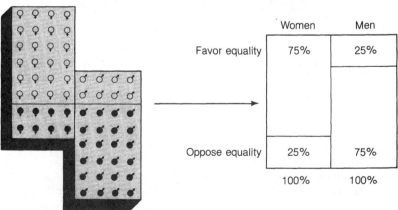

Sample

	Women	Men
Favor equality	75%	25%
Oppose equality	25%	75%
	100%	100%

selected a sample that gave us the results in Figure 16-3, we would look for a different explanation. Figure 16-4 illustrates that other explanation.

Notice that the sample selected in Figure 16-4 also shows a strong relationship between gender and equality, but this time the reason is quite different. We have selected a perfectly representative sample, but there is actually a strong relationship between the two variables in the population at large. In this latest figure, women are more likely to support equality than men in the population, and the sample reflects this.

In practice, of course, we never know what is true for the total population; that's why we select samples. If we selected a sample and found the strong relationship presented in Figures 16-3 and 16-4, we would need to decide

Figure 16-4   *A Representative Sample from a Population in Which the Variables Are Related*

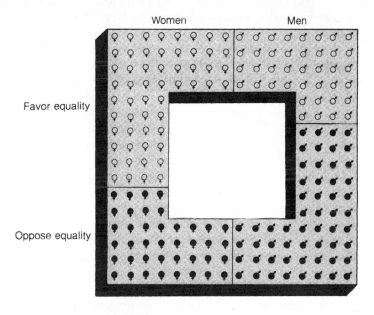

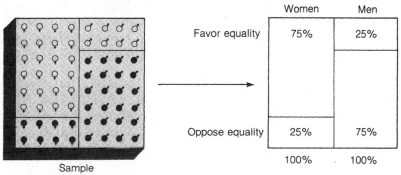

whether our finding accurately reflected the population or was simply a product of sampling error.

The Logic of Statistical Significance. The fundamental logic of tests of statistical significance, then, is the following. Faced with *any* discrepancy between the assumed independence of variables in a population and the observed distribution of sample elements, we can explain the discrepancy in either of two ways: (1) We can attribute it to an unrepresentative sample, or (2) we can reject the assumption of independence. The logic and statistics associated with probability sampling methods offer guidance relative to the varying probabilities of varying degrees of unrepresentativeness (expressed as sampling error). Most simply put, there is a *high* probability of a *small* degree of unrepresentativeness and a *low* probability of a *large* degree of unrepresentativeness.

The *statistical significance* of a relationship observed in a set of sample data, then, is always expressed in terms of probabilities. Significant at the .05 level ($p < .05$) simply means that the probability of a relationship as strong as the one observed being attributable to sampling error alone is no more than 5 in 100. Put somewhat differently, if two variables are independent of one another in the population and if 100 probability samples are selected from that population, no more than 5 of those samples should provide a relationship as strong as the one observed.

There is a corollary to confidence intervals in tests of significance, which represents the probability of the measured associations being due *only* to sampling error: the **level of significance.** Like confidence intervals, levels of significance are derived from a logical model in which several samples are drawn from a given population. In determining levels of significance, we assume that there is no association between the variables in the population and then ask what proportion of the samples drawn from that population would produce associations at least as great as those measured from the empirical data. Three levels of significance are frequently used in research reports: .05, .01, and .001. These figures mean, respectively, that the chances of obtaining the measured association as a result of sampling error are 5/100, 1/100, and 1/1,000.

Researchers who use tests of significance normally follow one of two patterns. Some specify in advance the level of significance they will regard as sufficient. If any measured association is statistically significant at that level, they will regard it as representing a genuine association between the two variables. In other words, they are willing to discount the possibility of its resulting from sampling error only.

Other researchers prefer to report the specific level of significance for each association, disregarding the conventions of .05, .01, and .001. Rather than reporting that a given association is significant at the .05 level, they might report significance at the .023 level, indicating that the chances of its having resulted from sampling error are 23 out of 1,000.

Chi Square.   Chi square (X^2) is a frequently used test of significance in social science. It is based on the **null hypothesis,** which is the assumption that there is no relationship between the two variables in the total population. Given the observed distribution of values on the two separate variables, we compute the conjoint distribution that would be expected if there were no relationship between the two variables. The result of this computation is a set of *expected frequencies* for all the cells in the contingency table. We then compare this expected distribution with the distribution of cases actually found in the sample data and determine the probability that the discovered discrepancy could have resulted from sampling error alone. An example will illustrate this procedure.

Assume that we are interested in the possible relationship between church attendance and gender for the members of a particular church. To test this relationship, we select a sample of 100 church members at random. We find that our sample is made up of forty men and sixty women and that 70 percent of our sample report having attended church during the preceding week while the remaining 30 percent say they did not.

If no relationship exists between sex and church attendance, then 70 percent of the men in the sample should have attended church during the preceding week and 30 percent should have stayed away. Moreover, women should have attended church in the same proportion. Table 16-7 (Part I) shows that, based on this model, twenty-eight men and forty-two women would have attended church, with twelve men and eighteen women not attending.

Table 16-7   *A Hypothetical Illustration of Chi Square*

I. Expected Cell Frequencies	Men	Women	Total
Attended church	28	42	70
Did not attend church	12	18	30
Total	40	60	100

II. Observed Cell Frequencies	Men	Women	Total
Attended church	20	50	70
Did not attend church	20	10	30
Total	40	60	100

III. (Observed − Expected)^2 ÷ Expected	Men	Women	
Attended church	2.29	1.52	$X^2 = 12.70$
Did not attend church	5.33	3.56	$p < .001$

Part II of Table 16-7 presents the observed attendance for the hypothetical sample of 100 church members. Note that twenty men report having attended church during the preceding week, while the remaining twenty say they did not. Among the women in the sample, fifty attended church and ten did not. Comparing the expected and observed frequencies (Parts I and II), we note that somewhat fewer men attended church than expected, while somewhat more women than expected attended.

Chi square is computed as follows. For each cell in the table, the researcher (1) subtracts the expected frequency for that cell from the observed frequency, (2) squares this quantity, and (3) divides the squared difference by the expected frequency. This procedure is carried out for each cell in the table, and the results are added together. (Part III of Table 16-7 presents the cell-by-cell computations.) The final sum is the value of chi square—12.70 in the present example.

This value is the overall discrepancy between the observed conjoint distribution in the sample and the distribution we should have expected if the two variables were unrelated to one another. Of course, the mere discovery of a discrepancy does not prove that the two variables are related; normal sampling error might produce discrepancies even when there is no relationship in the total population. The magnitude of the value of chi square, however, permits us to estimate the probability of that having happened.

To determine the statistical significance of the observed relationship, we must use a standard set of chi square values. That requires the computation of **the degrees of freedom**, which for chi square are computed as follows. The number of rows in the table of observed frequencies, minus one, is multiplied by the number of columns, minus one. This can be written as $(r - 1)(c - 1)$. In the present example, we have two rows and two columns (discounting the totals), so there is one degree of freedom.

Turning to a table of chi square values (see Appendix C), we find that for one degree of freedom and random sampling from a population in which there is no relationship between two variables, 10 percent of the time we should expect a chi square of at least 2.7. Thus, if we selected 100 samples from such a population, we should expect about ten of those samples to produce chi squares equal to or greater than 2.7. Moreover, we should expect chi square values of at least 6.6 in only 1 percent of the samples and chi square values of 7.9 in only .5 percent of the samples. The higher the chi square value, the less probable it is that the value could be attributed to sampling error alone.

In our example, the computed value of chi square is 12.70. If there were no relationship between sex and church attendance in the church member population and a large number of samples had been selected and studied, then we would expect a chi square of this magnitude in fewer than half of one percent of those samples. Thus, the probability of obtaining a chi square of this magnitude is less than .001, assuming random sampling has been used and there is no relationship in the population. We report this finding by saying that the relationship is statistically significant at the .001 level. Because it is so improbable that the observed relationship could have resulted from sampling error alone, we are likely to reject the null hypothesis and assume that a relationship exists between the two variables in the population of church members.

Most measures of association can be tested for statistical significance. Standard tables of values permit us to determine whether a given association is statistically significant and, if so, at what level. Any standard statistics textbook provides instructions on the use of such tables; we will not pursue the matter further here.

Review. Tests of significance provide an objective yardstick against which we can estimate the significance of associations between variables. They assist us in ruling out associations that might not represent genuine relationships in the population under study. A researcher who uses or reads reports of significance tests should remain wary of several dangers in their interpretation, however.

First, we have been discussing tests of *statistical* significance; no objective tests of substantive significance exist. Thus, we might be legitimately convinced that a given association is not due to sampling error, but we might be in the position of asserting without fear of contradiction that two variables are only slightly related to one another. Recall that sampling error is an inverse function of sample size: The larger the sample, the smaller the expected error. Thus, a correlation of, say, .1 might very well be significant (at a given level) if discovered in a large sample, whereas the same correlation between the same two variables would not be significant if found in a smaller sample. Of course, that makes perfectly good sense if you understand the basic logic of tests of significance. In the larger sample, there is less chance that the correlation could simply be the product of sampling error. In both samples, however, the correlation might represent a very weak and essentially zero correlation.

The distinction between statistical and substantive significance is perhaps best illustrated by cases in which there is absolute certainty that observed differences cannot be a result of sampling error. Such would be the case when we observe an entire population. Suppose we were able to learn the ages of every public official in the United States as well as the ages of every public official in the U.S.S.R. For argument's sake, assume further that the average age of American officials was found to be forty-five years old, compared to forty-six for Soviet officials. We would have gotten the ages of all officials, so there would be no question of sampling error. We know with certainty that the Soviet officials are older than their American counterparts. At the same time, we would obviously say that the difference was of no substantive significance. We would conclude, in fact, that officials in both countries were essentially the same age.

Second, lest you be misled by this hypothetical example, you should *not* calculate statistical significance on relationships observed in data collected from whole populations. Remember that tests of statistical significance measure the likelihood of relationships between variables being only a product of sampling error; if there is no sampling, there can be no sampling error.

Third, tests of significance are based on the same sampling assumptions used in the computation of confidence intervals. To the extent that these assumptions are not met by the actual sampling design, the tests of significance are not strictly legitimate.

As is the case in regard to most matters covered in this book, I have a personal prejudice. In this instance, it is against tests of significance. My objection is not to the statistical logic of such tests, since the logic is sound. Rather, I am concerned that such tests seem to mislead more than they enlighten. My principal reservations are the following:

1. Tests of significance make sampling assumptions that are virtually never satisfied by actual sampling designs.

2. They depend on the absence of nonsampling errors, a questionable assumption in most actual empirical measurements.

3. In practice, they are too often applied to measures of association that have been computed in violation of the assumptions made by those measures (for example, product-moment correlations computed from ordinal data).

4. Statistical significance is too easily misinterpreted as "strength of association," or substantive significance.

At the same time—perhaps paradoxically—I feel that tests of significance can be a valuable asset to the researcher as useful tools for the understanding of data. Although the above comments suggest an extremely conservative approach to tests of significance, specifically, that you should use them only when all assumptions are met, my general perspective is just the reverse. I encourage you to use any statistical technique (any measure of association or any test of significance) on any set of data if doing so will help you understand your data. If the computation of product-moment correlations among nominal variables and the testing of statistical significance in the context of uncontrolled sampling will meet this criterion, then I would encourage such activities. I say this in the spirit of what Hanan Selvin has referred to as data-dredging techniques. Anything goes if it ultimately leads to the understanding of data and of the social world under study.

The price of this radical freedom, however, is the giving up of strict, statistical interpretations. You will not be able to base the ultimate importance of your finding solely on a significant correlation at the .05 level. Whatever the avenue of discovery, empirical data must ultimately be presented in a legitimate manner, and their importance must be argued logically.

Summary

In this chapter, we have undertaken a wide-reaching overview of the world of social statistics. Though it has not been my intention to make you a statistical expert, you should understand the basic logic behind some of the more commonly used statistical techniques. We began by making a broad distinction between *descriptive* and *inferential* statistics.

Descriptive statistics are used to summarize data under study. Some descriptive statistics summarize the distribution of attributes on a single variable; others, called *measures of association,* summarize the associations between variables.

Inferential statistics are used to estimate the generalizability of findings arrived at through analysis of a sample to the larger population from which the sample has been selected. Some inferential statistics estimate the single-variable characteristics of the population (in terms of *confidence levels* and *confidence intervals*); others, called *tests of statistical significance,* estimate the relationships between variables in the population.

Many measures of association are based on a *proportionate reduction of error* (PRE) model. This model is based on a comparison of (a) the number of errors we would make in attempting to guess the attributes of a given variable for each case under study if we knew nothing but the distribution of attributes on that variable and (b) the number of errors we would make if we knew the joint distribution overall and were told for each case the attribute of one variable each time we were asked to guess the attribute of the other.

Lambda (λ) is an appropriate measure of association to be used in the analysis of two *nominal* variables. It also provides a clear illustration of the PRE model. Gamma (γ) is an appropriate measure of association to be used in the analysis of two *ordinal* variables. Pearson's *product-moment correlation (r)* is an appropriate measure of association to be used in the analysis of two *interval* or *ratio* variables.

Inferences about some characteristic of a population, such as the percentage of voters favoring Candidate A, must contain an indication of a *confidence interval* (the range within which the value is expected to be; for example, between 45 and 55 percent favor Candidate A) and an indication of the *confidence level* (the likelihood that the value does fall within that range, for example, 95 percent confidence). Computations of confidence levels and intervals are based on probability theory and assume that conventional probability sampling techniques have been employed in the study.

Inferences about the generalizability to a population of the associations discovered between variables in a sample involve *tests of statistical significance.* Most simply put, these tests estimate the likelihood that an association as large as the one observed could result from normal sampling error if no such association in fact exists between the variables in the larger population. Tests of statistical significance, then, are also based on probability theory and assume that conventional probability sampling techniques have been employed in the study. Statistical significance must not be confused with *substantive* significance; the latter means that an observed association is strong, important, meaningful, or worth writing home to your mother about.

The level of significance of an observed association is reported in the form of the probability that the association could have been produced merely by sampling error. To say that an association is significant at the .05 level is to say that an association as large as the one observed could not be expected to result from sampling error more than 5 times out of 100. Survey researchers tend to utilize a particular set of levels of significance in connection with tests of statistical significance: .05, .01, and .001. This is merely a convention, however.

Tests of statistical significance, strictly speaking, make assumptions about data and methods that are almost never completely satisfied by real survey research. Despite this disparity, the tests can serve a very useful function in the analysis and interpretation of data. You should, however, be wary of interpreting the "significance" of the test results too precisely.

Additional Readings

Hubert Blalock, *Social Statistics* (New York: McGraw-Hill, 1979).

Ramon E. Henkel, *Tests of Significance* (Beverly Hills, CA: Sage, 1976).

Margaret Platt Jendrek, *Through the Maze: Statistics with Computer Applications* (Belmont, CA: Wadsworth, 1985).

Leslie Kish, "Chance, Statistics, and Statisticians," *Journal of the American Statistical Association,* March 1978, vol. 73, no. 361, pp. 1–6.

Denton Morrison and Ramon Henkel, eds., *The Significance Test Controversy: A Reader* (Chicago: Aldine-Atherton, 1970).

Vicki Sharp, *Statistics for the Social Sciences* (Boston: Little, Brown, 1979).

17

Advanced Multivariate

Techniques

For the most part, this book has focused on rather rudimentary forms of data manipulation, such as the use of percentaged contingency tables. The elaboration model of analysis was presented in the form of contingency tables, as were the statistical techniques described in Chapter 16.

Now we move one step further and consider briefly a few more complex methods of data analysis and presentation. Where possible, each technique examined in this chapter will be presented from the logical perspective of the elaboration model. We will examine the following techniques: regression analysis, path analysis, factor analysis, analysis of variance, discriminant analysis, and log-linear models. These techniques are but a few of the many techniques available to the survey researcher.

My purpose in this chapter is to *introduce* you to these techniques. You won't come away from the chapter proficient in the use of the techniques, but you will be familiar with them when you encounter them in research reports. At the same time, you might want to learn more about how to use one or more of these advanced techniques; with this in mind, some references for further study have been given.

Regression Analysis

At several points in this text, I have referred to the general formula for describing the association between two variables: $Y = f(X)$. This formula is read "Y is a function of X," meaning that values of Y can be explained in terms of variations in the values of X. Stated more strongly, we might say that X causes Y, so the value of X determines the value of Y.

Regression analysis is a method of determining the specific function relating Y to X. There are several forms of regression analysis, depending on the complexity of the relationships being studied. We will begin with the simplest: **linear regression.**

Linear Regression

The regression model can be seen most clearly in the case of a perfect linear association between two variables. Figure 17-1 is a scattergram presenting in graphic form the conjoint values of X and Y as produced by a hypothetical study. It shows that for the four cases in our study the values of X and Y are identical. The relationship between the two variables in this instance is described by the equation $Y = X$; this equation is called the **regression equation.** Because all four points lie on a straight line, we could superimpose that line over the points; this line is the **regression line.**

The linear regression model has important descriptive uses. The regression line offers a graphic picture of the association between X and Y, and the regression equation is an efficient form for summarizing that association. The regression model has inferential value, as well. To the extent that the regression equation correctly describes the *general* association between the two variables, it can be used to predict other sets of values. If, for example, we know that a new case has a value of 3.5 on X, we can predict a value of 3.5 on Y as well. In practice, of course, studies are seldom limited to four cases, and the associations between variables are seldom as clear as the one presented in Figure 17-1.

A somewhat more realistic example is presented in Figure 17-2, which represents a hypothetical relationship between the number of hours spent studying and students' grade point averages. As was the case in our previous example, the values of Y (grade point averages) generally correspond to those

Figure 17-1 *Simple Scattergram of Values of X and Y*

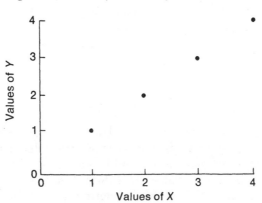

Figure 17-2 *A Scattergram of the Values of Two Variables with Regression Line Added (Hypothetical)*

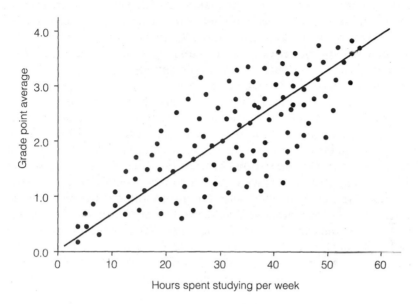

of X (hours spent studying); as values of X increase, so do values of Y. However, the association is not nearly as clear as it was for the case represented in Figure 17-1. It is not possible in Figure 17-2 to superimpose a straight line that will pass through all the points in the scattergram, but we can draw an approximate line showing the best possible linear representation of the several points. That line has been drawn on the graph.

If you have ever studied geometry, you'll know that any straight line on a graph can be represented by an equation of the form $Y = a + bX$, where X and Y are values of the two variables. In this equation, a equals the value of Y when X is 0, and b represents the *slope* of the line. If we know the values of a and b, we can calculate an estimate of Y for every value of X.

Regression analysis is a technique for establishing the regression equation representing the geometric line that comes closest to the distribution of points. This equation is valuable both descriptively and inferentially. First, the regression equation provides a mathematical description of the relationship between the variables. Second, the regression equation allows us to infer values of Y when we have values of X. Recalling Figure 17-2, we could estimate students' grade point averages if we knew how many hours they studied each week.

To improve the accuracy of your guessing, you construct a *regression line,* stated in the form of a regression equation that permits the estimation of values on one variable from values on the other. The general format for this equation is $Y' = a + b(X)$, where a and b are computed values, X is a given value on one variable, and Y' is the *estimated* value on the other. The values of a and b are computed so as to minimize the differences between actual values of Y and the corresponding estimates (Y') based on the known value of X. The sum of squared differences between actual and estimated values of Y is called the *unexplained variation* because it represents errors that still exist even when estimates are based on known values of X.

The *explained variation* is the difference between the total variation and the unexplained variation. Dividing the explained variation by the total variation produces a measure of the *proportionate reduction of error* corresponding to the similar quantity in the computation of lambda. In the present case, this quantity is the correlation squared, r^2. Thus, if $r = .7$, then $r^2 = .49$, meaning that about half the variation has been explained. In practice, we compute r rather than r^2, because the product-moment correlation can take either a positive or a negative sign, depending on the direction of the relationship between the two variables. (Computing r^2 and taking a square root would always produce a positive quantity.) Consult a standard statistics textbook for the method of computing r, though I anticipate that most readers using this measure will have access to computer programs designed to perform this function.

The preceding discussion may have been more statistical than you find comfortable, but social life is so complex that the simple linear regression model is often not sufficient to represent the true state of affairs. As we saw in Chapter 14, it is possible, using percentage tables, to analyze more than two variables. As the number of variables increases, such tables become increasingly complicated and harder to read. The regression model offers a useful alternative in such cases.

Multiple Regression

Survey researchers very often find that a given dependent variable is affected simultaneously by several independent variables. **Multiple regression** analysis provides a means of analyzing such situations. Beverly Yerg used such analysis when she set about studying teacher effectiveness in physical education, stating her expectations in the form of the following multiple regression equation:[1]

$$F = b_0 + b_1 I + b_2 X_1 + b_3 X_2 + b_4 X_3 + b_5 X_4 + e$$

---

[1] Adapted from Beverly J. Yerg, "Reflections on the Use of the RTE Model in Physical Education," *Research Quarterly for Exercise and Sport* (March 1981), p.42.

where

> F = Final pupil performance score
> I = Initial pupil performance score
> X_1 = Composite of guiding and supporting practice
> X_2 = Composite of teacher mastery of content
> X_3 = Composite of providing specific, task-related feedback
> X_4 = Composite of clear, concise task presentation
> b = Regression weight
> e = Residual

Notice that in place of the single X variable in a linear regression there are several X's; there are also several b's instead of just one. Also, Yerg has chosen to represent a as b_0 in this equation, with the same meaning as discussed above. Finally, the equation ends with a residual factor (e), which represents the square root of the variance in Y that is not accounted for by the X variables analyzed.

Beginning with this equation, Yerg calculated the values of the several b's in order to show the relative contributions of the various independent variables in determining final student performance scores. She also calculated the multiple-correlation coefficient as an indicator of the extent to which all six variables predicted the final scores. This procedure follows the same logic as the simple bivariate correlation discussed earlier, and it is traditionally reported as a capital R. In this case, $R = .877$, meaning that 77 percent of the variance (R^2)in final scores is explained by the six variables acting in concert.

Partial Regression

In explaining the elaboration model, we paid special attention to the relationship between two variables when a third test variable was held constant. Thus, we might examine the effect of education on prejudice with age held constant, testing the independent effect of education. To do this, we would compute the tabular relationship between education and prejudice separately for each age group.

Partial regressions are based on this same logical model. The equation summarizing the relationship between variables is computed on the basis of the test variables remaining constant. As in the case of the elaboration model, the result can then be compared with the uncontrolled relationship between the two variables in order to further clarify the overall relationship.

Curvilinear Regression

Up to now, we have been discussing the association among variables as represented by a straight line, though in more than two dimensions. The regression model is even more general than our discussion thus far implies.

If you have a knowledge of geometry, you will already know that curvilinear functions can also be represented by equations. For example, the equation $X^2 + Y^2 = 25$ describes a circle with a radius of 5. Raising variables to powers greater than 1 has the effect of producing curves rather than straight lines. In empirical research, there is no reason to assume that the relationship among all variables will be linear. In some cases, then, curvilinear regression analysis can provide a better understanding of empirical relationships than can any linear model.

Recall, however, that a regression line serves two functions. It describes a set of empirical observations, and it provides a *general* model for making inferences about the relationship between two variables in the general population represented by the observations. A very complex equation might produce an erratic line that would indeed pass through every individual point. In this sense, it would perfectly describe the empirical observations. There would be no guarantee, however, that such a line could adequately predict new observations or that it in any meaningful way represented the relationship between the two variables in general. Thus, it would have little or no inferential value.

Earlier in this book, we discussed the need for balancing detail and utility in data reduction. Ultimately, researchers attempt to provide the most faithful, yet also the simplest, representation of their data. This practice also applies to regression analysis. Data should be presented in the simplest fashion that best describes the actual data (thus, linear regressions are most frequently used). **Curvilinear regression** analysis presents the researcher with a new option in this regard, but it does not solve the problem altogether. Nothing does that.

Cautions in Regression Analysis

The use of regression analysis for statistical inferences is based on the same assumptions made in regard to correlation analysis: simple random sampling, the absence of nonsampling errors, and continuous interval data. Because social scientific research seldom completely satisfies these assumptions, you should use caution in assessing the results in regression analyses.

Also, regression lines—linear or curvilinear—can be useful for *interpolation* (estimating cases lying between those observed), but they are less trustworthy when used for *extrapolation* (estimating cases that lie beyond the range of observations). Recognizing this limitation on extrapolations is important for two reasons. First, you are likely to come across regression equations that seem to make illogical predictions. An equation linking hours spent studying and grade point averages, for example, might seem to suggest that students who study 100 hours a week should get a 7.0 grade point average (although the maximum is 4.0). This failure in predictive ability does not disqualify the equation but instead dramatizes the fact that its applicability is limited to a particular range of values. Second, researchers sometimes overstep this limitation, drawing inferences that lie outside their range of observation, and you would be right in criticizing them for that.

Path Analysis

Path analysis is a *causal* model for understanding relationships between variables. It is based on regression analysis, but it can provide a more useful graphic picture of relationships among several variables than is possible through other means. Path analysis assumes that the values on one variable are caused by the values on another, so it is essential that independent and dependent variables be distinguished. This requirement is not unique to path analysis, of course, but path analysis does provide a unique way of displaying explanatory results for interpretation.

Recall for a moment one of the ways the elaboration model was represented in Chapter 15. We might diagram the logic of interpretation as follows:

The logic of this presentation was that an independent variable had an impact on an intervening variable, which in turn had an impact on a dependent variable. The path analyst constructs similar patterns of relationships among variables, but the typical path diagram contains many more variables than shown above.

Besides diagramming a network of relationships among variables, path analysis shows the strengths of those several relationships. The strengths of the relationships are calculated from a regression analysis that produces numbers analogous to the partial relationships in the elaboration model. These **path coefficients,** or beta weights as they are also called, represent the strengths of the relationships between pairs of variables with the effects of all other variables in the model held constant.

The analysis in Figure 17-3, for example, focuses on the religious causes of anti-Semitism among Christian church members. The variables in the diagram are, from left to right, (1) orthodoxy, or the extent to which the subjects accept conventional beliefs about God, Jesus, the biblical miracles, and so forth; (2) particularism, the belief that one's religion is the "only true faith"; (3) acceptance of the view that the Jews crucified Jesus; (4) religious hostility toward contemporary Jews, such as believing that God is punishing them or that they will suffer damnation unless they convert to Christianity; and (5) secular anti-Semitism, such as believing that Jews cheat in business, are disloyal to their country, and so forth.

To start with, the researchers who conducted this analysis proposed that secular anti-Semitism was produced by moving through the five variables as follows: orthodoxy caused particularism, which caused the view of the histori-

Figure 17-3 Diagramming the Religious Sources of Anti-Semitism

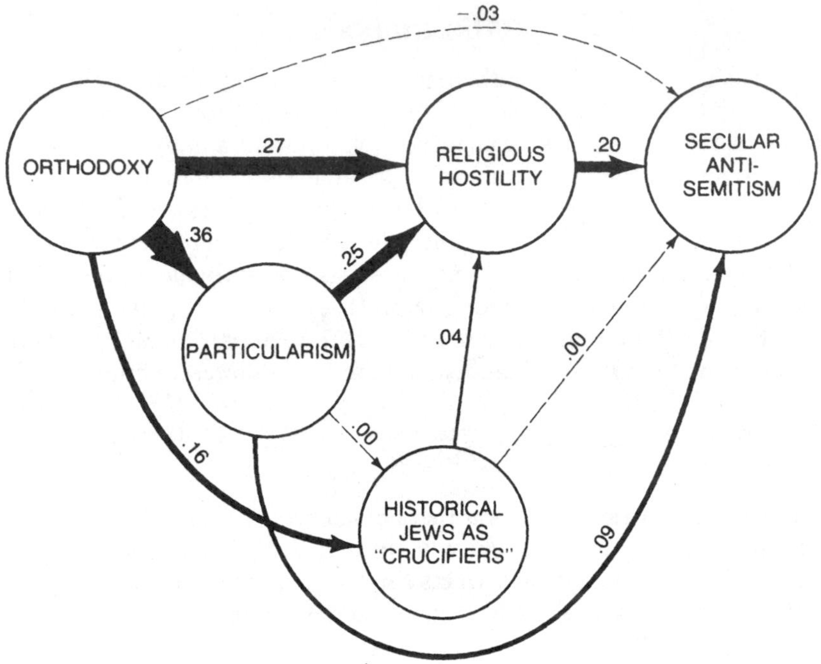

SOURCE: Rodney Stark, Bruce D. Foster, Charles Y. Glock, and Harold E. Quinley, *Wayward Shepherds—Prejudice & the Protestant Clergy.* Copyright © 1971 by Anti-Defamation League of B'nai Brith. Reprinted by permission of Harper & Row.

cal Jews as crucifiers, which caused religious hostility toward contemporary Jews, which resulted, finally, in secular anti-Semitism.

The path diagram tells a different story. The researchers found, for example, that belief in the historical role of Jews as the crucifiers of Jesus doesn't seem to matter in the process. And, although particularism is shown to be a part of one process resulting in secular anti-Semitism, the diagram also shows that anti-Semitism is reached more directly through orthodoxy and religious hostility. Orthodoxy produces religious hostility even without particularism, and religious hostility generates secular hostility in any event.

One last comment on path analysis is in order. Although it is an excellent way of handling complex causal chains and networks of variables, you must realize that path analysis itself does not tell the causal order of the variables. Nor was the path diagram generated by computer. The researcher decided the structure of relationships among the variables and used computer analysis merely to calculate the path coefficients that apply to the structure decided on.

Factor Analysis

Factor analysis is a different approach to multivariate analysis than regression analysis. Its statistical basis is complex enough and different enough from the foregoing discussions to suggest a very general discussion here. Factor analysis is used to discover patterns among the variations in values of several variables, essentially through the generation of artificial dimensions (factors) that correlate highly with several of the real variables. A computer must be used to perform this complex operation.

Suppose for the moment that a data file contains several indicators of subjects' prejudice. Each item should provide some indication of prejudice, but none will give a perfect indication. All the items, moreover, should be highly intercorrelated empirically. In a factor analysis of the data, the researcher would create an artificial dimension that would be highly correlated with each item measuring prejudice. Each subject would essentially receive a value on that artificial dimension, and the value assigned would be a good predictor of the observed attributes on each item.

Now suppose that the same study provided several indicators of subjects' alienation. It is likely that the factor analysis would also generate an artificial dimension highly correlated with each of those items.

The output of a factor analysis program consists of columns representing the different factors (artificial dimensions) generated from the observed relations among variables plus the correlations between each variable and each factor (called the **factor loadings**). In the above example, it is likely that one factor would more or less represent prejudice and another would more or less represent alienation. Data items measuring prejudice would have high loadings on (correlations with) the prejudice factor and low loadings on the alienation factor. Data items measuring alienation ability would have just the opposite pattern.

In practice, however, factor analysis does not proceed in this fashion. Rather, the variables are input to the program, and a series of factors with appropriate factor loadings are output. You must then determine the meaning of a given factor on the basis of those variables that load highly on it. The generation of factors, however, has no reference to the meaning of variables but only to their empirical associations. Two criteria are taken into account: (1) A factor must explain a relatively large portion of the variance found in the study variables, and (2) every factor must be more or less independent of every other factor. An example of the use of factor analysis follows.

Many survey researchers have studied the problem of delinquency. When you look deeply into the problem, however, you discover that there are many different types of delinquents. Morris Forslund, in a survey of high school students in a small Wyoming town, set out to create a typology of delinquency.[2] His questionnaire asked students to report whether they had committed a variety of delinquent acts. He then submitted their responses to factor analysis. The results are shown in Table 17-1. The various delinquent acts are listed on the left. The numbers shown in the body of the table are the factor loadings on the four factors constructed in the analysis. Notice that Forslund has labeled the dimensions. I have bracketed the items on each factor that led to his choice of labels. Forslund summarizes the results as follows:

> For the total sample four fairly distinct patterns of delinquent acts are apparent. In order of variance explained, they have been labeled: 1) Property Offenses, including both vandalism and theft; 2) Incorrigibility; 3) Drugs/Truancy; and 4) Fighting. It is interesting, and perhaps surprising, to find both vandalism and theft appear together in the same factor. It would seem that those high school students who engage in property offenses tend to be involved in both vandalism and theft. It is also interesting to note that drugs, alcohol and truancy fall in the same factor.[3]

Having determined this overall pattern, Forslund reran the factor analysis separately for boys and for girls. Essentially the same patterns emerged in both cases.

This example, I think, shows that factor analysis is an efficient method of discovering predominant patterns among a large number of variables. Instead of the researcher being forced to compare countless correlations—simple, partial, and multiple—to discover patterns, factor analysis can be used for this task. Incidentally, this is a good example of a helpful use of computers.

Factor analysis also presents data in a form that can be interpreted by the reader or researcher. For a given factor, the reader can easily discover the variables loading highly on it, thus noting clusters of variables, or the reader can easily discover which factors a given variable is or is not loaded highly on.

But factor analysis also has disadvantages. First, as noted above, factors are generated without any regard to substantive meaning. Researchers will often find factors producing very high loadings for a group of substantively disparate variables. They might find, for example, that prejudice and religiosity have high positive loadings on a given factor, while education has an equally high negative loading. Surely the three variables are highly correlated, but what does the factor represent? All too often, inexperienced researchers will be led into naming such factors as "religio-prejudicial lack of education" or something similarly nonsensical.

---

[2] Morris A. Forslund, "Patterns of Delinquency Involvement: An Empirical Typology," paper presented to the Annual Meeting of the Western Association of Sociologists and Anthropologists, Lethbridge, Alberta, February 8, 1980.
[3] Ibid., p. 4.

Table 17-1   *Factor Analysis: Delinquent Acts, Whites*

Delinquent Act	Property Offenses Factor I	Incorrigibility Factor II	Drugs/ Truancy Factor III	Fighting Factor IV
Broke street light, etc.	.669	.126	.119	.167
Broke windows	.637	.093	.077	.215
Broke down fences, clothes lines, etc.	.621	.186	.186	.186
Taken things worth $2 to $50	.616	.187	.233	.068
Let air out of tires	.587	.243	.054	.156
Taken things worth over $50	.548	−.017	.276	.034
Thrown eggs, garbage, etc.	.526	.339	−.023	.266
Taken things worth under $2	.486	.393	.143	.077
Taken things from desks, etc., at school	.464	.232	−.002	.027
Taken car without owner's permission	.461	.172	.080	.040
Put paint on something	.451	.237	.071	.250
Disobeyed parents	.054	.642	.209	.039
Marked on desk, wall, etc.	.236	.550	−.061	.021
Said mean things to get even	.134	.537	.045	.100
Disobeyed teacher, school official	.240	.497	.223	.195
Defied parents to their face	.232	.458	.305	.058
Made anonymous telephone calls	.373	.446	.029	.135
Smoked marijuana	.054	.064	.755	−.028
Used other drugs for kicks	.137	.016	.669	.004
Signed name to school excuse	.246	.249	.395	.189
Drank alcohol, parents absent	.049	.247	.358	.175
Skipped school	.101	.252	.319	.181
Beat up someone in a fight	.309	.088	.181	.843
Fought—hit or wrestled	.242	.266	.070	.602
Percent of variance	67.2	13.4	10.9	8.4

SOURCE: Morris A. Forslund, "Patterns of Delinquency Involvement: An Empirical Typology." Paper presented to the Annual Meeting of the Western Association of Sociologists and Anthropologists. Lethbridge, Alberta, February 8, 1980. The table above is adapted from page 10.

Second, factor analysis is often criticized on basic philosophical grounds. Recall an earlier statement that to be legitimate a hypothesis must be disconfirmable: You must be able to specify the conditions under which the hypothesis would be disproved. No matter what data are input, factor analysis produces a solution in the form of factors. Thus, if the researcher were asking "Are there any patterns among these variables?" the answer would always be yes. This is not the case for all factor analysis, but you should be warned that when factor analysis is used for exploratory purposes, the factors produced are not necessarily meaningful.

My personal view of factor analysis is the same as my view of other complex modes of analysis. It can be an extremely useful tool for the survey researcher, and its use should be encouraged whenever such activity can assist researchers in understanding a body of data. As in all cases, however, you must remain aware that such tools are only tools and never magical solutions.

Analysis of Variance

Analysis of variance, abbreviated **ANOVA,** applies the logic of statistical significance discussed earlier. Fundamentally, the cases under study are combined into groups representing an independent variable, and the extent to which the groups differ from one another is analyzed in terms of some dependent variable. The extent to which the groups do differ is compared with the standard of random distribution: Could we expect to obtain such differences if we had assigned cases to the various groups through random selection?

We will look briefly now at two common forms of ANOVA: *one-way* analysis of variance and *two-way* analysis of variance.

One-Way Analysis of Variance

Suppose we want to compare income levels of Republicans and Democrats to see if Republicans are really richer. We select a sample of individuals for our study, and in our questionnaire we ask the sample (1) which political party they identify with and (2) their total income for last year. We calculate the mean or median incomes for each group, finding that the Republicans in our sample have a mean income of $21,000, compared to $19,000 for the Democrats. Clearly, our Republicans are richer than our Democrats, but is the difference "significant"? Would we have been likely to get a $2,000 difference if we had created two groups by way of random selection?

ANOVA answers this question through the use of *variance.* Most simply put, the variance of a distribution (of incomes, for example) is a measurement of the extent to which the values in a set of values are clustered close to the mean or range very high or low away from it.

Figure 17-4 *Two Distribution Patterns of the Incomes of Republicans and Democrats*

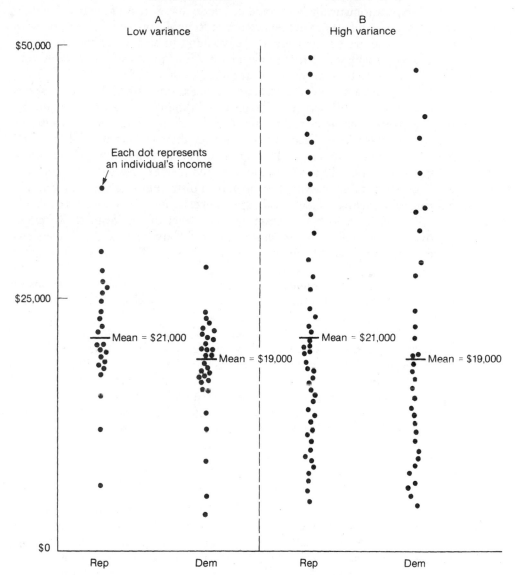

Figure 17-4 illustrates these two possibilities. Notice that in both distributions Republicans have a mean income of $21,000 and Democrats $19,000. In Part A, most Republicans have incomes relatively close to the mean of $21,000, and most Democrats have incomes close to their party's mean of $19,000. Part B, however, presents quite a different picture. Although the group means are the same as in Part A, both Republicans and Democrats have incomes ranging from very high to very low, with considerable overlap in the parties' distributions. In technical terms, there is a higher degree of variance in Part B than in Part A.

Impressionistically, we would conclude that Part A of Figure 17-4 indicates a genuine difference in the incomes of Republicans and Democrats. With data like those presented in Part B, we wouldn't be so sure; in this case, there seems more likelihood that the normal variations produced by random sampling error could have produced means of $21,000 and $19,000.

In an actual ANOVA, statistical calculations rather than impressions are used to make this decision. The observed difference in means is expressed as standardized multiples and fractions of the observed variance. Because the variance in Part A of Figure 17-4 is smaller than in Part B, $2,000 would represent a larger difference in Part A than in Part B. The resulting difference of means—standardized by the variance—would then be checked against a standard statistical table showing the theoretical distribution of such values, as in our earlier discussion of statistical significance. Ultimately, we would conclude that the difference was significant at some level of significance. We might discover, for example, that sampling error would have produced a difference as large as the one observed only one time in a thousand. Thus, we would say that difference was "significant at the .001 level."

In the above example, I have glossed over the actual calculations in favor of the basic logic of the procedure. In practice, such calculations are now typically performed by computer programs.

This simplest case is often referred to as a "*t*-test" for the difference between two means. With more than two groups, the calculations become more complex, because more comparisons must be made. Basically, it is necessary to compare the group means with one another and examine the variation among the values in each group. The end result of the analysis, as discussed in the simplest case, is expressed in terms of statistical significance—the likelihood of the observed differences resulting from sampling error in random selection.

Two-Way Analysis of Variance

One-way ANOVA represents a form of bivariate analysis (political party and income were the two variables in our example). As we have seen, however, survey researchers often engage in multivariate analysis. Two-way ANOVA permits the simultaneous examination of more than two variables. Suppose, for example, that we suspect that the income differences between Republicans and Democrats are a function of education. Our hypothesis is that Republicans are better educated than Democrats and that educated people—regardless of party—earn more, on the average, than people with less education. A two-way ANOVA would sort out the effects of the two explanatory variables, in a manner similar to that of elaboration and of partial correlations and regressions, discussed earlier.

Discriminant Analysis

Discriminant analysis offers an interesting twist on several of the techniques we have already examined in this chapter. Its logic is similar to that of multiple regression, except that the dependent variable can be nominal; regression, you will recall, requires interval variables. We will look at a simple example for purposes of illustration.

Figure 17-5 represents six writers. Three of the writers do their writing by hand (with a pen), and three write on computers. (Yes, I know there are other alternatives, but I said this was going to be simple.) Our task is to account for the difference in writing method. Can we find a way of predicting whether a given writer uses a pen or a computer?

Figure 17-6 explores two variables we might logically expect to have an impact on how the writers write. *Age* might make a difference, because the older writers might have grown accustomed to writing by hand and might have difficulty adapting to the new technology, while the younger writers would have grown up with computers. *Income* could make a difference, because computers cost more than pens. Figure 17-6 plots each writer on the graph on the basis of his or her age and income. See if you can reach any conclusion from this graph about what might account for the difference in writing method.

Figure 17-7 further clarifies the conclusion you might have drawn. Income alone seems an adequate predictor, at least as far as these six writers are concerned. Writers earning $30,000 or less all use pens, and those earning $55,000 or more all use computers.[4]

Life is seldom that simple, however, even in simplified illustrations. So let's muddy the water a bit. Figure 17-8 presents the six hypothetical writers in a somewhat more complicated configuration in terms of their ages and incomes. Notice that we cannot draw a line that would separate the pens from the computers, using either age or income.

Figure 17-5 *Six Writers: Three Who Write by Hand and Three Who Use Computers*

---

[4] If you said, "Ah, but the relationship might go in the opposite direction—how you write determines how much you earn—give yourself a pat on the back for an excellent insight, and then set it aside for purposes of this illustration. For now, let's assume that income causes writing method rather than the other way around.

Figure 17-6 *Plotting the Six Writers in Terms of Age and Income*

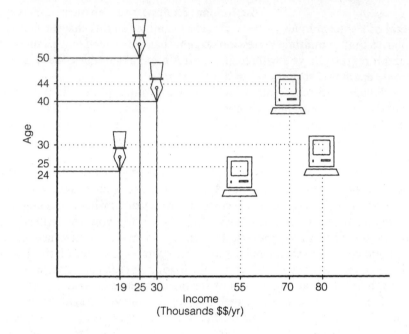

Figure 17-7 *Income Alone Is Sufficient to Predict Writing Method*

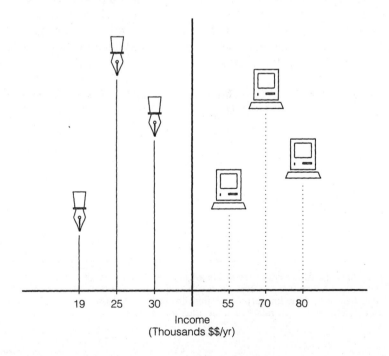

Figure 17-8   A Slightly More Complicated Pattern

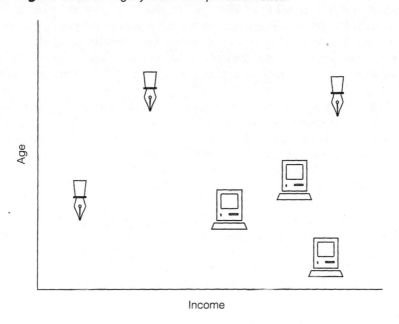

Figure 17-9   Separating the Pens from the Computers

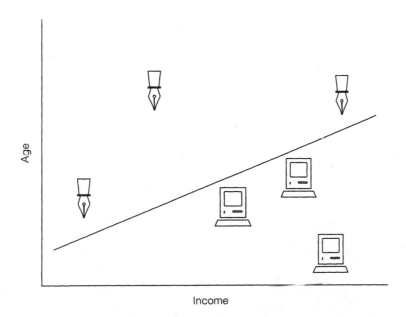

If you study Figure 17-8 a little more carefully, however, you will discover that it *is* possible to draw a line that separates the pens from the computers. It is just not perpendicular to either axis of the graph. Figure 17-9 shows the line that achieves our aim. To take advantage of the line that separates the pens from the computers, we need to find a way of predicting which writers fall on which side of that line. Figure 17-10 illustrates how this is done.

By constructing a new line perpendicular to the dividing line, we can calculate where each writer would fall on the new, composite dimension. This calculation would take a form similar to the regression equations discussed earlier. The equation would look something like the following:

New dimension = a + (b × Age) + (c × Income)

A discriminant analysis computer program would be able to take the values of age and income, examine their relationship to writing method, and

Figure 17-10 *Plotting the Six Writers on a New Dimension*

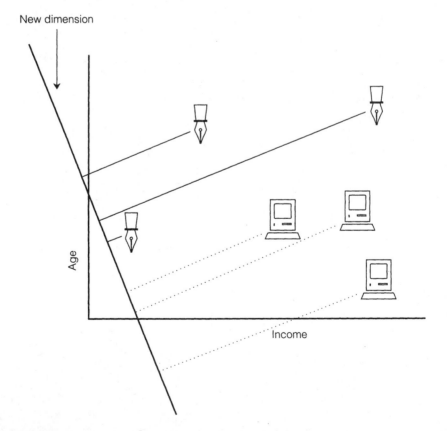

New dimension

Age

Income

generate an equation that would allow you to predict the writing method of additional writers based on their ages and incomes.

Log-Linear Models

Suppose we want to know whether political orientation is related to party affiliation. Are liberals, for example, more likely than conservatives to be Democrats? By dividing our sample into two groups—liberal and conservative—we can calculate the percentage of Democrats in each group. If we find a higher percentage among the liberals, we conclude that political orientation and party affiliation are indeed related.

In this example, and in the tabular analyses of Chapters 14 and 15, all the dependent variables analyzed were dichotomous, that is, comprised of two attributes. When the dependent variable is not dichotomous, however, matters become more complex. Suppose that besides Democrats and Republicans our sample includes Independents, Socialists, and Libertarians. It no longer makes sense to examine the percentage of liberals and conservatives who are Democrats, any more than it makes sense to look only at the percentages affiliated with any one of the other groups. Looking at each group independently would result in more tables than could be easily interpreted.

The complexity of this situation is increased if the explanatory variable is not dichotomous. Suppose we add "moderates" to the liberals and conservatives, or suppose we examine the interactive effects of other explanatory variables such as race and religion on the political equation. As you can imagine, the resulting percentage tables would become incredibly complicated.

Log-linear models offer a potential solution to this complexity. This technique, which involves rather elaborate logarithmic calculations, is based on specifying models that describe the interrelationships among variables and then comparing expected and observed table-cell frequencies. (The logic here is similar to that for *chi square,* discussed earlier.) H. T. Reynolds describes the process:

> At the outset of log-linear analysis, as in most statistical procedures, the investigator proposes a model that he feels might fit the data. The model is a tentative statement about how a set of variables are interrelated. After choosing the model, he next estimates the frequencies expected in a sample of the given size *if the model were true.* He then compares these estimates, *F*, with the observed values.[5]

In specifying the models to be tested in a log-linear analysis, the researcher will consider direct relationships between the dependent variable and each independent variable, relationships between pairs of independent

---

[5] H. T. Reynolds, *Analysis of Nominal Data* (Beverly Hills, CA: Sage, 1977), pp. 76–77.

variables, and three-variable (and more, depending on the total number of variables) relationships similar to those already discussed in the elaboration model (Chapter 15). We will consider a three-variable case taken from the preceding example.

We might suspect that a person's political party affiliation ("party") is a function of political orientation ("philosophy") and race. The components of this model, then, include (1) the direct effect of philosophy on party, (2) the direct effect of race on party, (3) the effect of race on philosophy, (4) the effect of race on the relationship between philosophy and party (as in the elaboration model), and (5) the effect of philosophy on the relationship between race and party. Though each of these components will have *some* explanatory power, log-linear analysis provides a means of identifying which are the most important and which can, as a practical matter, be ignored. Although the calculations involved in log-linear analysis are many and complex, computer programs can perform them all handily.

Log-linear analysis seems to be growing in popularity in the social sciences, though it is still not used as much as some of the other methods discussed in this chapter. It is a relatively new technique, and not many social scientists have been trained in its use.

Log-linear analysis has two main shortcomings. First, its logic makes certain mathematical assumptions that might not be satisfied by a particular set of data, though this issue is far too complex to be pursued here. Second, as with other summary techniques discussed, the results of log-linear analysis do not permit the immediate, intuitive grasp possible in simple comparisons of percentages or means. Because of this, log-linear methods would not be appropriate—even if statistically justified—in cases where the analysis can be managed through simple percentage tables. It is best reserved for complex situations in which tabular analyses are not powerful enough.

This completes our discussion of some of the analytical techniques commonly used by social scientists. I have only brushed the surface of each, and there are many other techniques that I haven't touched on at all. My purpose has been to give you a preview of some of the techniques you might want to study in more depth later on and to familiarize you with them in the event you run across them in reading the research reports of others.

Summary

In this chapter, we have probed a little deeper into the possibilities of complex, quantitative analysis of survey research data. I have not tried to train you to use these techniques, but it is useful for you to have some familiarity with them. Perhaps you will find some that you want to learn about more fully. You should at least be able to react a little more effectively to research reports that use these techniques. The techniques described are only a few of the more popular techniques survey researchers use today, but I think they offer a useful overview of possibilities.

Regression analysis represents the relationships between variables in the form of equations, which can be used to predict the values of a dependent variable on the basis of values of one or more independent variables.

The basic regression equation—for a simple *linear regression*—is of the form $Y = a + bX$. Y in this case is the value (estimated) of the dependent variable; a is some constant value; b is another numerical value, which is multiplied by X, the value of the independent variable.

Regression equations are computed on the basis of a *regression line,* that geometric line that represents, with the least amount of discrepancy, the actual location of points in a scattergram. A *multiple regression* analysis results in a regression equation, which estimates the values of a dependent variable from the values of several independent variables. A *partial regression* analysis examines the effects of several independent variables, with each independent variable's effect expressed separately while the effects of all others are held constant.

A *curvilinear regression* analysis permits the "best-fitting" regression line to be something other than a straight line. The curvature of the regression line is achieved by permitting the values of the independent variables to be raised to powers greater than 1, for example, squared, cubed, and so forth.

Path analysis is a method of presenting graphically the networks of causal relationships among several variables. It illustrates graphically the primary "paths" of variables through which independent variables cause dependent ones. *Path coefficients* are standardized regression coefficients that represent the partial relationships between variables.

Factor analysis, feasible only with a computer, is an analytical method of discovering the general dimensions represented by a collection of actual variables. These general dimensions, or *factors,* are calculated hypothetical dimensions that are not perfectly represented by any of the empirical variables under study but are highly associated with groups of empirical variables. A *factor loading* indicates the degree of association between a given empirical variable and a given factor.

Analysis of variance (ANOVA) is based on comparing variations between and within groups and determining whether between-group differences could reasonably have occurred in simple random sampling or whether it is more reasonable to conclude that they represent a genuine relationship between the variables involved.

Discriminant analysis seeks to account for variation in some dependent variable by finding a hypothetical, composite dimension that separates categories of the dependent variable. It results in an equation that scores people on the basis of that hypothetical dimension and allows us to predict their values on the dependent variable.

Log-linear models offer a method for analyzing complex relationships among several nominal variables having more than two attributes each.

Additional Readings

Ramon E. Henkel, *Tests of Significance* (Beverly Hills, CA: Sage, 1976).

David K. Hildebrand, James D. Laing, and Howard Rosenthal, *Analysis of Ordinal Data* (Beverly Hills, CA: Sage, 1977).

Gudmund R. Iversen and Helmut Norpoth, *Analysis of Variance* (Beverly Hills, CA: Sage, 1976).

Jae-On Kim and Charles W. Mueller, *Introduction to Factor Analysis* (Beverly Hills, CA: Sage, 1978).

David Knoke and Peter J. Burke, *Log-Linear Models* (Beverly Hills, CA: Sage, 1980).

Michael S. Lewis-Beck, *Applied Regression: An Introduction* (Beverly Hills, CA: Sage, 1980).

H. T. Reynolds, *Analysis of Nominal Data* (Beverly Hills, CA: Sage, 1977).

18

The Reporting of

Survey Research

So far in this book, we have considered the activities that comprise the *doing* of a survey. Now we turn to an often neglected subject: reporting the survey to others. Unless the survey is properly communicated, all the efforts devoted to conducting it will go for naught.

Before proceeding further on this topic, I should suggest one absolutely basic guideline. Good social scientific reporting requires good English (unless you are writing in a foreign language). Whenever we ask the "figures to speak for themselves," they tend to remain mute. Whenever we use unduly complex terminology or construction, communication is reduced. I most strongly advise that you read (and reread at approximately three-month intervals) an excellent small book by William Strunk, Jr., and E. B. White, *The Elements of Style*. If you do this faithfully, and if even 10 percent of the contents rub off, you stand a rather good chance of making yourself understood and perhaps having your findings appreciated.

Scientific reporting has several functions, and it is a good idea to keep these functions in mind. First, the report communicates to an audience a body of specific data and ideas. The report should provide those specifics clearly and with sufficient detail to permit an informed evaluation. Second, the scientific report should be viewed as a contribution to the general body of scientific knowledge. While remaining appropriately humble, you should always regard your research report as an addition to what is known about social behavior. Finally, the report should stimulate and direct further inquiry.

Some Basic Considerations

Within the general guidelines mentioned above, different reports serve different purposes. A report appropriate for one purpose might be wholly inappropriate for another. This section of the chapter deals with some of the basic considerations in regard to writing suitable reports.

Audience

Before drafting your report, you must ask yourself who you hope will read it. Normally, you should make a distinction between fellow scientists and laypersons. If writing for the former, you can make certain assumptions as to their existing knowledge and can perhaps summarize certain points rather than explaining them in detail. Similarly, you may appropriately use more technical language than you could for an audience of laypersons. A marketing researcher writing to a client, for example, should explain things in more detail than would be necessary in addressing an audience of other marketing researchers.

At the same time, you should always remain aware that any science is composed of factions or cults. Terms and assumptions acceptable to your immediate colleagues might only confuse other scientists. This constraint applies with regard to substance as well as techniques.

Form and Length of Report

I should begin this subsection by saying that my comments apply to both written and verbal reports. The nature of the report, however, is affected by which form you are using.

It is useful to think about the variety of reports that might result from a research project. To begin, you might want to prepare a short *research note* for publication in an academic or technical journal. Such reports should be approximately one to five pages in length (double-spaced, typed) and should be concise and direct. In a short amount of space, you will not be able to present the state of the field in any detail, and your methodological notes will have to be somewhat abbreviated, as well. Basically, you should tell why you feel a brief note is justified by your findings and then tell what those findings are.

Often, researchers must prepare reports for the sponsors of their research. These vary greatly in length, of course. In preparing such a report, however, you should consider the audience for the report—scientific or lay— and the reasons for sponsoring the project in the first place. It is both bad politics and bad manners to bore sponsors with research findings that have no interest or value to them. At the same time, it might be useful to summarize the ways in which the research has advanced basic scientific knowledge (if indeed it has).

Working papers or monographs are another form of research reporting. Especially if you are engaged in a large and complex project, it will be useful for you to obtain comments on your analysis and on the interpretation of your data. A working paper constitutes a tentative presentation with an implicit request for comments. Working papers can also vary in length, and they can present all the research findings of the project or only a portion of them. Your professional reputation is not at stake in a working paper, so you should feel free to present tentative interpretations that you cannot altogether justify, identifying them as such and asking for evaluations.

Many research projects result in papers to be delivered at professional meetings. Often, these serve the same purpose as working papers. You are able to present findings and ideas of possible interest to your colleagues and ask for their comments. Although the length of professional papers will vary depending on the organization of the meetings, you are strongly encouraged to say too little rather than too much. Whereas a working paper may ramble somewhat through a variety of tentative conclusions, conference participants should not be forced to sit through a verbal unveiling of the same. Interested listeners can always ask for more details later, and uninterested listeners can gracefully escape.

Probably the most popular research report is an article published in an academic journal. Again, lengths vary; you should examine the lengths of articles previously published by the journal in question. As a rough guide, however, twenty-five typed pages is as good a length as any. The section of this chapter on the organization of the report is based primarily on the structure of a journal article, so I will say no more at this point except to indicate that student term papers should be written on this model. As a general rule, a term paper that would make a good journal article would also make a good term paper.

A book, of course, represents the most prestigious form of research report. It has all the advantages of the working paper—length and detail—but should be a more polished document. Because the publication of research findings as a book gives those findings an appearance of greater substance and worth, you have a special obligation to your audience. Although you will still hope to receive comments from your colleagues, possibly leading you to revise your ideas, you must realize that other, less sophisticated readers might be led to accept your findings uncritically.

Aim of the Report

Earlier in this book, we considered the different purposes of survey research projects. In preparing your report, you should keep these same differences in mind.

Some reports might focus primarily on the *exploration* of a topic of interest. Inherent in this aim is the tentativeness and incompleteness of the conclusions. You should clearly indicate to your audience the exploratory aim of the study and point to the shortcomings of the particular project. An important aspect of an exploratory report is its utility in pointing the way to more refined research on the topic.

Most studies have a *descriptive* purpose, and the research reports from such studies will have a descriptive element. You should carefully distinguish those descriptions that apply only to the sample from those that are inferred to apply to the population. Whenever inferential descriptions are made, you should give your audience some indication of the probable range of error in those descriptions.

Many reports have an *explanatory* aim; you want to point to causal relationships among variables. Depending on the probable audience for your report, you should carefully delineate the rules of explanation that lie behind your computations and conclusions; as in the case of description, you must give your readers some guide to the relative certainty of your conclusions.

Finally, some research reports might have the aim of *proposing action*. In a study of the roots of prejudice, for example, you might want to suggest ways in which prejudice can be reduced on the basis of your research findings. This aim often presents knotty problems, as your own values and orientations could interfere with your proposals. It is perfectly legitimate for your proposals to be motivated by personal values, but you must ensure that the specific actions you propose are warranted by your data. Thus, you should be especially careful to spell out the logic by which you have moved from empirical data to proposed action.

Organization of the Report

Although the organization of reports differs somewhat on the basis of form and purpose, it is possible to suggest a general format for presenting research data. The following comments apply most directly to a journal article, but with some modification they apply to most other forms of research reports, as well.

Purpose and Overview

Beginning with a brief statement of the purpose of the study and the main findings of the analysis is always helpful to your readers. In a journal article, this can sometimes be accomplished in the form of an abstract.

You might find summarizing difficult to do, especially if your analysis involved considerable detective work with important findings revealing themselves only as a result of imaginative deduction and data manipulation. In such a case, you might want to lead readers through the same exciting process, chronicling the discovery process with a degree of suspense and surprise. To the extent that this form of reporting gives an accurate picture of the research process, I believe it has considerable instructional value. Nevertheless, some of your readers might not be interested in following the entire research account, and not knowing the purpose and general conclusions in advance could make it difficult for them to understand the significance of the study.

An old forensic dictum says, "Tell them what you're going to tell them, tell them, and tell them what you told them." You would do well to follow this dictum in the preparation of research reports.

Review of the Literature

Because every research report should be placed in the context of the general body of scientific knowledge, you should always indicate where your report fits in that picture. After presenting the general purpose of your study, you should proceed to bring readers up to date on the previous research in the area, pointing to general agreements and disagreements among earlier researchers.

In some cases, you might want to challenge previously accepted ideas. You should carefully review the studies that have led to the acceptance of those ideas and then indicate the factors that have not been previously considered or the logical fallacies present in the previous research.

When you are concerned with resolving a disagreement among previous researchers, you should organize your review of the literature around the opposing points of view. You should summarize the research that supports one view, then summarize the research that supports the other, and finally suggest the reasons for the disagreement.

To an extent, your review of the literature serves a bibliographic function for readers, indexing the prior research on a given topic. Review can be overdone, however, and you should avoid an opening paragraph that runs three pages and mentions every previous study in the field. The comprehensive bibliographic function can best be served by a bibliography at the end of the report; the review of the literature should focus only on studies that have direct relevance to the present study.

Avoiding Plagiarism

Whenever you are reporting on the work of others, you must be clear about who said what. It is essential that you avoid plagiarism, which is the theft of another's words and/or ideas—whether intentional or accidental—and the presentation of those words and ideas as your own. Because this is a common and sometimes unclear problem, we will take the time to examine it in some detail.

Following are the main ground rules regarding plagiarism:

- You cannot use another writer's exact words without using quotation marks and giving a complete citation that indicates the source of the quotation in such a way that your reader could locate that quotation in its original context. As a rule of thumb, give a citation for any passage of eight or more words.
- It is not acceptable to *edit* or *paraphrase* another's words and present the revised version as your own work.
- Finally, it is not even acceptable to present another's *ideas* as your own—even if you use totally different words to express those ideas.

An example should clarify what is or is not acceptable in the use of another's work. Consider the following original passage.

Systems are like babies: once you get one, you have it. They don't go away. On the contrary, they display the most remarkable persistence. They not only persist; they grow. And as they grow, they encroach. The growth potential of systems was explored in a tentative, preliminary way by Parkinson, who concluded that administrative systems maintain an average growth of 5 to 6 percent per annum regardless of the work to be done. Parkinson was right so far as he goes, and we must give him full honors for initiating the serious study of this important topic.[1]

Now we will look at some *acceptable* ways you might make use of Gall's work in a term paper.

Acceptable: John Gall, in his work on *Systemantics,* draws a humorous parallel between systems and infants: "Systems are like babies: once you get one, you have it. They don't go away. On the contrary, they display the most remarkable persistence. They not only persist; they grow."[2]
Acceptable: John Gall warns that systems are like babies. Create a system and it sticks around. Worse yet, Gall notes, systems keep growing larger and larger.[3]
Acceptable: It has also been suggested that systems have a natural tendency to persist, even grow and encroach (Gall 1975: 33).[4]

Here, now, are some *unacceptable* uses of the same material, reflecting some common errors.

Unacceptable: In this paper, I want to look at some of the characteristics of the social systems we create in our organizations. First, systems are like babies: once you get one, you have it. They don't go away. On the contrary, they display the most remarkable persistence. They not only persist; they grow.[5]
Unacceptable: In this paper, I want to look at some of the characteristics of the social systems we create in our organizations. First, systems are a lot like children: once you get one, it's yours. They don't go away; they persist. They not only persist, in fact: they grow.[6]
Unacceptable: In this paper, I want to look at some of the characteristics of the social systems we create in our organizations. One thing I've noticed is that once you create a system, it never seems to go away. Just the opposite, in fact: they have a tendency to grow. You might say systems are a lot like children in that respect.[7]

Each unacceptable example above is an example of plagiarism, and each represents a serious offense. Admittedly, there are some "gray areas." Some ideas are more or less in the public domain, not "belonging" to any one person.

---

[1] John Gall, *Systemantics: How Systems Work and Especially How They Fail* (New York: Pocket Books, 1975), pp. 33–34. [Note: Gall previously gave a full citation for Parkinson.]
[2] This would be accompanied by the footnote: John Gall, *Systemantics: How Systems Work and Especially How They Fail* (New York: Pocket Books, 1975), p. 33.
[3] This would be accompanied by the footnote: John Gall, *Systemantics: How Systems Work and Especially How They Fail* (New York: Pocket Books, 1975), p. 33.
[4] *Note*: This format requires that you give a complete citation in your bibliography.
[5] It is unacceptable to directly quote someone else's materials without using quotation marks and giving a full citation.
[6] It is unacceptable to edit another's work and present it as your own.
[7] It is unacceptable to paraphrase someone else's ideas and present them as your own.

You might hold an idea on your own that someone else has already put in writing. If you have a question about a specific situation, discuss it with your instructor in advance.

I have discussed this topic in some detail because it is important that you place your research in the context of what others have done and said, yet the improper use of others' materials is a serious offense. Mastering the matter of proper attribution of ideas, however, is a part of your "coming of age" as a scholar.

Study Design and Execution

A research report that contains interesting findings and conclusions can be a cause of frustration when you are unable to determine the methodological design and execution of the study. The worth of all scientific findings depends heavily on the manner in which the data were collected and analyzed.

In reporting the design and execution of a survey research project, you should always include the following: the population, the sampling frame, the sampling method, the sample size, the data collection method, the completion rate, and the method of data processing and analysis. With experience, you will be able to report these details in a rather short space without omitting anything required for your readers' evaluation of the study.

Analysis and Interpretation

Having set the study in the perspective of previous research and having described its design and execution, you should proceed to the presentation of your data. The section on guidelines for reporting analyses will provide further information in this regard. For now, a few general comments are in order.

The presentation of data, the manipulations of those data, and your interpretations should be integrated into a logical whole. It is frustrating to readers to discover a collection of seemingly unrelated analyses and findings with a promise that all the loose ends will be tied together later in the report. Every step in the analysis should make sense at the time it is reported. You should present your rationale for a particular analysis, present the data relevant to it, interpret the results, and then indicate where that result leads next.

Summary and Conclusions

Following the forensic dictum mentioned earlier, you must be sure to summarize the research report. You should avoid reviewing every specific finding, but you should review all the significant ones, pointing out once more their general significance.

The report should conclude with a statement of what you have discovered about your subject matter and where future research might be directed. A quick review of recent journal articles would probably indicate a very high frequency of occurrence of the concluding statement "It is clear that much more research is needed." This is probably always a true conclusion, but it is of little value unless you can offer pertinent suggestions as to what the nature of that future research should be. You should review the particular shortcomings of your own study and suggest ways in which those shortcomings might be avoided by future researchers.

Guidelines for Reporting Analyses

The presentation of data analyses should provide a maximum of detail without being cluttered. You can accomplish this best by continually examining the aims described below.

Data should be presented in such a way as to permit recomputations by readers. In the case of percentage tables, for example, readers should be able to collapse categories and recompute the percentages. They should be given sufficient information to permit them to percentage the table in the opposite direction from your presentation.

All aspects of the analysis should be described in sufficient detail to permit a secondary analyst to replicate the analysis from the same body of data. This means being able to create the same indexes and scales, produce the same tables, arrive at the same regression equations, obtain the same factors and factor loadings, and so forth. This re-creation will seldom be done, of course, but if the report is presented in such a manner as to make it possible, your readers will be far better equipped to evaluate it.

A final guide to the reporting of methodological details is that your readers should be in a position to completely replicate the entire study independently. They should be able to identify the same population, select the same size sample in the same manner, collect data in the same manner, execute the same analyses, and see if they arrive at the same conclusions. It should be recalled from an earlier discussion that replicability is an essential norm of science in general. A single study does not prove a point; only a series of studies can begin to do this. Unless studies can be replicated, no meaningful series of studies can take place.

I have previously mentioned the importance of integrating data, analysis, and interpretations in your report. A more specific guideline can be offered in this regard. Tables, charts, and figures should be integrated into the text of the report, appearing near that portion of the text that discusses them. Students sometimes describe their analyses in the body of the report and place all the tables in an appendix at the end.[8] This procedure greatly impedes your readers.

---

[8] I should point out that some publishers prefer to receive materials in this fashion, because different people process textual and graphic materials in the publication process. With modern word processing, however, it is not difficult to create two versions of a paper—one for readers and another for publication.

As a general rule, it is best to (1) describe the purpose for presenting a table, (2) present it, and (3) review and interpret it.

Be explicit in drawing conclusions. Although research is typically conducted for the purpose of drawing general conclusions, you should carefully note the specific basis for such conclusions. Otherwise, you might lead your reader into accepting unwarranted conclusions.

Point out any qualifications or conditions warranted in the evaluation of your conclusions. Typically, the person who actually conducted a survey is in the best position to know its shortcomings and the tentativeness of the conclusions, and you should give your reader the advantage of that knowledge. Failing to do so can misdirect future research and result in the waste of research funds.

I will conclude with a point already made at the outset of this chapter, as it is extremely important. Research reports should be written in the best possible literary style. Writing clearly is easier for some people than for others, and it is always harder than writing poorly. You are again referred to the Strunk and White volume. You would do well to follow this procedure: Write. Read Strunk and White. Revise. Reread Strunk and White. Revise again. This endeavor will be difficult and time-consuming, but so is science.

Summary

A perfectly designed, carefully executed, and brilliantly analyzed survey will be altogether worthless unless you are able to communicate your findings to others. This chapter has attempted to provide some general and specific guidelines toward achieving that end. Your best guides are logic, clarity, and honesty. Ultimately, there is probably no substitute for practice.

Additional Readings

Pauline Bart and Linda Frankel, *The Student Sociologist's Handbook* (Glenview, IL: Scott, Foresman, 1981).

Leonard Becker and Clair Gustafson, *Encounter with Sociology: The Term Paper* (San Francisco: Boyd and Fraser, 1976).

H. W. Fowler, *A Dictionary of Modern English Usage* (New York: Oxford University Press, 1965).

James Gruber, Judith Pryor, and Patricia Berge, *Materials and Methods for Sociology Research* (New York: Neal-Schuman, 1980).

Tze-chung Li, *Social Science Reference Sources: A Practical Guide* (Westport, CT: Greenwood Press, 1980).

William Strunk, Jr., and E. B. White, *The Elements of Style* (New York: Macmillan, 1959).

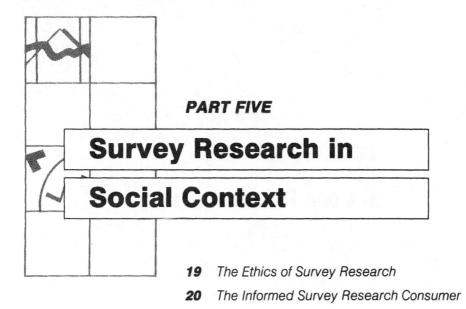

For the most part, this book has addressed the skills of doing survey research. Every attempt has been made to place this particular research method within the context of science in general. The concluding chapters reflect my view that even this perspective is not sufficiently broad.

The prefatory remarks that opened the book noted the ways in which survey research affects the lives of us all, whether we do surveys or not. It is not sufficient, therefore, to regard survey research solely as a neutral, scientific technique. Rather, you should be sensitive to the social context in which surveys are conducted.

Chapter 19 is devoted to a discussion of research ethics as they relate to survey methods. A discussion of general ethical guidelines in survey research is followed by a series of illustrations of research situations in which ethical issues are involved. Even though few of the situations can be unequivocally regarded as ethical or unethical, I believe that a considered examination of them will make you more sensitive to the knotty ethical problems often faced by survey researchers.

Chapter 20 speaks to you in a somewhat different fashion. You might not conduct a lot of surveys after you leave this course, but you will certainly read about them and, in some cases, have to make decisions based on them. My purpose in the final chapter, therefore, is to empower you as an effective consumer of survey research.

19

The Ethics of

Survey Research

A theoretically oriented textbook on survey research methods would provide the student only with an ideal image of how surveys should be conducted. In previous portions of this book, I have sought to impress upon you the fact that a variety of administrative and practical concerns impinge upon the research process, so that living up to the ideal model is not always possible. The enlightened survey researcher should be aware of those additional constraints and be able to balance administrative and scientific factors in order to arrive at the best possible compromise.

This chapter addresses another nonscientific constraint—*ethical* issues in survey research. These ethical concerns are not a part of the scientific method.[1] Nevertheless, they comprise a set of norms that scientists in most disciplines are obliged to follow. In many instances, these ethical norms directly conflict with scientific procedures, just as administrative concerns sometimes do. You should be aware of the possible conflicts so that you can ultimately conduct the most scientific, ethical research.

Science in and of itself is amoral. The law of gravity and the correlation between education and prejudice are neither moral nor immoral. Scientists, however, are not amoral, nor are those who might use the results of scientific inquiry. Thus, scientific research can be conducted and/or used for either moral or immoral purposes. Of course, one man's morality is another man's immoral-

---

[1] When the book was first published in 1973, a number of reviewers objected to the inclusion of a chapter on research ethics. In response to my comments on "the rights of respondents," I recall one reviewer retorting, "What about the rights of science?" In the years since, however, it is my sense that no social research textbook could be published without some substantial treatment of ethical issues. In my view, that's a healthy sign.

ity. These are lessons that nuclear physicists learned a quarter of a century ago and that social scientists are still learning today.

There is no way to ensure that all scientists will always be motivated by ethical concerns when they engage in scientific research, nor is there any way to ensure that scientific findings will be used only for ethical purposes. We can, however, point out a set of more or less agreed upon ethical norms relating to the execution of research. This chapter presents some of the more common ethical problems that appear in survey research and suggests ethical solutions to them that do not seriously endanger the "scientific" quality of the research itself.

Voluntary Participation

Survey research almost always represents an intrusion into people's lives. The interviewer's knock on the door or the arrival of a questionnaire in the mail signals the beginning of an activity that the respondent has not requested and that might require a significant amount of time and energy. Moreover, the survey often asks the respondent to reveal personal information—attitudes and personal characteristics—that might be unknown to his or her friends and associates. Yet survey research requires that the respondent reveal such information to a complete stranger.

Other professionals, such as physicians and lawyers, also require such information. Their requests, however, can be justified on the grounds that the information is required for them to serve the personal interests of clients from whom they are requesting the information. Survey researchers can seldom make this claim. Like medical researchers, we can only argue that our research effort might ultimately help people in general.

A major tenet of medical research ethics is that experimental participation must be *voluntary*. The same norm applies to survey research. No one should be forced to participate in a survey. This norm is far easier to accept in theory than to apply in practice, however. Again, medical research provides a useful parallel. Many experimental drugs are tested on prisoners. In the most rigorously ethical cases, the prisoners are told the nature—and the possible dangers—of the experiment, they are told that participation is completely voluntary, and they are further instructed that they can expect no special rewards (such as early parole) for participation. Even under these conditions, volunteers are often clearly motivated by the belief that they will personally benefit from their cooperation.

When instructors in introductory sociology classes ask their students to fill out questionnaires that they hope to analyze and publish, they should always impress upon the students that participation in the survey is completely voluntary. Even so, many students will clearly fear that nonparticipation will somehow affect the grades they receive in the course. In such a case, instructors should be especially sensitive to the implied sanctions and make special provisions to obviate them. Instructors might leave the room while the questionnaires are completed and dropped in a box, or they might ask students to return the questionnaires by mail or put them in a box near the door upon arriving at the next meeting of the class.

In this particular situation, instructors should be encouraged, if appropriate, to involve their students in the processing and analysis of the data, thereby providing students with a learning experience personally valuable to them. Students should, of course, receive appropriate credit in any published report on the study. (This technique should not be used as a method of obtaining cheap labor.)

The idea that participation in surveys must be voluntary goes directly against a scientific concern. If statistical techniques are to be used legitimately in the analysis of survey data, then every member of the random sample should participate. Even with less rigorous statistical plans, you will desire a high completion rate in order to ensure a reasonably representative sample. Thus, while you cannot ethically require participation, you will usually do everything possible to obtain it.

In an interview survey, interviewers will typically be trained to persuade uncertain respondents. "It will only take a few minutes," "The results of the study will be valuable to people living in this area," and "May I come back at a more convenient time?" are all important phrases in the interviewer's vocabulary. The fortunate research director will discover some interviewers who are more persuasive than others, and these interviewers will probably be assigned to follow up on sample members who refused the initial request. (Experience has shown that an eight-months-pregnant interviewer is more successful than the average in this regard.) Clearly, the line between ethical persuasion and coercion is a fuzzy one.

In a mail survey, you will normally send follow-up mailings to members of the sample who have not responded to earlier requests. Usually, the follow-up mailings are accompanied by special appeals for cooperation; again, the line between persuasion and coercion is not always clear.

No Harm to Respondents

Survey research should never injure the respondents who have volunteered to cooperate with you. Perhaps the clearest instance of this norm in practice concerns the revealing of information that would embarrass respondents or endanger their home life, friendships, jobs, and so forth. This situation is discussed more fully in the section on anonymity and confidentiality.

It is also possible for respondents to be harmed in the course of an interview, however, and you should be aware of this possibility and guard against it. Even with the most professional (that is, neutral) interviewer, respondents might feel extremely uncomfortable about reporting deviant behavior and attitudes they think are unpopular, or about acknowledging low income, the receipt of welfare payments, and the like.

Surveys often force respondents to face aspects of themselves that they do not normally consider. In retrospect, certain past behaviors might appear to them as unjust or immoral, and they might now regret having exhibited them. The survey, then, could be the source of a continuing personal agony for respondents. They might begin thinking that they are not as moral and ethical as they ought to be, and this thought might continue to bother them.

There is no way to ensure against all these possibilities. At the same time, some questionnaire items are more likely to produce such reactions than others, and you should be sensitive to this possibility. If a given item seems likely to produce unpleasant reactions for the respondent, you should have the firmest scientific grounds for asking it. Unless it is vital to your research aims, you should remove it from the questionnaire. If the item is both essential and sensitive, then you will find yourself in an ethical netherworld, and you yourself might be forced to do some personal agonizing.

Although the fact often goes unrecognized, respondents can be harmed by the analysis and reporting of data. Every now and then, respondents read the books published from the surveys they participated in. Reasonably sophisticated respondents will be able to locate themselves in the various indexes and tables. Having done this, they might find themselves characterized—though not identified by name—as bigoted, unpatriotic, irreligious, and so forth. At the very least, this discovery is likely to trouble them and threaten their self-images. Yet the whole purpose of the research project might be to explain why some people are prejudiced while others are not.

I once conducted a survey of Episcopal churchwomen in Northern California. Ministers in a sample of churches were asked to distribute questionnaires to a specified sample of members, collect them, and return them to the research office. One minister read through the questionnaires from the sample before returning them and then proceeded to deliver a hellfire and brimstone sermon to the congregation, saying that many among them were atheists and were going to hell. Even though the minister could not know or identify the respondents who gave particular responses, it seems certain that many respondents were personally harmed by the action.

Like voluntary participation, not harming respondents is an easy norm to accept in theory but is often difficult to ensure in practice. Sensitivity to the issue and experience with its applications, however, should improve your batting average.

Anonymity and Confidentiality

The clearest situation involving the protection of respondents' interests and well-being concerns the protection of their identity. If revealing their survey responses would injure them in any way, adherence to this norm becomes even more important. Two techniques—*anonymity* and *confidentiality*—assist you in protecting respondents' identity, although the two are often confused.

Anonymity

A respondent can be considered **anonymous** when you yourself cannot identify a given response with a given respondent. This means that an interview survey respondent can never be considered anonymous, because an interviewer collects the information from an identifiable respondent. (This assumes that standard sampling methods are followed.) An example of anonymity is a mail survey in which no identification numbers are put on the questionnaire prior to its return to the research office.

Of course, anonymity complicates any follow-up plans for increasing response rates. If you do not know who among your sample failed to reply, you cannot contact only the nonrespondents. As an alternative, however, you could mail the questionnaire again to all members of the original sample, asking those who had already replied to ignore the second appeal, or you could employ the postcard technique discussed in Chapter 9.

Despite the difficulties attendant upon ensuring anonymity, in some situations you might be well advised to pay the necessary price. In studies of drug use among university students, for example, you might choose specifically not to know the identity of respondents, for two reasons. First, providing respondents with anonymity might increase the likelihood and honesty of responses. Second, you might not want to face the possibility of being asked for the names of drug offenders. In the latter case, an anonymous survey will spare you from having to choose whether to (1) violate your promise to respondents, (2) risk going to jail for withholding the names, or (3) pray that the questionnaires will accidentally catch fire and be destroyed.[2]

Confidentiality

In a **confidential** survey, you are able to identify a given person's responses but essentially promise that you will not. In an interview survey, for example, you would be in a position to make public the income reported by a given respondent, but you assure respondents that this will not be done.

A number of techniques can help you better ensure your adherence to this guarantee. To begin, interviewers and others with access to respondent identifications should be trained in their ethical responsibilities. As soon as possible, all names and/or addresses should be removed from questionnaires and replaced with identification numbers. A master identification file should be created linking numbers to names to permit the later correction of missing or contradictory information, but this file should not be available except for legitimate purposes.

---

[2] Note: Always have the questionnaires in a safe place outdoors when you pray for a fire. Otherwise, it takes forever to get the smoke smell out of your office.

Whenever a survey is confidential rather than anonymous, you are responsible for making that fact clear to the respondent. The use of the term "anonymous" to mean "confidential" should never be tolerated.

Inferred Identity

Even in a truly anonymous survey, identifying a given respondent is sometimes possible, particularly with open-ended questions. If a respondent lists his or her occupation as "President of ABC Company," the cat is out of the bag. Sometimes, the multivariate analysis of closed-ended questions will permit the identification of a given respondent.

This possibility always exists in any survey, anonymous or not; you cannot rule it out altogether. You should never attempt to make such identifications, however, and you should make sure that your research workers do not make such attempts. Moreover, you should never report aggregated data in such a way that readers will be able to make such identifications. For this reason, the United States Census Bureau will not report aggregated data containing fewer than fifteen cases per cell in a table.

Hidden Identification

Occasionally, some researchers conduct surveys in which respondents are ensured anonymity when, in fact, they are identifiable. Sometimes the return address of the research office contains a "box number" with a different number for each respondent. On occasion, researchers have entered identification numbers under the stamps placed on return envelopes. Probably some researchers have written numbers with lemon juice.

I suspect that in virtually all such cases—and they are probably few— the researchers have attempted to maintain the confidentiality of the data and have made no attempt to harm respondents. (It is my impression that those researchers most convinced of their personal "morality" are the most willing to engage in such practices.) Whatever the motivations or scientific value, such practices are unacceptable. If a survey is confidential rather than anonymous, say so. Don't lie. Finally, all surveys should be at least confidential.

Identifying Purpose and Sponsor

Often you face the dilemma that respondents' knowledge of your survey's purpose and/or sponsor might affect the answers they will provide. In more extreme cases, it might affect the likelihood of cooperation. If the interviewer introduces herself as being engaged in a study of prejudice, it stands to reason that the respondent will be rather careful not to sound prejudiced. Or if the study is identified as being sponsored by the welfare office, a recipient's responses are likely to be modified somewhat to guard against the loss of benefits. If the local university is in particularly bad repute among the population—say, as a result of student demonstrations—a university-sponsored study could face severe opposition and produce a low completion rate.

Almost any specification of purpose and/or sponsor will have some effect both on completion rates and on the answers given by those who participate. These considerations, of course, affect the scientific quality of the data and the conclusions to be drawn from their analysis. Yet deceiving the respondent as to the study's purpose and/or sponsor raises an ethical issue.

My general feeling is that you are obliged to be honest with your respondents with regard to both the sponsor and the auspices of the study. If the study is being conducted by a university research office on behalf of the state government, this should be spelled out in the interviewer's introduction (or in the introductory letter accompanying a self-administered questionnaire). Moreover, I am especially wary of fictitious organizations—research office names made up for purposes of a particular study.

It should be noted that honesty in this regard is not only more ethical but also has a practical value. Whenever you attempt to deceive your respondents as to the sponsor or auspices of a study, you run the risk that the truth will become known during the study. In such a case, there is a good chance that the resulting notoriety will cause the study to suffer more than would a study openly conducted by an unpopular research agency or sponsor.

Honesty with regard to purpose is more difficult, because a given survey will often serve a variety of analytical purposes. Thus, you cannot be completely honest with respondents, since you cannot fully anticipate the uses that the survey will serve. Of course, a precise statement of the primary purposes of the study will probably affect responses even more than will disclosure of the sponsor.

In view of these concerns, I tend to be somewhat more liberal regarding a survey's purpose than in the case of sponsorship. The following guidelines seem appropriate: (1) You should reveal nothing about the purpose of the study that is likely to affect the reliability of responses. (2) At the same time, you should tell respondents whatever you can about purposes when such information is not likely to affect responses. (3) Explanations of purpose should be general rather than specific. (4) You should never offer fictitious reasons for the study.

Analysis and Reporting

Just as you have ethical obligations to respondents, you also have ethical obligations to your readers in the scientific community.

These latter norms are not often considered in the ethical realm, so a few comments are in order.

In conducting any rigorous survey, you should be more familiar with the technical shortcomings of the study than anyone else, and you should make these shortcomings known to your readers. If, near the conclusion of the study, you discover that a particular subset of the population was omitted from sampling, you should disclose this fact. If you believe that several respondents misunderstood a questionnaire item, you should apprise readers of this fact. Any defect in the study design or analysis that will have any possible effect on the conclusions drawn should be noted openly.

Negative findings should be reported if they are at all related to the analysis being reported. An unfortunate myth persists that in scientific reporting only positive discoveries are worth reporting (journal editors are sometimes guilty of this, as well). From the standpoint of the scientific community, however, knowing that two variables are not related to each other is often as important as knowing that they are. Sometimes, as in the Chapter 15 discussion of *The American Soldier*, the lack of expected correlations can be even more useful. There is certainly no reason to be embarrassed by the discovery that two variables are not related to each other.

Similarly, you should take care not to fall victim to the temptation to "save face" by describing empirical findings as the products of a preplanned analytical strategy when this is not the case. It is simply a fact of life that many findings arrive unexpectedly, even though they might seem patently obvious in retrospect. Embroidering such events with descriptions of fictitious hypotheses not only is dishonest but also tends to mislead inexperienced researchers into thinking that all scientific inquiry is preplanned.

If an unexpected association between variables appears, it should be presented as unexpected. If the entire analytical strategy was radically restructured during the course of the study, your reader should be let in on the secret. This revelation is especially beneficial in that other researchers will be made aware of the fact that another, seemingly functional, strategy is not appropriate to the subject matter. Science generally progresses through honesty and is retarded by ego-based deception.

A Professional Code of Ethics

Because ethical issues in social research are both important and ambiguous, most professional associations have created and published formal codes of conduct that describe what is considered acceptable and unacceptable professional behavior. As an illustration, I have presented the code of conduct of the American Association for Public Opinion Research; AAPOR is an interdisciplinary association of both academic and commercial survey researchers (see Figure 19-1).

Figure 19-1 *Code of Professional Ethics and Practices*

We, the members of the American Association for Public Opinion Research, subscribe to the principles expressed in the following code. Our goals are to support sound and ethical practice in the conduct of public opinion research and in the use of such research for policy and decision-making in the public and private sectors, as well as to improve public understanding of opinion research methods and the proper use of opinion research results.

We pledge ourselves to maintain high standards of scientific competence and integrity in conducting, analyzing, and reporting our work in our relations with survey respondents, with our clients, with those who eventually use the research for decision-making purposes, and with the general public. We further pledge ourselves to reject all tasks or assignments that would require activities inconsistent with the principles of this code.

THE CODE

I. Principles of Professional Practice in the Conduct of Our Work

  A. We shall exercise due care in developing research designs and survey instruments, and in collecting, processing, and analyzing data, taking all reasonable steps to assure the reliability and validity of results.

    1. We shall recommend and employ only those tools and methods of analysis which, in our professional judgment, are well suited to the research problem at hand.

    2. We shall not select research tools and methods of analysis because of their capacity to yield misleading conclusions.

    3. We shall not knowingly make interpretations of research results, nor shall we tacitly permit interpretations that are inconsistent with the data available.

    4. We shall not knowingly imply that interpretations should be accorded greater confidence than the data actually warrant.

  B. We shall describe our methods and findings accurately and in appropriate detail in all research reports, adhering to the standards for minimal disclosure specified in Section III.

  C. If any of our work becomes the subject of a formal investigation of an alleged violation of this Code, undertaken with the approval of the AAPOR Executive Council, we shall provide additional information on the survey in such detail that a fellow survey practitioner would be able to conduct a professional evaluation of the survey.

II. Principles of Professional Responsibility in Our Dealings With People

  A. The Public:

    1. If we become aware of the appearance in public of serious distortions of our research, we shall publicly disclose what is required to correct these distortions, including, as appropriate, a statement to the public media, legislative body, regulatory agency, or other appropriate group, in or before which the distorted findings were presented.

Figure 19-1   *Code of Professional Ethics and Practices (Continued)*

B. Clients or Sponsors:

1. When undertaking work for a private client, we shall hold confidential all propri-
etary information obtained about the client and about the conduct and findings
of the research undertaken for the client, except when the dissemination of the
information is expressly authorized by the client, or when disclosure becomes
necessary under terms of Section I-C or II-A of this Code.

2. We shall be mindful of the limitations of our techniques and capabilities and
shall accept only those research assignments which we can reasonably expect to accomplish within these limitations.

C. The Profession:

1. We recognize our responsibility to contribute to the science of public opinion
research and to disseminate as freely as possible the ideas and findings which
emerge from our research.

2. We shall not cite our membership in the Association as evidence of professional competence, since the Association does not so certify any persons or
organizations.

D. The Respondent:

1. We shall strive to avoid the use of practices or methods that may harm, humili-
ate, or seriously mislead survey respondents.

2. Unless the respondent waives confidentiality for specified uses, we shall hold
as privileged and confidential all information that might identify a respondent
with his or her responses. We shall also not disclose or use the names of respondents for non-research purposes unless the respondents grant us permission to do so.

III. Standard for Minimal Disclosure

Good professional practice imposes the obligation upon all public opinion researchers to include, in any report of research results, or to make available when that
report is released, certain essential information about how the research was conducted. At a minimum, the following items should be disclosed:

1. Who sponsored the survey, and who conducted it.

2. The exact wording of questions asked, including the text of any preceding instruction or explanation to the interviewer or respondents that might reasonably
be expected to affect the response.

3. A definition of the population under study, and a description of the sampling
frame used to identify this population.

4. A description of the sample selection procedure, giving a clear indication of the
method by which the respondents were selected by the researcher, or whether
the respondents were entirely self-selected.

5. Size of samples and, if applicable, completion rates and information on eligibility
criteria and screening procedures.

(continued)

Figure 19-1 Code of Professional Ethics and Practices (Continued)

6. A discussion of the precision of the findings, including, if appropriate, estimates of sampling error, and a description of any weighting or estimating procedures used.

7. Which results are based on parts of the sample, rather than on the total sample.

8. Method, location, and dates of data collection.

SOURCE: American Association for Public Opinion Research, 1989. Used with permission.

Ethics—Relevant Illustrations

The ethics of survey research, or of any scientific research, are not clear-cut. In this chapter, I have indicated relatively few firm guidelines, and most of those are subject to debate because they represent my own personal orientation.

My primary concern has been to make you more *sensitive* to ethical issues in survey research. Being able to recognize ethical considerations in real research situations is far more important than simply memorizing a set of ethical norms. With this in mind, the chapter concludes with the description of several research situations, most of them real, some hypothetical. For the most part, these situations can be honestly debated with integrity. You should identify the ethical issues contained in them and consider steps you might take to ensure that the research is as ethical as possible.

1. As an instructor in market research, you ask your students to complete questionnaires that you will then use for the analysis of a research problem of interest to you.

2. In a proposed study of attitudes among new law school graduates, you make an agreement with the state bar association to include a questionnaire in the bar exam materials. Completion of the questionnaire will be a requirement for licensing.

3. You agree to conduct a survey for the local city government. Because the city government is in general disfavor with the public, you instruct interviewers to say only that they are doing the study for university researchers.[3]

4. Your analysis of your data has produced so many surprises that your initial hypotheses have been wholly displaced by the findings that have appeared in your hectic and often confused analysis. The final conclusions are such that you are ashamed of not having

---

[3] Recall from Chapter 9 that methodological research has shown that university sponsorship increases response rates.

begun with hypotheses appropriate to them. To save face, you write your report as though you had.

5. After an interview study of deviant behavior, law enforcement officials order you to identify for them those respondents who reported being participants in looting during a recent riot. Rather than acting as an accomplice after the fact, you comply with the order.

6. At the completion of your analysis, you discover that twenty-five of the 2,000 interviews were falsified by the interviewers. You choose to ignore this fact in your report.

7. A person who was not selected in the sample for a survey contacts you and insists on being interviewed. You conduct an interview and then throw the questionnaire away.

8. You obtain a list of *National Review* subscribers for a study of consumer behavior. Questionnaires are sent to this group with the explanation that they have been "selected at random."

9. Race matching is considered essential for a study of racial prejudice. This means, however, that in the large city being studied your minority interviewers will be working under generally poorer conditions than white interviewers.

10. In a survey in higher education, you want to examine the effect of various background factors on academic achievement. To measure achievement, you obtain students' grade point averages from the university administration.

11. Respondents are assured that a questionnaire they are asked to complete is anonymous. In fact, a serial number has been placed inconspicuously on the questionnaire to permit the analysis of other information collected about the respondents from other sources.

12. In a study of sexual behavior, you want to overcome respondents' reluctance to report what they might consider deviant behavior. Thus, you use the following item: "Everyone masturbates now and then; about how often do you masturbate?"

13. Respondents are told that a survey is being conducted simply to determine how people felt about a series of public issues. In fact, you are interested in determining sources of opposition to a particular issue.

14. You discover that 85 percent of the university student body smoke marijuana regularly. Publication of this finding will probably create a furor in the community. Because you are not planning to analyze drug use in depth, you decide to ignore the finding.

15. You are given a contract to conduct a study and prepare a report

for a sponsor regarding a particular topic. You find that the data provide an opportunity to examine a related issue, although it is one that the sponsor is not concerned with. You use project funds to cover the costs of analysis and preparation of a paper, which you then deliver to a professional association meeting.

16. To test the extent to which respondents might try to save face by expressing attitudes on matters about which they are wholly uninformed, you ask for attitudes regarding a fictitious issue.

17. A research questionnaire is circulated among students as part of their university registration packet. Although students are not told they must complete the questionnaire, the hope is that they will believe they must, thus ensuring a higher completion rate.

Summary

The requirements of *scientific* research sorely tax the researcher's imagination and ingenuity. Practical, administrative constraints often further complicate matters by ruling out ideal research procedures. Ethical concerns are likely to place an additional burden on you. Having discovered that you have neither the time nor the money to execute the best possible study, you might arrive at a brilliant compromise that is both administratively feasible and scientifically sound only to discover that it violates ethical concerns.

As I have repeated throughout this book, good scientific research is often difficult to achieve. Ethical scientific research might be harder yet, but you cannot afford to give up on any of these concerns. You must conduct research that is scientifically sound, administratively feasible, and ethically defensible. You must not violate people's rights and well-being in an attempt to help them.

Additional Readings

Tom L. Beauchamp, Ruth R. Faden, Jay Wallace, Jr., and LeRoy Walters, *Ethical Issues in Social Science Research* (Baltimore: The Johns Hopkins Press, 1982).

Edward Diener and Rick Crandall, *Ethics in Social and Behavioral Research* (Chicago: University of Chicago Press, 1978).

Albert E. Gollin, "In Search of a Useable Past," *Public Opinion Quarterly*, vol. 49 (Fall 1985), pp. 414–424.

Duncan MacRae, Jr., *The Social Function of Social Science* (New Haven, CT: Yale University Press, 1976).

Steven McGuire, "A Feminist Ethic for Science," *Humanity and Society* (November 1984), pp. 461–467.

Maurice Richter, Jr., *Science as a Cultural Process* (Cambridge, MA: Schenkman, 1972).

Burns W. Roper, "Some Things That Concern Me," *Public Opinion Quarterly,* vol. 48 (Fall 1983), pp. 303–309.

Gideon Sjoberg, *Ethics, Politics, and Social Research* (Cambridge, MA: Schenkman, 1967).

20

The Informed Survey

Research Consumer

There may have been times during this course and this book that you felt the material wasn't really appropriate for you. You might find it hard to believe right now that you will ever become a professional survey researcher. You're probably right. On a per capita basis, there just aren't all that many survey researchers in the world.

Frankly, the chances are slight ($p < .001$) that you will ever be responsible for designing and running a multistage, area probability survey of the nation's work force for the purpose of determining whether a statistically significant relationship exists between gender and income among matched subsets of wage earners. But notice that you know basically what that means and could say something about the various steps involved in such a study: questionnaire design, sampling, data collection, calculating mean incomes for those employed full-time or part-time, and so on. Even if you needed to refer to this and other books to refresh your memory on some details, professionals do the same thing all the time.

You might never conduct a major survey of your own, but you will be able to evaluate the research of others. You have become what is sometimes called the "informed consumer." This is not an insignificant status, because you will be exposed to survey research nearly every day for the rest of your life. You will read reports of surveys in newspapers and magazines, you will hear about survey research on television and radio, and politicians and manufacturers will tell you what "research shows." It is very important, therefore, that you be able to protect yourself; you need to be able to separate the wheat from the chaff even if you never grow wheat on your own.

In this final chapter of the book, I offer some consumer tips for you to use in reviewing reports of survey research in the years to come. In part, these guidelines serve as a review of much of the territory covered in this book. Feel free to use this section as a review of the book, but realize that my primary purpose here is to strengthen your ability to evaluate survey research rather than your ability to design and execute it. As a consequence, I have skipped over several topics included in the book and have focused only on topics where consumer skills are particularly germane. I will organize the consumer's guide along pretty much the same lines as the rest of the book, however.

Whenever you are confronted with what a "survey shows," you might want to ask the questions listed below. Not all questions apply to all surveys, but this listing should give you enough to choose from.

Research Design

- What was the purpose of the study: exploration, description, explanation, or a combination? Was the research design appropriate to that purpose?

- Who conducted the research? Who paid for it? What motivated the study? If the study's conclusions happen to correspond to the interests of the sponsor or researcher, that doesn't disqualify the conclusions, but you want to be especially wary.

- What was the unit of analysis? Was it appropriate to the purpose of the study? Are the conclusions drawn from the research appropriate to the unit of analysis? For example, have the researchers surveyed households and ended up making assertions about individuals, or vice versa?

- Is this a cross-sectional or longitudinal study? Be especially wary of longitudinal assertions that are made on the basis of cross-sectional observations.

- If longitudinal data have been collected, be sure that comparable measurements have been made at each point in time. Have the same questions been asked each time?

- If a panel study has been conducted, check the number of people who dropped out ("panel attrition") over the course of the study.

Measurement

- What are the names of the concepts under study?

- Have the researchers delineated different dimensions of those variables where appropriate? Have those distinctions been kept straight in the analysis and reporting?

- What indicators have been chosen as measures of those dimensions and concepts? Is each indicator a valid measure of what it is intended to measure? What else could the indicator be a measure of? Is it a reliable measure? Has the reliability been tested?

- Do the measurements of variables correspond with the way previous researchers have measured them? If not, do the present researchers have good reasons for innovating?

- Exactly how are questions in the questionnaire worded? Be wary of researchers who merely paraphrase the questions.

- Are all the questions asked clear and unambiguous? Could they have been misinterpreted by respondents? If so, what might the given answers mean other than what the researchers have assumed?

- Are respondents capable of answering the questions asked? If not, they might answer anyway, but their answers might not mean anything.

- Are any of the questions double-barreled? Look for conjunctions like *and* and *or*. Are respondents being asked to agree or disagree with two ideas, when they might prefer to agree with one and disagree with the other?

- Do the questions contain negative terms? If so, respondents might have misunderstood the questions and answered inappropriately.

- Is there a danger of social desirability in any of the questions? Is any answer so right or so wrong that respondents might have answered on the basis of what people would think of them?

- As a general rule, it is a good idea to test all questionnaire items by asking them of yourself to see how you would answer. Any difficulty you might have in answering might also apply to others. Then try to assume different points of view (e.g., liberal versus conservative, religious versus irreligious) and ask how the questions might sound to someone with that point of view.

- What are the levels of measurement—nominal, ordinal, interval, or ratio—for the several variables? Are they the appropriate levels?

- If closed-ended questions have been asked, were the answer categories provided appropriate, exhaustive, and mutually exclusive?

- If open-ended questions were asked, how have the answers been categorized? Have the researchers guarded against their own bias creeping in during the coding of open-ended responses?

- Have composite measurements been used, for example, indexes, scales, or typologies? If so, are they appropriate to the purpose of the study? Have they been constructed correctly?

Sampling

- Was studying a sample appropriate, or should all elements have been studied?

- Has a probability sample been selected? If not, be wary of all conclusions drawn from the study, especially any descriptive assertions.

- What is the population the researchers want to draw conclusions about?
- What sampling frame has been used? Is it an appropriate representation of the population that interests the researcher? What elements of the population have been omitted from the sampling frame, and what extraneous elements have been included?
- What specific sampling techniques have been employed: simple random sampling, systematic sampling, or cluster sampling? Have the researchers stratified the sampling frame prior to sampling? Have the stratification variables been chosen wisely? That is, are they relevant to the variables under study?
- How large a sample was selected? What was the completion rate? That is, what percent of the sample responded? Are there any likely differences between those who responded and those who didn't?
- Even assuming that the respondents are representative of those selected in the sample, what is the sampling error to be expected from a sample of this size? Have the researchers reported the sampling error?
- Have the researchers tested for representativeness, comparing the sex distribution of the population and of respondents, for example, or their ages, ethnicity, education, or income?
- Ultimately, do the individuals (or other unit of analysis) studied represent the larger population from which they were chosen? That is, do conclusions drawn about the sample tell us anything about some meaningful population, including human beings in general?

Data Analysis

- What statistical techniques have been used in the analysis of data? Are they appropriate to the levels of measurement of the variables involved?
- Have the researchers undertaken all relevant analyses? Have all appropriate variables been examined? Would it have been important, for example, to replicate the results separately among men and women?
- Is it possible that the correlation observed between two variables might have been caused by a third, antecedent variable, making the observed relationship spurious?
- Have tests of statistical significance been used? If so, have they been interpreted correctly? Has statistical significance been confused with substantive significance?
- Does a particular research finding really make a difference? Does it matter? Is an observed difference between subgroups, for example, a

large or meaningful one? Are there any implications for action? Not all findings need to have policy implications, but it is one yardstick to use in reviewing research.

- Have the researchers gone too far beyond their actual findings in drawing conclusions and implications?
- Are there logical flaws in the analysis and interpretation of data?

Data Reporting

- Have the researchers placed this particular project in the context of previous research on the topic? Does their research add to, modify, replicate, or contradict previous studies?
- In general, have the researchers reported the details of the study design and execution fully? Are there parts of the report that seem particularly vague or incomplete in the reporting of details?
- Have the researchers reported any flaws or shortcomings in the study design or execution? Are there any suggestions for improved research on the topic in the future?

I trust that this short checklist will assist you in reviewing and evaluating the surveys you come across in your daily activities. I hope you have profited from this course and that the textbook has supported you in the learning process. I have a special fondness for survey research, and it has been my pleasure to share this powerful research technique with you.

Additional Readings

Earl Babbie, *Social Research for Consumers* (Belmont, CA: Wadsworth, 1982).

Phillip Fellin, Tony Tripodi, and Henry J. Meyer, eds., *Exemplars of Social Research* (Itasca, IL: F. E. Peacock Publishers, 1969).

Albert E. Gollin, "Polling and the News Media," *Public Opinion Quarterly*, vol. 51 (Winter 1987), pp. S86–S94.

Robert M. Groves, "Research on Survey Data Quality," *Public Opinion Quarterly*, vol. 51 (Winter 1987), pp. S156–S172.

Darrell Huff, *How to Lie with Statistics* (New York: W. W. Norton, 1954).

A. J. Jaffe and Herbert F. Spirer, *Misused Statistics: Straight Talk for Twisted Numbers* (New York: Marcel Dekker, 1987).

Paul F. Lazarsfeld, William H. Sewell, and Harold L. Wilensky, eds., *The Uses of Sociology* (New York: Basic Books, 1967).

Seymour Sudman and Norman Bradburn, "The Organizational Growth of Public Opinion Research in the United States," *Public Opinion Quarterly*, vol. 51 (Winter 1987), pp. S67–S78.

Tony Tripodi, Phillip Fellin, and Henry J. Meyer, *The Assessment of Social Research* (Itasca, IL: F. E. Peacock Publishers, 1969).

Appendixes

A

Table of Random Numbers

10480	15011	01536	02011	81647	91646	69179	14194	62590	36207	20969	99570	91291	90700
22368	46573	25595	85393	30995	89198	27982	53402	93965	34095	52666	19174	39615	99505
24130	48360	22527	97265	76393	64809	15179	24830	49340	32081	30680	19655	63348	58629
42167	93093	06243	61680	07856	16376	39440	53537	71341	57004	00849	74917	97758	16379
37570	39975	81837	16656	06121	91782	60468	81305	49684	60672	14110	06927	01263	54613
77921	06907	11008	42751	27756	53498	18602	70659	90655	15053	21916	81825	44394	42880
99562	72905	56420	69994	98872	31016	71194	18738	44013	48840	63213	21069	10634	12952
96301	91977	05463	07972	18876	20922	94595	56869	69014	60045	18425	84903	42508	32307
89579	14342	63661	10281	17453	18103	57740	84378	25331	12566	58678	44947	05585	56941
85475	36857	53342	53988	53060	59533	38867	62300	08158	17983	16439	11458	18593	64952
28918	69578	88231	33276	70997	79936	56865	05859	90106	31595	01547	85590	91610	78188
63553	40961	48235	03427	49626	69445	18663	72695	52180	20847	12234	90511	33703	90322
09429	93969	52636	92737	88974	33488	36320	17617	30015	08272	84115	27156	30613	74952
10365	61129	87529	85689	48237	52267	67689	93394	01511	26358	85104	20285	29975	89868
07119	97336	71048	08178	77233	13916	47564	81056	97735	85977	29372	74461	28551	90707
51085	12765	51821	51259	77452	16308	60756	92144	49442	53900	70960	63990	75601	40719
02368	21382	52404	60268	89368	19885	55322	44819	01188	65255	64835	44919	05944	55157
01011	54092	33362	94904	31273	04146	18594	29852	71585	85030	51132	01915	92747	64951
52162	53916	46369	58586	23216	14513	83149	98736	23495	64350	94738	17752	35156	35749
07056	97628	33787	09998	42698	06691	76988	13602	51851	46104	88916	19509	25625	58104

SOURCE: Abridged from *Handbook of Tables for Probability and Statistics*, Second edition, edited by William H. Beyer (Cleveland: The Chemical Rubber Company, 1968). Reproduced by permission of the publishers, The Chemical Rubber Company.

(continued)

48663	91245	85528	14346	09172	30168	90229	04734	59193	22178	30421	61666	99904	32812
54164	58492	22421	74103	47070	25306	76468	26384	58151	06646	21524	15227	96909	44592
32639	32363	05597	24200	13363	38005	94342	28728	35806	06912	17012	64161	18296	22851
29334	27001	87637	87308	58731	00256	45834	15398	46557	41135	10367	07684	36188	18510
02438	33062	28834	07351	19731	92420	60952	61280	50001	67658	32586	86679	50720	94953
81525	72295	04839	96423	24878	82651	66566	14778	76797	14780	13300	87074	79666	95725
29676	20591	68086	26432	46901	20849	89768	81536	86645	12659	92259	57102	80428	25280
00742	57392	39064	66432	84673	40027	32832	61362	98947	96067	64760	64584	96096	98253
05366	04213	25669	26422	44407	44048	37937	63904	45766	66134	75470	66520	34693	90449
91921	26418	64117	94305	26766	25940	39972	22209	71500	64568	91402	42416	07844	69618
00582	04711	87917	77341	42206	35126	74087	99547	81817	42607	43808	76655	62028	76630
00725	69884	62797	56170	86324	88072	76222	36086	84637	93161	76038	65855	77919	88006
69011	65795	95876	55293	18988	27354	25575	08625	40801	59920	29841	80150	12777	48501
25976	57948	29888	88604	67917	48708	13912	82271	65424	69774	33611	54262	85963	03547
09763	83473	73577	12908	30883	18317	28290	35797	05998	41688	34952	37888	38917	88050
91567	42595	27958	30134	04024	36385	29880	99730	55536	84855	29080	09250	79656	73211
17955	56349	90999	49127	20044	59931	06115	20542	18059	02008	73708	83517	36103	42791
46503	18584	18845	49618	02304	51038	20655	58727	28168	15475	56942	53389	20562	87338
92157	89634	94824	78171	84610	82834	09922	25417	44137	48413	25555	21246	35509	20468
14577	62765	35605	81263	39667	47358	56873	56307	61607	49518	89656	20103	77490	18062
98427	07523	33362	64270	01638	92477	66969	98420	04880	45585	46565	04102	46880	45709
34914	63976	88720	82765	34476	17032	87589	40836	32427	70002	70663	88863	77775	69348
70060	28277	39475	46473	23219	53416	94970	25832	69975	94884	19661	72828	00102	66794
53976	54914	06990	67245	68350	82948	11398	42878	80287	88267	47363	46634	06541	97809
76072	29515	40980	07391	58745	25774	22987	80059	39911	96189	41151	14222	60697	59583
90725	52210	83974	29992	65831	33857	50490	83765	55657	14361	31720	57375	56228	41546
64364	67412	33339	31926	14883	24413	59744	92351	97473	89286	35931	04110	23726	51900
08962	00358	31662	25388	61642	34072	81249	35648	56891	69352	48373	45578	78547	81788
95012	68379	93526	70765	10592	04542	76463	54328	02349	17247	28865	14777	62730	92277
15664	10493	20492	38391	91132	21999	59516	81652	27195	48223	46751	22923	32261	85653

(continued)

(continued)

16408	81899	04153	53381	79401	21438	83035	92350	36693	31238	59649	91754	72772	02338
18629	81953	05520	91962	04739	13092	97662	24822	94730	06496	35090	04822	86774	98289
73115	35101	47498	87637	99016	71060	88824	71013	18735	20286	23153	72924	35165	43040
57491	16703	23167	49323	45021	33132	12544	41035	80780	45393	44812	12515	99831	91202
30405	83946	23792	14422	15059	45799	22716	19792	09983	74353	68668	30429	70735	25499
16631	35006	85900	98275	32388	52390	16815	69298	82732	38480	73817	32523	41961	44437
96773	20206	42559	78985	05300	22164	24369	54224	35083	19687	11052	91491	60383	19746
38935	64202	14349	82674	66523	44133	00697	35552	35970	19124	63318	29686	03387	59846
31624	76384	17403	53363	44167	64486	64758	75366	76554	31601	12614	33072	60332	92325
78919	19474	23632	27889	47914	02584	37680	20801	72152	39339	34806	08930	85001	87820
03931	33309	57047	74211	63445	17361	62825	39908	05607	91284	68833	25570	38818	46920
74426	33278	43972	10119	89917	15665	52872	73823	73144	88662	88970	74492	51805	99378
09066	00903	20795	95452	92648	45454	09552	88815	16553	51125	79375	97596	16296	66092
42238	12426	87025	14267	20979	04508	64535	31355	86064	29472	47689	05974	52468	16834
16153	08002	26504	41744	81959	65642	74240	56302	00033	67107	77510	70625	28725	34191
21457	40742	29820	96783	29400	21840	15035	34537	33310	06116	95240	15957	16572	06004
21581	57802	02050	89728	17937	37621	47075	42080	97403	48626	68995	43805	33386	21597
55612	78095	83197	33732	05810	24813	86902	60397	16489	03264	88525	42786	05269	92532
44657	66999	99324	51281	84463	60563	79312	93454	68876	25471	93911	25650	12682	73572
91340	84979	46949	81973	37949	61023	43997	15263	80644	43942	89203	71795	99533	50501
91227	21199	31935	27022	84067	05462	35216	14486	29891	68607	41867	14951	91696	85065
50001	38140	66321	19924	72163	09538	12151	06878	91903	18749	34405	56087	82790	70925
65390	05224	72958	28609	81406	39147	25549	48542	42627	45233	57202	94617	23772	07896
27504	96131	83944	41575	10573	08619	64482	73923	36152	05184	94142	25299	84387	34925
37169	94851	39117	89632	00959	16487	65536	49071	39782	17095	02330	74301	00275	48280
11508	70225	51111	38351	19444	66499	71945	05422	13442	78675	84081	66938	93654	59894
37449	30362	06694	54690	04052	53115	62757	95348	78662	11163	81651	50245	34971	52924
46515	70331	85922	38329	57015	15765	97161	17869	45349	61796	66345	81073	49106	79860
30986	81223	42416	58353	21532	30502	32305	86482	05174	07901	54339	58861	74818	46942
63798	64995	46583	09785	44160	78128	83991	42865	92520	83531	80377	35909	81250	54238

```
82486  84846  99254  67632  43218  50076  21361  64816  51202  88124  41870  52689  51275  83556
21885  32906  92431  09060  64297  51674  64126  62570  26123  05155  59194  52799  28225  85762
60336  98782  07408  53458  13564  59089  26445  29789  85205  41001  12535  12133  14645  23541
43937  46891  24010  25560  86355  33941  25786  54990  71899  15475  95434  98227  21824  19585
97656  63175  89303  16275  07100  92063  21942  18611  47348  20203  18534  03862  78095  50136

03299  01221  05418  38982  55758  92237  26759  86367  21216  98442  08303  56613  91511  75928
79626  06486  03574  17668  07785  76020  79924  25651  83325  88428  85076  72811  22717  50585
85636  68335  47539  03129  65651  11977  02510  26113  99447  68645  34327  15152  55230  93448
18039  14367  61337  06177  12143  46609  32989  74014  64708  00533  35398  58408  13261  47908
08362  15656  60627  36478  65648  16764  53412  09013  07832  41574  17639  82163  60859  75567

79556  29068  04142  16268  15387  12856  66227  38358  22478  73373  88732  09443  82558  05250
92608  82674  27072  32534  17075  27698  98204  63863  11951  34648  88022  56148  34925  57031
23982  25835  40055  67006  12293  02753  14827  23235  35071  99704  37543  11601  35503  85171
09915  96306  05908  97901  28395  14186  00821  80703  70426  75647  76310  88717  37890  40129
59037  33300  26695  62247  69927  76123  50842  43834  86654  70959  79725  93872  28117  19233

42483  78077  69882  61657  34136  79180  97526  43092  04098  73571  80799  76536  71255  64239
46764  86273  63003  93017  31204  36692  40202  35275  57306  55543  53203  18098  47625  88684
03237  45430  55417  63282  90816  17349  88298  90183  36600  78406  06216  95787  42579  90730
86591  81482  52667  61582  14972  90053  89534  76036  49199  43716  97548  04379  46370  28672
38534  01715  94964  87288  65680  43772  35560  12918  85537  62738  19636  51132  25739  56947
```

B

Estimated Sampling

Error for a Binomial

(95% Confidence Level)

How to use this table: Find the intersection between the sample size and the approximate percentage distribution of the binomial in the sample. The number appearing at this intersection represents the estimated sampling error, at the 95% confidence level, expressed in percentage points (plus or minus).

Example: In a sample of 400 respondents, 60% answer yes and 40% answer no. The sampling error is estimated at plus or minus 4.9 percentage points. The confidence interval, then, is between 55.1% and 64.9%. We would estimate (95% confidence) that the proportion of the total population who would say yes is somewhere within that interval.

Sample Size	Binomial Percentage Distribution				
	50/50	60/40	70/30	80/20	90/10
100	10	9.8	9.2	8	6
200	7.1	6.9	6.5	5.7	4.2
300	5.8	5.7	5.3	4.6	3.5
400	5	4.9	4.6	4	3
500	4.5	4.4	4.1	3.6	2.7
600	4.1	4	3.7	3.3	2.4
700	3.8	3.7	3.5	3	2.3
800	3.5	3.5	3.2	2.8	2.1
900	3.3	3.3	3.1	2.7	2
1000	3.2	3.1	2.9	2.5	1.9
1100	3	3	2.8	2.4	1.8
1200	2.9	2.8	2.6	2.3	1.7
1300	2.8	2.7	2.5	2.2	1.7
1400	2.7	2.6	2.4	2.1	1.6
1500	2.6	2.5	2.4	2.1	1.5
1600	2.5	2.4	2.3	2	1.5
1700	2.4	2.4	2.2	1.9	1.5
1800	2.4	2.3	2.2	1.9	1.4
1900	2.3	2.2	2.1	1.8	1.4
2000	2.2	2.2	2	1.8	1.3

C

Distribution of

Chi Square

				Probability			
df	.99	.98	.95	.90	.80	.70	.50
1	.0³157	.0³628	.00393	.0158	.0642	.148	.455
2	.0201	.0404	.103	.211	.446	.713	1.386
3	.115	.185	.352	.584	1.005	1.424	2.366
4	.297	.429	.711	1.064	1.649	2.195	3.357
5	.554	.752	1.145	1.610	2.343	3.000	4.351
6	.872	1.134	1.635	2.204	3.070	3.828	5.348
7	1.239	1.564	2.167	2.833	3.822	4.671	6.346
8	1.646	2.032	2.733	3.490	4.594	5.527	7.344
9	2.088	2.532	3.325	4.168	5.380	6.393	8.343
10	2.558	3.059	3.940	4.865	6.179	7.267	9.342
11	3.053	3.609	4.575	5.578	6.989	8.148	10.341
12	3.571	4.178	5.226	6.304	7.807	9.034	11.340
13	4.107	4.765	5.892	7.042	8.634	9.926	12.340
14	4.660	5.368	6.571	7.790	9.467	10.821	13.339
15	5.229	5.985	7.261	8.547	10.307	11.721	14.339
16	5.812	6.614	7.962	9.312	11.152	12.624	15.338
17	6.408	7.255	8.672	10.085	12.002	13.531	16.338
18	7.015	7.906	9.390	10.865	12.857	14.440	17.338
19	7.633	8.567	10.117	11.651	13.716	15.352	18.338
20	8.260	9.237	10.851	12.443	14.578	16.266	19.337
21	8.897	9.915	11.591	13.240	15.445	17.182	20.337
22	9.542	10.600	12.338	14.041	16.314	18.101	21.337
23	10.196	11.293	13.091	14.848	17.187	19.021	22.337
24	10.856	11.992	13.848	15.659	18.062	19.943	23.337
25	11.524	12.697	14.611	16.473	18.940	20.867	24.337
26	12.198	13.409	15.379	17.292	19.820	21.792	25.336
27	12.879	14.125	16.151	18.114	20.703	22.719	26.336
28	13.565	14.847	16.928	18.939	21.588	23.647	27.336
29	14.256	15.574	17.708	19.768	22.475	24.577	28.336
30	14.953	16.306	18.493	20.599	23.364	25.508	29.336

(continued)

For larger values of df, the expression $\sqrt{2\chi^2} - \sqrt{2df - 1}$ may be used as a normal deviate with unit variance, remembering that the probability of χ^2 corresponds with that of a single tail of the normal curve.

SOURCE: I am grateful to the Literary Executor of the late Sir Ronald A. Fisher, F.R.S., to Dr. Frank Yates, F.R.S., and to Longman Group Ltd., London, for permission to reprint Table IV from their book *Statistical Tables for Biological, Agricultural, and Medical Research* (6th ed., 1974).

Probability

df	.30	.20	.10	.05	.02	.01	.001
1	1.074	1.642	2.706	3.841	5.412	6.635	10.827
2	2.408	3.219	4.605	5.991	7.824	9.210	13.815
3	3.665	4.642	6.251	7.815	9.837	11.341	16.268
4	4.878	5.989	7.779	9.488	11.668	13.277	18.465
5	6.064	7.289	9.236	11.070	13.388	15.086	20.517
6	7.231	8.558	10.645	12.592	15.033	16.812	22.457
7	8.383	9.803	12.017	14.067	16.622	18.475	24.322
8	9.524	11.030	13.362	15.507	18.168	20.090	26.125
9	10.656	12.242	14.684	16.919	19.679	21.666	27.877
10	11.781	13.442	15.987	18.307	21.161	23.209	29.588
11	12.899	14.631	17.275	19.675	22.618	24.725	31.264
12	14.011	15.812	18.549	21.026	24.054	26.217	32.909
13	15.119	16.985	19.812	22.362	25.472	27.688	34.528
14	16.222	18.151	21.064	23.685	26.873	29.141	36.123
15	17.322	19.311	22.307	24.000	28.259	30.578	37.697
16	18.841	20.465	23.542	26.296	29.633	32.000	39.252
17	15.511	21.615	24.769	27.587	30.995	33.409	40.790
18	20.601	22.760	25.989	28.869	32.346	34.805	42.312
19	21.689	23.900	27.204	30.144	33.687	36.191	43.820
20	22.775	25.038	28.412	31.410	35.020	37.566	45.315
21	23.858	26.171	29.615	32.671	36.343	38.932	46.797
22	24.939	27.301	30.813	33.924	37.659	40.289	48.268
23	26.018	28.429	32.007	35.172	38.968	41.638	49.728
24	27.096	29.553	33.196	36.415	40.270	42.980	51.179
25	28.172	30.675	34.382	37.652	41.566	44.314	52.620
26	29.246	31.795	35.563	38.885	42.856	45.642	54.052
27	30.319	32.912	36.741	40.113	44.140	46.963	55.476
28	31.391	34.027	37.916	41.337	45.419	48.278	56.893
29	32.461	35.139	39.087	42.557	46.693	49.588	58.302
30	35.530	36.250	40.256	43.773	47.962	50.892	59.703

D

Normal Curve Areas

z	.00	.01	.02	.03	.04	.05	.06	.07	.08	.09
0.0	.0000	.0040	.0080	.0120	.0160	.0199	.0239	.0279	.0319	.0359
0.1	.0398	.0438	.0478	.0517	.0557	.0596	.0636	.0675	.0714	.0753
0.2	.0793	.0832	.0871	.0910	.0948	.0987	.1026	.1064	.1103	.1141
0.3	.1179	.1217	.1255	.1293	.1331	.1368	.1406	.1443	.1480	.1517
0.4	.1554	.1591	.1628	.1664	.1700	.1736	.1772	.1808	.1844	.1879
0.5	.1915	.1950	.1985	.2019	.2054	.2088	.2123	.2157	.2190	.2224
0.6	.2257	.2291	.2324	.2357	.2389	.2422	.2454	.2486	.2517	.2549
0.7	.2580	.2611	.2642	.2673	.2704	.2734	.2764	.2794	.2823	.2852
0.8	.2881	.2910	.2939	.2967	.2995	.3023	.3051	.3078	.3106	.3133
0.9	.3159	.3186	.3212	.3238	.3264	.3289	.3315	.3340	.3365	.3389
1.0	.3413	.3438	.3461	.3485	.3508	.3531	.3554	.3577	.3599	.3621
1.1	.3643	.3665	.3686	.3708	.3729	.3749	.3770	.3790	.3810	.3830
1.2	.3849	.3869	.3888	.3907	.3925	.3944	.3962	.3980	.3997	.4015
1.3	.4032	.4049	.4066	.4082	.4099	.4115	.4131	.4147	.4162	.4177
1.4	.4192	.4207	.4222	.4236	.4251	.4265	.4279	.4292	.4306	.4319
1.5	.4332	.4345	.4357	.4370	.4382	.4394	.4406	.4418	.4429	.4441
1.6	.4452	.4463	.4474	.4484	.4495	.4505	.4515	.4525	.4535	.4545
1.7	.4554	.4564	.4573	.4582	.4591	.4599	.4608	.4616	.4625	.4633
1.8	.4641	.4649	.4656	.4664	.4671	.4678	.4686	.4693	.4699	.4706
1.9	.4713	.4719	.4726	.4732	.4738	.4744	.4750	.4756	.4761	.4767
2.0	.4772	.4778	.4783	.4788	.4793	.4798	.4803	.4808	.4812	.4817
2.1	.4821	.4826	.4830	.4834	.4838	.4842	.4846	.4850	.4854	.4857
2.2	.4861	.4864	.4868	.4871	.4875	.4878	.4881	.4884	.4887	.4890
2.3	.4893	.4896	.4898	.4901	.4904	.4906	.4909	.4911	.4913	.4916
2.4	.4918	.4920	.4922	.4925	.4927	.4929	.4931	.4932	.4934	.4936
2.5	.4938	.4940	.4941	.4943	.4945	.4946	.4948	.4949	.4951	.4952
2.6	.4953	.4955	.4956	.4957	.4959	.4960	.4961	.4962	.4963	.4964
2.7	.4965	.4966	.4967	.4968	.4969	.4970	.4971	.4972	.4973	.4974
2.8	.4974	.4975	.4976	.4977	.4977	.4978	.4979	.4979	.4980	.4981
2.9	.4981	.4982	.4982	.4983	.4984	.4984	.4985	.4985	.4986	.4986
3.0	.4987	.4987	.4987	.4988	.4988	.4989	.4989	.4989	.4990	.4990

Source: Abridged from Table I of *Statistical Tables and Formulas*, by A. Hald (New York: John Wiley & Sons, 1952). Used by permission of John Wiley & Sons, Inc.

Glossary

analysis of variance (ANOVA) A form of data analysis in which the variance of a *dependent variable* is examined for the whole *sample* and for separate subgroups created on the basis of some *independent variable*(s). See Chapter 17.

anonymous Describes a survey in which the identity of the *respondent* cannot be determined by anyone, including the survey director. *Note:* Interview surveys are rarely anonymous, since the interviewer usually knows the identity of the respondent. See Chapter 19 and *confidential*.

area probability sample A form of *multistage cluster sample* in which geographical areas such as census blocks or tracts serve as the first-stage *sampling unit*. Units selected in the first stage of sampling are then listed—all the households on each selected block would be written down after a trip to the block—and such lists are subsampled. See Chapter 5.

attributes Characteristics of persons or things. See *variables* and Chapter 7.

average An ambiguous term generally suggesting typical or normal. The *mean, median,* and *mode* are specific examples of mathematical averages. See Chapter 14.

bias That quality of a measurement device that tends to result in a misrepresentation of what is being measured in a particular direction. For example, the *questionnaire* item "Don't you agree that the president is doing a good job?" would be biased in that it would generally encourage more favorable responses. See Chapter 7 for more on this topic. OR: The thing inside you that makes other people or groups seem consistently better or worse than they really are. OR: What a nail looks like after you hit it crooked. (If you drink, don't drive.)

binomial variable A *variable* that has only two *attributes* is binomial. Sex is an example, having the attributes male and female. OR: The advertising slogan used by the Nomial Widget Company. See Chapter 5.

bivariate analysis The analysis of two *variables* simultaneously for the purpose of determining the empirical relationship between them. The construction of a simple percentage table and the computation of a simple correlation coefficient are examples of bivariate analyses. See Chapter 14 for more on this topic.

Bogardus social distance scale A measurement technique for determining the willingness of people to participate in social relations—of varying degrees of closeness—with other kinds of people. It is an especially efficient technique in that several discrete answers can be summarized without any of the original details of the data being lost. This technique is described in Chapter 8.

CATI See *computer-assisted telephone interviewing*.

causal relationship Relationship between two *variables* in which one variable causes the other. For example, we might say that increasing education causes a decline in prejudice. See Chapter 1.

census An enumeration of the characteristics of some *population*. A census is often similar to a survey, with the difference that the census collects data from all members of the population while the survey is limited to a *sample*. See Chapters 3 and 5.

chi square A test of *statistical significance* appropriate for *nominal* and *ordinal* variables. See Chapter 16.

closed-ended *Questionnaire* items in which the *respondent* is provided with standardized answers to choose from. For example:

   What is your gender?
     [ ] Male
     [ ] Female
See Chapter 7.

cluster Natural grouping of units, used in *multistage cluster sampling*. For example, colleges provide clusters of students. City blocks provide clusters of households. See Chapter 5.

cluster sampling See *multistage cluster sampling*.

codebook The document used in data processing and analysis that tells the location of different *variables* in a data file and the meaning of the codes used to represent different *attributes* of those variables. See Chapter 11 for more discussion and illustrations. OR: The document that cost you thirty-eight boxtops just to learn that Captain Marvelous wanted you to brush your teeth and always tell the truth. OR: The document that allows CIA agents to learn that Captain Marvelous wants them to brush their teeth.

coding The process whereby raw data are transformed into standardized form suitable for machine processing and analysis. See Chapter 11.

coefficient of reproducibility A measure of the extent to which a *scale* score allows you to reconstruct accurately the specific data that went into the construction of the scale. See Chapter 8 for a fuller description and an illustration. OR: Fecundity.

cohort study A study in which some specific group is studied over time although data may be collected from different members in each set of observations. A study of the occupational history of the class of 1970, in which *questionnaires* were sent every five years, for example, would be a cohort study. See Chapter 4 for more on this topic.

computer-assisted telephone interviewing (CATI) An adaptation of computers to structure and facilitate telephone interviewing, including sample selection, *questionnaire* presentation, data recording, and analysis. See Chapter 10.

concept An idea or mental image used to summarize and stand for a set of objects, experiences, or thoughts. For example, the concept "chair" stands for a varied set of objects. "Upper class" and "alienated" are concepts typical of the social sciences. See Chapter 1.

conceptualization The mental process whereby fuzzy and imprecise notions (*concepts*) are made more specific and precise. Suppose you want to study prejudice. What do you mean by prejudice? Are there different kinds of prejudice? What are they? See Chapter 7, which is all about conceptualization and its pal *operationalization.*

confidence interval The range of values within which a population *parameter* is estimated to fall. A survey, for example, may show 40 percent of a *sample* favoring Candidate A (poor devil). Although the best estimate of the support existing among all voters would also be 40 percent, we would not expect it to be exactly that. We might, therefore, compute a confidence interval (e.g., from 35 to 45 percent) within which the actual percentage of the population probably lies. Note that it is necessary to specify a *confidence level* in connection with every confidence interval. See Chapters 5 and 16. OR: How close to an alligator you dare to get.

confidence level The estimated probability that a population *parameter* lies within a given *confidence interval.* Thus, we might be 95 percent confident that between 35 and 45 percent of all voters favor Candidate A. See Chapters 5 and 16.

confidential Describes a survey in which the research workers know or are capable of knowing the responses given by specific *respondents* but guarantee to keep those responses secret. See Chapter 19 and *anonymous.*

construct validity The degree to which a measure relates to other *variables* as expected within a system of theoretical relationships. See Chapter 7.

content analysis A research method based on the coding of recorded communications such as books, newspapers, speeches, television commercials, and so on. See Chapters 2 and 5.

content validity The degree to which a measure covers the range of meanings included within the *concept.* See Chapter 7.

contextual study A survey in which data are collected to describe the condi-

tions within which *respondents* function. Thus, you might survey students in a state and collect data about the schools they attend as well as about the students themselves. See Chapter 4.

contingency question A survey question that is to be asked of only some *respondents,* as determined by their responses to some other question. For example, all respondents might be asked whether they belong to the Cosa Nostra, and only those who said yes would be asked how often they go to company meetings and picnics. The latter would be a contingency question. See Chapter 7 for illustrations.

contingency table A format for presenting the relationships among *variables* in the form of percentage distributions. See Chapter 14 for several illustrations and for guides to creating one. OR: The card table you keep around in case your guests bring their seven kids with them to dinner.

control group In experimentation, a group of subjects to whom no experimental stimulus is administered and who should resemble the *experimental group* in all other respects. The comparison of the control group and the experimental group at the end of the *experiment* points to the effect of the experimental *stimulus.* See Chapter 2.

control variable A *variable* that is held constant in an attempt to further clarify the relationship between two other variables. Having discovered a relationship between education and prejudice, for example, we might hold sex constant by examining the relationship between education and prejudice among men only and then among women only. In this example, sex would be the control variable. See Chapter 15 to find out how important the proper use of control variables is in analysis.

correlation An ambiguous term used to indicate a co-relationship or correspondence between *variables*; as the values of one variable change, the values of the other variable typically change in a patterned fashion. We speak of a positive correlation when the values of both variables increase or decrease together and a negative correlation when the values of one variable increase while values of the other decrease. The term is also used with a more specific meaning in reference to Pearson's *r* product-moment correlation. See Chapters 1 and 16.

correlation matrix A tabular format for presenting the bivariate *correlations* among several *variables*. The list of variables is presented across the top of the table and down the side. Any given cell of the table contains the correlation between the two variables whose intersection defines that cell. See Chapter 16.

criterion-related validity The degree to which a measure relates to some external criterion. For example, the validity of the college board is shown in its ability to predict the college success of students. See Chapter 7.

cross-sectional study A study based on observations that represent a single point in time, contrasted with a *longitudinal survey*. See Chapter 4.

curvilinear regression A form of *regression analysis* that allows for nonlinear relationships among variables. See Chapter 17.

deduction The logical model in which specific expectations or *hypotheses* are developed on the basis of general principles. Starting from the general principle that all deans are meanies, you might anticipate that this one won't let you change courses. That anticipation would be the result of deduction. See also *induction* and Chapter 1. OR: What the Internal Revenue Service said your good-for-nothing moocher of a brother-in-law technically isn't. OR: Of a duck.

degrees of freedom A quantity needed to assess the significance of a statistical relationship. In a set of values, it is the number of values we would need to know in order to calculate the rest. If you knew that two people had ten dollars between them, knowing how much money one had would also tell you how much the other had. In this case, we would say that there was one degree of freedom, in the sense that only one value could vary. If there were three people in this example, we'd have two degrees of freedom. See Chapter 16. OR: Being able to choose between Biology 101 and Chemistry 101 to satisfy your natural sciences requirement.

dependent variable That *variable* that is assumed to depend on or be caused by another (called the *independent variable*). If you find that income is partly a function of amount of formal education, income is being treated as a dependent variable. See Chapter 15. OR: A wimpy variable.

descriptive statistics Statistical computations describing either the characteristics of a *sample* or the relationship among *variables* in a sample. Descriptive statistics merely summarize a set of sample observations, whereas *inferential statistics* move beyond the description of specific observations to make inferences about the larger *population* from which the sample observations were drawn. See Chapter 16.

dimension An aspect of a *variable*; for example, we might speak of the "belief dimension" of the variable religiosity or the "foreign policy" dimension of the variable political orientation. See Chapter 7.

disconfirmability The possibility of specifying conditions under which a *hypothesis* would be deemed false. Unless a hypothesis is disconfirmable, it cannot be genuinely tested. See Chapter 1.

discriminant analysis An analytical technique adapting the logic of *regression analysis* to *nominal*-level *variables*. See Chapter 17.

dispersion The distribution of values around some central value, such as an *average*. The *range* is a simple example of a measure of dispersion. Thus, we might report that the *mean* age of a group is 37.9 and that the range is from 12 to 89. See Chapter 14.

distorter variable In the elaboration model, a control variable that interacts with the independent and dependent variables so as to make the relationship between IV and DV appear the opposite of what it is. See Chapter 15.

ecological fallacy Erroneously drawing conclusions about individuals based solely on the observation of groups. See Chapter 2.

elaboration model A logical method for understanding causal relationships among *variables*. The relationship between two variables is examined among subsets created by a *control variable*. See Chapters 3 and 15. OR: A model who talks too much.

element In sampling, the elements are the units that the population is composed of, e.g., individuals or voters. See Chapter 5.

EPSEM sample Equal-probability-of-selection method. A *sample* design in which each member of a *population* has the same chance of being selected into the sample. See Chapter 5.

experiment A research method involving the controlled manipulation of an *independent variable* (the *stimulus*) to determine its impact on a *dependent variable*. Usually involves the use of a *control group* and *pre-* and *posttesting*. See Chapter 1.

experimental group In an *experiment*, the subjects who receive no treatment, providing a comparison for the experimental subjects. See Chapter 2. OR: A group that likes to mess around.

explanation An elaboration model outcome in which a relationship between two variables is "explained away" by

controlling for an antecedent variable. In such a situation, we say the original relationship was "spurious," not genuine. See Chapter 15.

explanatory study A study designed to uncover *causal relationships* among *variables*. See Chapter 4.

external validation The process of testing the *validity* of a measure, such as an *index* or *scale*, by examining its relationship to other presumed indicators of the same *variable*. If the index really measures prejudice, for example, it should correlate with other indicators of prejudice. See Chapter 8 for a fuller discussion and for illustrations. OR: Checking your house number for accuracy.

face validity That quality of an indicator that makes it seem a reasonable measure of some *variable*. That the frequency of church attendance is some indication of a person's religiosity seems to make sense without a lot of explanation. It has face validity. See Chapters 7 and 8. OR: When your face looks like your driver's license photo (rare).

factor analysis A complex algebraic method for determining the general dimensions or factors that exist within a set of concrete observations. See Chapter 17 for more details.

factor loading In *factor analysis*, measures of the *correlation* between an item and one of the calculated factors. See Chapter 17. OR: Eating a bunch of factors before the big race.

focus groups Small groups brought together for guided discussions of a particular subject, such as a new consumer product, a political issue, and so on. They can provide an in-depth examination of attitudes and preferences, but the groups are typically not rigorously representative. See Chapter 5.

frequency distribution A description of the number of times the various *attributes* of a *variable* are observed in a *sample*. The report that 53 percent of a sample were men and 47 percent were women would be a simple example of a frequency distribution. Another example would be the report that fifteen of the cities studied had populations under 10,000, twenty-three had populations between 10,000 and 25,000, and so forth. See Chapter 14.

gamma A measure of association appropriate to *ordinal* variables. See Chapter 16. OR: Gampa's wife.

Guttman scale A type of composite measure used to summarize several discrete observations and to represent some more general *variable*. See Chapter 8.

homogeneity Quality of sameness. A group of people are homogeneous to the degree that they are similar to one another, or alike. See Chapter 5.

hypothesis An expectation about the nature of things derived from a theory. It is a statement of something that ought to be observed in the real world if the theory is correct. See *deduction* and Chapter 1. OR: A graduate-student paper explaining why hypopotamuses are the way they are.

hypothesis testing The determination of whether the expectations that a *hypothesis* represents are, indeed, found to exist in the real world. See Chapter 1. OR: An oral examination centering around a graduate-student paper explaining why hypopotamuses are the way they are.

independent variable A *variable* whose values are not problematical in an analysis but are taken as simply given. An independent variable is presumed to cause or determine the value of a *dependent variable*. If we discover that religiosity is partly a function of sex—women are more religious than men—sex is the independent variable and religiosity is the dependent variable. Note that any given variable might be treated as independent in one part of an analysis and dependent in another part of the analysis. Religios-

ity might become an independent variable in the explanation of crime. See Chapter 15. OR: A hard-to-control variable. OR: A variable with money.

index A type of composite measure that summarizes several specific observations and represents some more general dimension. Contrasted with *scale*. See Chapter 8. OR: The location of playing cards.

induction The logical model in which general principles are developed from specific observations. Having noted that Jews and Catholics are more likely to vote Democratic than Protestants are, you might conclude that religious minorities in the United States are more affiliated with the Democratic party and explain why. That would be an example of induction. See also *deduction* and Chapter 1. OR: The culinary art of stuffing ducks.

inferential statistics The body of statistical computations relevant to making inferences from findings based on sample observations to some larger *population*. See also *descriptive statistics* and Chapter 16. Not to be confused with infernal statistics, which have something to do with the population of Hell.

interchangeability of indexes A term coined by Paul Lazarsfeld referring to the logical proposition that if some general *variable* is related to another variable, then all indicators of the variable should have that relationship. See Chapter 13 for a fuller description of this topic and a graphic illustration.

internal validation The process whereby the individual items composing a composite measure are correlated with the measure itself. This provides one test of the wisdom of including all the items in the composite measure. See also *external validation* and Chapter 8.

interpretation A technical term used in connection with the *elaboration model*. It represents the research outcome in which a *control variable* is discovered to be the mediating factor through which an *independent variable* has its effect on a *dependent variable*. See Chapter 15. OR: Explaining to your parents that a 1.0 grade point average means you are the best student in your class.

intersubjective Characterized by agreement among observers. When we say that something is "objectively" true, we usually mean that different observers agree it is true. See Chapter 1.

interval measure A level of measurement describing a *variable* whose *attributes* are rank-ordered and have equal distances between adjacent attributes. The Fahrenheit temperature scale is an example of this, since the distance between 17° and 18° is the same as that between 89° and 90°. See also *nominal measure, ordinal measure,* and *ratio measure* and Chapter 7.

interview A data-collection encounter in which one person (an interviewer) asks questions of another (a *respondent*). Interviews can be conducted face-to-face or by telephone. See Chapter 10. OR: Between views.

judgmental sample A type of *non-probability sample* in which you select the units to be observed on the basis of your own judgment about which ones will be the most useful or representative. Another name for this is *purposive sample*. See Chapter 5 for more details. OR: A sample of opinionated people.

lambda A measure of association appropriate to nominal variables. See Chapter 16. OR: Child of a eweda and a ramda.

law In theory, a universal generalization about a set of facts, sometimes also called a principle. The law of gravity is an example. See Chapter 1.

level of significance In the context of *tests of statistical significance,* the degree of likelihood that an observed, empirical

relationship could be attributable to sampling error. A relationship is significant at the .05 level if the likelihood of its being only a function of sampling error is no greater than 5 out of 100. See Chapter 16. OR: An important tool for carpenters.

Likert scale A type of composite measure developed by Rensis Likert in an attempt to improve the levels of measurement in social research through the use of standardized response categories in survey *questionnaires*. Likert items are those utilizing such response categories as strongly agree, agree, disagree, and strongly disagree. Such items can be used in the construction of true Likert scales or in the construction of other types of composite measures. See Chapters 7 and 8.

linear regression A regression model that tests whether the relationship between two *variables* can be represented by a straight line. See Chapter 17. OR: Recalling your childhood, year by year, for a therapist.

logic A system of inference making. The rules by which you may derive *hypotheses* or conclusions from a set of observations or starting assumptions. While some systems of logic are rigorously explicated in philosophy, we all operate within implicit logical systems (see *rational,* for example). See Chapter 1.

log-linear models A form of data analysis that uses logarithmic calculations to simplify the analysis of complex multivariate cross-tabulations. See Chapter 17.

longitudinal survey A study design involving the collection of data at different points in time, as contrasted with a *cross-sectional study*. See also Chapter 4 and *trend study, cohort study,* and *panel study*.

matrix questions A *questionnaire* format in which *respondents* are presented with a table of items, each of which is to be answered from among a standardized set of responses, such as "strongly agree," "agree," and so on. See Chapter 7. OR: "Would you be willing to have your sister marry a matrix?" "What's your favorite matrix?"

mean An *average* computed by summing the values of several observations and dividing by the number of observations. If you now have a grade point average of 4.0 based on ten courses and you get an F in this course, your new grade point (mean) average will be 3.6. See Chapter 14. OR: The quality of the thoughts you might have if your instructor did that to you.

measurement Fundamental research process in which the *units of analysis* are identified with specific *attributes* on the *variables* under study. For example, the process of deciding that some voters are liberal, others conservative, and so forth. See Chapter 2.

median Another *average*, representing the value of the "middle" case in a rank-ordered set of observations. If the ages of five men are 16, 17, 20, 54, and 88, the median would be 20. (The *mean* would be 39.) See Chapter 14. OR: The dividing line between safe driving and exciting driving.

mode Still another *average*, representing the most frequently observed value or *attribute*. If a *sample* contains 1,000 Protestants, 275 Catholics, and 33 Jews, Protestant is the modal category. See Chapter 14 for more thrilling disclosures about averages. OR: Better than apple pie á la median.

multiple regression A regression model in which an equation is developed for the purpose of estimating values on a *dependent variable* by knowing the values on several *independent variables*. See Chapter 17. OR: Recalling several past lives for a New Age therapist.

multistage cluster sample A multistage sample in which natural groups (*clusters*) are sampled initially, with the members of each selected group subsampled afterward. For example, you might select a *sample* of U.S. colleges and universities from a directory, get lists of the students at all the selected schools, and then draw samples of students from each. This procedure is discussed in Chapter 5. See also *area probability sample*. OR: Pawing around in a box of macadamia nut clusters to take all the big ones for yourself.

multivariate analysis The analysis of the simultaneous relationships among several *variables*. Examining simultaneously the effects of age, sex, and social class on religiosity would be an example of multivariate analysis. See Chapters 4, 15, and 17.

mutually exclusive A quality desired in the set of responses offered with a closed-ended question—that no more than one response could apply to any given *respondent*. See Chapter 1.

nominal measure A level of measurement describing a *variable* whose many *attributes* are only different, as distinguished from *ordinal, interval,* or *ratio measures*. Sex is an example of a nominal measure. See Chapter 7.

nonprobability sample A *sample* selected in some fashion other than those suggested by probability theory. Examples include *judgmental* (purposive), *quota*, and *snowball samples*. See Chapter 5.

nonsampling error Those imperfections of data quality that are a result of factors other than sampling error. Examples include misunderstanding of questions by *respondents,* erroneous recordings by interviewers and coders, keypunch errors, and so forth. See Chapter 16. OR: The mistake you made in deciding to interview everyone rather than selecting a sample.

null hypothesis In connection with *hypothesis testing* and *tests of statistical significance,* that hypothesis that suggests there is no relationship between the *variables* under study. You might conclude that the two variables are related after having statistically rejected the null hypothesis. See Chapter 16. OR: A hypothesis about unconsummated marriages.

objectivity Doesn't exist. See *intersubjectivity* and Chapter 1.

observation unit The unit from which data are collected. It is typically the *unit of analysis,* but the two can differ, as when households (observation unit) are interviewed and data are collected about the individual members (the units of analysis). See Chapter 5.

open-ended *Questionnaire* items that ask *respondents* to supply their own answers in their own words. See Chapters 3 and 7 and *closed-ended.*

operational definition The concrete and specific definition of something in terms of the operations by which observations are to be categorized. The operational definition of "earning an A in this course" might be "correctly answering at least 90 percent of the final exam questions." See Chapter 7.

operationalization One step beyond *conceptualization.* Operationalization is the process of developing *operational definitions.* See Chapters 1 and 7. OR: A fancy word for surgery.

ordinal measure A level of measurement describing a *variable* whose *attributes* may be rank-ordered along some dimension. An example would be socioeconomic status as composed of the attributes high, medium, and low. See also *nominal measure, interval measure,* and *ratio measure* and Chapter 7.

ordinal variable A variable whose attributes differ from one another by some measure of magnitude, such as ranging

from low to high, weak to strong, and so on. See Chapter 7.

panel attrition In a panel survey, the failure of some *respondents* to participate in all waves of the study. See Chapter 4.

panel study A type of *longitudinal study* in which data are collected from the same *sample* (the panel) at several points in time. See Chapters 3 and 4. OR: Put oak veneer sheets on the walls of your reading room at home.

paradigm A model or frame of reference for observing and understanding. In the social sciences, three major paradigms are functionalism, interactionism, and conflict theory. See Chapter 1.

parallel samples A research design technique in which *samples* from related *populations* are selected to provide for a *contextual* analysis. For example, we might select a sample of college students and also select their professors for study. See Chapter 4.

parameter A characteristic of a *population.* See Chapter 5 and *statistic.*

partial regression A *regression* model in which one *dependent variable* is estimated on the basis of one *independent variable,* while a third *variable* is held constant. See Chapter 17. OR: Eating only half a chocolate cake after losing ten pounds on your new diet.

participant observation A research technique in which researchers directly participate in the events being studied, either revealing or concealing their identities as researchers. See Chapter 2.

path analysis A form of *multivariate analysis* in which the *causal relationships* among *variables* are presented in graphic format. See Chapter 17. OR: Trying to choose the trail without bears and poison ivy.

path coefficient A measure of the association between two *variables* in a *path analysis.* See Chapter 17.

periodicity A cyclical pattern within a list of *sampling units,* such as the same number of apartments on every floor, the same number of houses on each subdivision block, and so forth. If the period of such a cycle corresponds to the *sampling interval* in a *systematic sample,* you might select an unrepresentative sample, such as only apartments by the trash chute or only corner houses. See Chapter 5.

pilot study A miniaturized run-through of a study for the purpose of testing all aspects of the study design. See Chapter 12. OR: A survey of airline captains. OR: Checking kitchen stove igniters.

population A specification of the *universe* to be sampled. For example, we might specify the universe "Americans" as "all adult Americans living in households on July 1, 1990. See Chapter 5.

posttest In *experiments,* measurements of the *dependent variable* subsequent to the administration of the *stimulus* to the *experimental group.* See Chapter 2. OR: Checking for termites along the pasture fence.

pretest (experiments) In *experiments,* measurements of the *dependent variable* prior to the administration of the *stimulus* to the *experimental group.* See Chapter 2.

pretest (survey research) Testing elements of a research design, such as *questionnaire* items, sampling techniques, and so on. See Chapter 12.

probabilistic Abiding by the laws of probability in shaping a descriptive pattern or association among *variables.* Probability sampling, for example, yields *samples* that provide close—though not perfect—estimates of characteristics of the *population* from which the samples were selected. Or we might observe that children from broken homes are "more likely" to be delinquent than others, even though the rela-

tionship is not perfect. See Chapter 1. OR: The likelihood that the bullet embedded in your Adolf Hitler wood carving will have distinctive markings on it.

probability proportionate to size (PPS) A type of *multistage cluster sample* in which *clusters* are selected not with equal probabilities (see *EPSEM sample*) but with probabilities proportionate to their sizes, as measured by the number of units to be subsampled. See Chapter 5. OR: Method for picking prize-fight winners.

probability sample The general term for a *sample* selected in accord with probability theory, typically involving some *random-selection* mechanism. Specific types of probability samples include *area probability sample, EPSEM sample, PPS, simple random sample,* and *systematic sample.* See Chapter 5.

probe A technique employed in interviewing to solicit a more complete answer to a question. It is a nondirective phrase or question used to encourage a *respondent* to elaborate on an answer. Examples include "Anything more?" and "How is that?" See Chapter 10 for a discussion of interviewing.

product-moment correlation See *correlation* and Chapter 16.

proportionate reduction of error (PRE) A logical model for conceptualizing the associations among *variables.* If we were to guess the political party affiliations of a *sample* of people, for example, we would make many errors. However, we would make fewer errors if we knew their parents' party affiliations and guessed that they were the same. The reduction of errors, as a percentage, is the PRE. See Chapter 16. OR: Guessing the longest answers in a multiple-choice exam.

proposition Conclusion drawn logically from a set of starting assumptions (axioms) in a theoretical system. See Chapter 1. OR: An offer of *rapport* that should never be made to an undercover police officer.

purposive sample See *judgmental sample* and Chapter 5. OR: A no-nonsense sample.

quantification The conversion of data to a quantitative or numerical format through *coding*. See Chapters 3 and 11. OR: The conversion of airliners for service to and from Australia.

quantitative analysis The numerical representation and manipulation of observations for the purpose of describing and explaining the phenomena that those observations reflect. See Chapter 3.

questionnaire A document containing questions and other types of items designed to solicit information appropriate to analysis. Questionnaires are used primarily in *survey research* and also in *experiments,* field research, and other modes of observation. See Chapters 3 and 9.

quota sample A type of *nonprobability sample* in which units are selected into the *sample* on the basis of prespecified characteristics so that the total sample will have the same distribution of characteristics as is assumed to exist in the *population* being studied. See Chapter 5. OR: Australian expression of admiration for a properly selected sample, as in "That's quota sample you've picked, mate."

random digit dialing An aspect of *CATI.* Computers can be used to generate sets of seven-digit numbers at random, which are then called as the survey *sample*—rather than selecting the names and numbers of actual telephone subscribers. See Chapters 5 and 10. OR: A desperation method for getting dates.

random selection The selection of *samples* through the use of probability methods. Every *element* in the *population* sampled must have a known

nonzero chance of selection. See Chapters 2 and 5.

range A measure of *dispersion* composed of the highest and lowest values of a *variable* in some set of observations. In your class, for example, the range of ages might be from 17 to 37. See Chapter 16. OR: Deer and buffalo residence.

rapport The easy, relaxed, and trusting relationship established between an interviewer and a respondent that makes it more likely that the respondent will answer questions accurately and fully. See Chapter 3.

ratio measure A level of measurement describing a *variable* whose *attributes* have all the qualities of *nominal, ordinal,* and *interval measures* and in addition are based on a "true zero" point. Age is an example of a ratio measure. See Chapter 7.

rational Reasonable; making sense. Rationality is a *paradigm* within which science operates and which is implicit in much of daily life. It is important to realize that life is not rational; rather, rationality is a framework we superimpose over our experiences and observations in an effort to deal with life effectively. See Chapter 1.

regression analysis A method of data analysis in which the relationships among *variables* are represented in the form of an equation, called a *regression equation.* See Chapter 17 for a discussion of the different forms of regression analysis. OR: What seems to happen to your knowledge of social research methods just before an exam.

regression equation An equation generated to explain and predict the values of an independent variable on the basis of the values of one or more dependent variables. If Y is always twice X, the regression equation would be $Y = 2X$. Most regression equations are a bit more complex, however. See Chapter 17.

regression line The graphic representation of a *regression equation*. It is the line drawn through a set of points that comes closest to them (measured as the distance between points and the line) and, in the process, provides a graphic picture of the relationship between the *variables* that the points are plotted against. See Chapter 17. OR: A type of ice-breaker used at New Age parties, e.g., "Hey, baby, want to hear about my past lives?"

reliability That quality of measurement methods that suggests that the same data would have been collected each time in repeated observations of the same phenomenon. In the context of a survey, we would expect that the question "Did you attend church last week?" would have higher reliability than the question "About how many times have you attended church in your life?" This is not to be confused with *validity*. See Chapter 7.

replication Generally, the duplication of an *experiment* to expose or reduce error. It is also a technical term used in connection with the *elaboration model,* referring to the *elaboration outcome* in which the initially observed relationship between two *variables* persists when a *control variable* is held constant. See Chapter 15. OR: An application for replacement parts.

representativeness That quality whereby a *sample* has the same distribution of characteristics as the *population* from which it was selected. By implication, descriptions and *explanations* derived from an analysis of the sample may be assumed to represent similar ones in the population. Representativeness is enhanced by *probability sampling* and provides for generalizability and the use of *inferential statistics*. See Chapter 5. OR: A noticeable quality in the presentation-of-self of some members of the U.S. Congress.

respondent A person who provides data for analysis by responding to a survey *questionnaire*. See Chapter 10.

response rate The number of persons participating in a survey divided by the number selected in the *sample,* given in the form of a percentage. This is also called the completion rate or, in self-administered surveys, the return rate—the percentage of *questionnaires* sent out that are returned. See Chapter 9.

sample A set of *respondents* selected for study in such a manner as to ensure that whatever is learned about those comprising the sample would also be true of the *population* from which they were selected. See Chapters 3 and 5.

sampling distribution Since characteristics of a *sample* will approximate, but not necessarily match perfectly, the characteristics of the *population* from which they are selected, this refers to the distribution of approximations that would be produced if we were to select a large number of samples, each providing an estimate of the characteristic in question within the whole population. See Chapter 5.

sampling frame That list or quasi list of units composing a *population* from which a *sample* is selected. If the sample is to be representative of the population, it is essential that the sampling frame include all (or nearly all) members of the population. See Chapter 5. OR: A device used for drying samples.

sampling interval The standard distance between *elements* selected from a *population* in a *systematic sample*. See Chapter 5.

sampling ratio The proportion of *elements* in the *population* that is selected to be in a *sample*. See Chapter 5.

sampling units The units (e.g., individuals, households, and so on) that are sampled. (Is this a tidy definition or what?) The sampling unit is usually, though not necessarily, the same as the *observation unit.* For example, we might sample households but then interview (observe) each individual in the selected households. See Chapter 5.

scale A type of composite measure composed of several items that have a logical or empirical structure among them. Examples of scales include *Bogardus social distance, Guttman, Likert,* and *Thurstone scales.* Contrasted with *index.* See Chapter 8. OR: One of the less appetizing parts of a fish.

secondary analysis A form of research in which the data collected and processed by one researcher are reanalyzed—often for a different purpose—by another. This is especially appropriate in the case of survey data. Data archives are repositories or libraries for the storage and distribution of data for secondary analysis. See Chapter 3. OR: Estimating the weight and speed of an opposing team's defensive backfield.

selected precinct sampling A *nonprobability sampling* technique sometimes used in political polling. Based on past voting patterns, the research establishes a set of precincts that, taken together, seem to be representative of voters in the whole jurisdiction under study. The danger is that the sample precincts and/or the whole jurisdiction may have changed sufficiently that the *sample* is no longer representative of the whole. See Chapter 5.

simple random sample A type of *probability sample* in which the units composing a *population* are assigned numbers, a set of random numbers is then generated, and the units with those numbers are included in the *sample.* Although probability theory and the calculations it provides assume this basic sampling method, it is seldom used, for practical reasons. An equivalent alternative is the *systematic sample* (with a ran-

dom start). See Chapter 5. OR: A random sample with a low IQ.

snowball sample A *nonprobability sampling* method often employed in field research. Each person interviewed may be asked to suggest additional people for interviewing. See Chapter 5. OR: Picking the icy ones to throw at your methods instructor.

sociometric study An examination of relationship networks, such as friendships, decision making, and so forth. See Chapter 4.

specification (elaboration model) Generally, the process through which *concepts* are made more specific. It is also a technical term used in connection with the *elaboration model,* representing the elaboration outcome in which an initially observed relationship between two *variables* is replicated among some subgroups created by the *control variable* and not among others. In such a situation, you will have specified the conditions under which the original relationship exists, e.g., among men but not among women. See Chapter 15.

specifications (questionnaire construction) A set of detailed instructions on the proper handling of ambiguous or complex survey situations. Whenever an interviewer doesn't know how to deal with a particular situation or response, the solution should be found in the specifications. See Chapter 10.

standard error The standard deviation of a *sampling distribution.* See Chapter 5. OR: A common mistake.

statistic Although this term is used in general reference to numerical data, it has a more specific meaning in connection with survey sampling: a numerical description of a *sample,* such as "45 percent male." It corresponds to a *parameter,* which is a numerical description of a whole *population.* In sampling, we calculate sample statistics as estimates of population parameters. See Chapter 5. OR: Annoying noise in your radio.

stimulus In experimentation, the event whose impact is being tested. In medical research, it might be a new drug. In criminology, it could be a counseling program. The experimental stimulus corresponds to the *independent variable* of survey analysis. See Chapter 2.

stratification The grouping of the units composing a *population* into homogeneous groups (or strata) before sampling. This procedure, which may be used in conjunction with *simple random, systematic,* or *cluster sampling,* improves the representativeness of a *sample,* at least in terms of the stratification *variables.* See Chapter 5.

stratum A relatively homogeneous grouping of *sampling units* from which *samples* are selected. In sampling college students, for example, we might sample separately from among freshmen, sophomores, juniors, and seniors. Each class would be a stratum. All are relatively homogeneous in that they share the same class standing and are more or less similar in terms of characteristics related to class, such as age. Stratified sampling involves sampling from within strata for the purpose of achieving greater *representativeness* in the overall sample. See Chapter 5.

suppressor variable In the elaboration model, a control variable that interacts with the independent and dependent variables so as to conceal the relationship between IV and DV. See Chapter 15.

survey population The specific list or quasi list of *elements* from which a survey sample is selected. For example, our *population* might be "Chicago voters in 1990," but our survey population (from which we actually select our *sample*) might be "Chicago voters listed on the registered-voter list on July 15, 1990." Ultimately, the samples we select are representative of the survey populations from which they are selected. See Chapter 5. OR: A retirement community of survey researchers.

survey research See Chapters 1–20.

syllogism A model, sometimes used for reasoning in classical logic, wherein a set of assertions leads logically to a conclusion. For example: (1) Textbooks contain words. (2) This book is a textbook. (3) Therefore, this book must contain words. (Elementary, my dear Watson.) See Chapter 1.

systematic sample A type of *probability sample* in which every *k*th unit in a list is selected for inclusion in the *sample,* e.g., every twenty-fifth student in the college directory of students. *k* is computed by dividing the size of the *population* by the desired sample size and is called the *sampling interval.* Within certain constraints, systematic sampling is a functional equivalent of *simple random sampling* and is usually easier to do. Typically, the first unit is selected at random. See Chapter 5 and *snowball sample.* OR: Picking every third one whether it's icy or not.

tests of statistical significance A class of statistical computations that indicate the likelihood that the relationship observed between *variables* in a *sample* can be attributed to sampling error only. See *inferential statistics* and Chapter 16. OR: A determination of how important statistics have been in improving humankind's lot in life. OR: An examination that can radically affect your grade in this course and your grade point average as well.

theory A comprehensive explanation for some sector of existence, including (1) definitions of the *elements* making up what is to be explained, (2) a set of assumptions and axioms that will be taken as the starting point for the theory, and (3) a set of interrelated statements about relationships among the elements. See Chapter 1. OR: An implicit assumption about things, as in "He had a theory that the garbage truck would stop for a new Mercedes."

Thurstone scale A type of composite measure constructed in accord with the weights assigned by "judges" to various indicators of some *variables*. See Chapter 8.

trend study A type of *longitudinal study* in which a given characteristic of some *population* is monitored over time. An example is the series of Gallup polls showing the political-candidate preferences of the electorate over the course of a campaign, even though different *samples* are interviewed at each point. See Chapter 4. OR: Analyzing the ups and downs of women's hemlines and men's hair.

typology The classification (typically *nominal*) of observations in terms of their *attributes* on two or more *variables*. The classification of newspapers as liberal-urban, liberal-rural, conservative-urban, or conservative-rural would be an example. See Chapter 8.

units of analysis The "what" or "who" being studied. In social science research, the most typical units of analysis are individual people. See Chapter 4.

univariate analysis The analysis of a single *variable* for purposes of description. *Frequency distributions, averages,* and measures of *dispersion* are examples of univariate analysis, as distinguished from *bivariate* and *multivariate analysis.* See Chapter 14.

universe The general, abstract mass of people about whom we wish to draw conclusions, e.g., "Americans." In order to select *samples* for study, however, we need to be more specific in identifying whom we have in mind. See *population* and *survey population* and Chapter 5.

validity A term used to describe a measure that accurately reflects the *concept* it is intended to measure. For example, your IQ would seem a more valid measure of your intelligence than would the number of hours you spend in the library. It is important to realize that the ultimate validity of a measure can never be proved. Yet we may agree as to its relative validity on the basis of *face validity, criterion validity, content validity, construct validity, internal validation,* and *external validation.* This must not be confused with *reliability.* See Chapter 7.

variables Logical groupings of *attributes.* The variable "sex" is made up of the attributes "male" and "female." See Chapters 5 and 7.

weighting A procedure employed in connection with sampling whereby selected units with unequal probabilities are assigned weights in such a manner as to make the *sample* representative of the *population* from which it was selected. See Chapter 5. OR: Olde English for hanging around for somebody who never gets there on time.

Bibliography

Gabriel Almond and Sidney Verba, *The Civic Culture* (Princeton, NJ: Princeton University Press, 1963).

Barbara A. Anderson, Brian D. Silver, and Paul R. Abramson, "The Effects of the Race of the Interviewer on Race-Related Attitudes of Black Respondents in SRC/CPS National Election Studies, *Public Opinion Quarterly,* vol. 52 (Fall 1988), pp. 289–324.

Earl Babbie, *Observing Ourselves: Essays in Social Research* (Belmont, CA: Wadsworth, 1986).

Earl Babbie, *The Practice of Social Research* (Belmont, CA: Wadsworth, 1989).

Earl R. Babbie, *Science and Morality in Medicine* (Berkeley: University of California Press, 1970).

Earl Babbie, *Social Research for Consumers* (Belmont, CA: Wadsworth, 1982).

William Sims Bainbridge, *Survey Research: A Computer-Assisted Introduction* (Belmont, CA: Wadsworth, 1989).

Pauline Bart and Linda Frankel, *The Student Sociologist's Handbook* (Glenview, IL: Scott, Foresman, 1981).

Tom L. Beauchamp, Ruth R. Faden, Jay Wallace, Jr., and LeRoy Walters, *Ethical Issues in Social Science Research* (Baltimore: The Johns Hopkins Press, 1982).

Leonard Becker and Clair Gustafson, *Encounter with Sociology: The Term Paper* (San Francisco: Boyd and Fraser, 1976).

W. I. B. Beveridge, *The Art of Scientific Investigation* (New York: Vintage Books, 1950).

Jacques Billiet and Geert Loosveldt, "Improvement of the Quality of Responses to Factual Survey Questions by Interviewers," *Public Opinion Quarterly,* vol. 52 (Summer 1988), pp. 190–211.

George F. Bishop, Alfred J. Tuchfarber, and Robert W. Oldendick, "Opinions on Fictitious Issues: The Pressure to Answer Survey Questions," *Public Opinion Quarterly,* vol 50 (Summer 1986), pp. 240–250.

Hubert Blalock, *Social Statistics* (New York: McGraw-Hill, 1979).

Norman M. Bradburn, Seymour Sudman, and Associates, *Improving Interview Method and Questionnaire Design* (San Francisco: Jossey-Bass, 1979).

K. A. Brownlee, "A Note on the Effects of Nonresponse on Surveys," *Journal of the American Statistical Association,* vol. 52, no. 277 (1957), pp. 29–32.

Herbert Butterfield, *The Origins of Modern Science* (New York: Macmillan, 1960).

Edward G. Carmines and Richard A. Zeller, *Reliability and Validity Assessment* (Beverly Hills, CA: Sage, 1979).

William G. Cochran, *Sampling Techniques* (New York: John Wiley & Sons, 1963).

Stephen Cole, *The Sociological Method* (Chicago: Markham, 1972).

Thomas D. Cook and Donald T. Campbell, *Quasi-Experimentation: Design and Analysis Issues for Field Settings* (Chicago: Rand McNally, 1979).

Paul C. Cozby, *Using Computers in the Behavioral Sciences* (Palo Alto, CA: Mayfield, 1984).

James A. Davis, *Elementary Survey Analysis* (Englewood Cliffs, NJ: Prentice-Hall, 1971).

Edward Diener and Rick Crandall, *Ethics in Social and Behavioral Research* (Chicago: University of Chicago Press, 1978).

Don A. Dillman, *Mail and Telephone Surveys: The Total Design Method* (New York: John Wiley & Sons, 1978).

Marjorie N. Donald, "Implications of Nonresponse for the Interpretation of Mail Questionnaire Data," *Public Opinion Quarterly,* vol. 24, no. 1 (1960), pp. 99–114.

Emile Durkheim, *The Rules of Sociological Method,* translated by Sarah Solovay and John Mueller, edited by George Catlin (New York: Free Press, 1962).

Emile Durkheim, *Suicide: A Study in Sociology,* translated by George Simpsen (New York: Free Press, 1951).

Perry Edwards and Bruce Broadwell, *Data Processing,* 2d ed. (Belmont, CA: Wadsworth, 1982).

Phillip Fellin, Tony Tripodi, and Henry J. Meyer, eds., *Examples of Social Research* (Itasca, IL: F. E. Peacock Publishers, 1969).

Morris A. Forslund, "Patterns of Delinquency Involvement: An Empirical Typology," paper presented to the Annual Meeting of the Western Association of Sociologists and Anthropologists, Lethbridge, Alberta, February 8, 1980.

H. W. Fowler, *A Dictionary of Modern English Usage* (New York: Oxford University Press, 1965).

Richard J. Fox, Melvin R. Crask, and Jonghoon Kim, "Mail Survey Response Rates," *Public Opinion Quarterly,* vol. 52 (Winter 1988), pp. 467–491.

Martin R. Frankel and Lester R. Frankel, "Fifty Years of Survey Sampling in the United States," *Public Opinion Quarterly,* vol. 51 (Winter 1987), pp. S127–S138.

Linton Freeman, *Elementary Applied Statistics* (New York: John Wiley & Sons, 1968).

John Gall, *Systemantics: How Systems Work and Especially How They Fail* (New York: Pocket Books, 1975).

Charles Y. Glock, ed., *Survey Research in the Social Sciences* (New York: Russell Sage Foundation, 1967).

Charles Y. Glock, Benjamin B. Ringer, and Earl R. Babbie, *To Comfort and to Challenge* (Berkeley: University of California Press, 1967).

Charles Y. Glock and Rodney Stark, *Christian Beliefs and Anti-Semitism* (New York: Harper & Row, 1966).

Charles Y. Glock and Rodney Stark, *Religion and Society in Tension* (Chicago: Rand McNally, 1965).

M. Patricia Golden, ed., *The Research Experience* (Itasca, IL: F. E. Peacock, 1976).

Albert E. Gollin, "Polling and the News Media," *Public Opinion Quarterly*, vol. 51 (Winter 1987), pp. S86–S94.

Albert E. Gollin, "In Search of a Useable Past," *Public Opinion Quarterly*, vol. 49 (Fall 1985), pp. 414–424.

Raymond L. Gorden, *Interviewing: Strategy, Techniques, and Tactics* (Homewood, IL: Dorsey Press, 1969).

Julius Gould and William Kolb, *A Dictionary of the Social Sciences* (New York: Free Press, 1964).

John Goyder, "Face-to-Face Interviews and Mailed Questionnaires: The Net Difference in Response Rate," *Public Opinion Quarterly*, vol. 49 (Summer 1985), pp. 234–252.

Robert M. Groves, "Research on Survey Data Quality," *Public Opinion Quarterly*, vol. 51 (Winter 1987), pp. S156–S172.

Robert M. Groves and Nancy A. Mathiowetz, "Computer Assisted Telephone Interviewing: Effects on Interviewers and Respondents," *Public Opinion Quarterly*, vol. 48 (Spring 1984), pp. 356–369.

James Gruber, Judith Pryor, and Patricia Berge, *Materials and Methods for Sociology Research* (New York: Neal-Schuman, 1980).

Phillip Hammond, ed., *Sociologists at Work* (New York: Basic Books, 1964).

M. H. Hansen, W. N. Hurwitz, and W. G. Madow, *Sample Survey Methods and Theory*, 2 vols. (New York: John Wiley & Sons, 1953).

David Heise, ed., *Microcomputers in Social Research* (Beverly Hills, CA: Sage, 1981).

Ramon E. Henkel, *Tests of Significance* (Beverly Hills, CA: Sage, 1976).

David K. Hildebrand, James D. Laing, and Howard Rosenthal, *Analysis of Ordinal Data* (Beverly Hills, CA: Sage, 1977).

Travis Hirschi and Hanan Selvin, *Principles of Survey Analysis* (New York: Free Press, 1973).

Ole Holsti, *Content Analysis for the Social Sciences and Humanities* (Reading, MA: Addison-Wesley, 1969).

Jacob Hornik and Shmuel Ellis, "Strategies to Secure Compliance for a Mall Intercept Interview," *Public Opinion Quarterly*, vol. 52 (Winter 1988), pp. 539–551.

Darrell Huff, *How to Lie with Statistics* (New York: W. W. Norton, 1954).

Morton Hunt, *Profiles of Social Research: The Scientific Study of Human Interactions* (New York: Basic Books, 1985).

Herbert Hyman, *Survey Design and Analysis* (New York: Free Press, 1955).

William Irvine, *Apes, Angels, and Victorians* (New York: Meridian Books, 1959).

Gudmund R. Iversen and Helmut Norpoth, *Analysis of Variance* (Beverly Hills, CA: Sage, 1976).

A. J. Jaffe and Herbert F. Spirer, *Misused Statistics: Straight Talk for Twisted Numbers* (New York: Marcel Dekker, 1987).

Margaret Platt Jendrek, *Through the Maze: Statistics with Computer Applications* (Belmont, CA: Wadsworth, 1985).

Robert L. Kahn and Charles F. Cannell, *The Dynamics of Interviewing* (New York: John Wiley & Sons, 1967).

Abraham Kaplan, *The Conduct of Inquiry* (San Francisco: Chandler Publishing Co., 1964).

Jae-On Kim and Charles W. Mueller, *Introduction to Factor Analysis* (Beverly Hills, CA: Sage, 1978).

Leslie Kish, "Chance, Statistics, and Statisticians," *Journal of the American Statistical Association,* vol. 73, no. 361 (March 1978), pp. 1–6.

Leslie Kish, *Survey Sampling* (New York: John Wiley & Sons, 1965).

David Knoke and Peter J. Burke, *Log-Linear Models* (Beverly Hills, CA: Sage, 1980).

Jon A. Krosnick and Duane F. Alwin, "A Test of the Form-resistant Correlation Hypothesis: Ratings, Rankings, and the Measurement of Values," *Public Opinion Quarterly,* vol. 52 (Winter 1988), pp. 526–538.

Thomas S. Kuhn, *The Structure of Scientific Revolutions* (Chicago: University of Chicago Press, 1970).

Sanford Labovitz and Robert Hagedorn, *Introduction to Social Research* (New York: McGraw-Hill, 1971).

Paul J. Lavrakas, *Telephone Survey Methods: Sampling, Selection, and Supervision* (Beverly Hills, CA: Sage, 1987).

Paul F. Lazarsfeld, Ann K. Pasanella, and Morris Rosenberg, eds., *Continuities in "The Language of Social Research"* (New York: Free Press, 1972), secs. III–IV.

Paul F. Lazarsfeld and Morris Rosenberg, eds., *The Language of Social Research* (New York: Free Press, 1955), pp. 15–108.

Paul F. Lazarsfeld, William H. Sewell, and Harold L. Wilensky, eds., *The Uses of Sociology* (New York: Basic Books, 1967).

Michael S. Lewis-Beck, *Applied Regression: An Introduction* (Beverly Hills, CA: Sage, 1980).

Tze-chung Li, *Social Science Reference Sources: A Practical Guide* (Westport, CT: Greenwood Press, 1980).

John Lofland, *Analyzing Social Settings* (Belmont, CA: Wadsworth, 1984).

Helena Znaniecki Lopata, "Widowhood and Husband Sanctification," *Journal of Marriage and the Family* (May, 1981), pp. 439–450.

Duncan MacRae, Jr., *The Social Function of Social Science* (New Haven, CT: Yale University Press, 1976).

George McCall and J. L. Simmons, eds., *Issues in Participant Observation: A Text and Reader* (Reading, MA: Addison-Wesley, 1969).

Steven McGuire, "A Feminist Ethic for Science," *Humanity and Society,* vol. 8 (November 1984), pp. 461–467.

Robert K. Merton, ed., *Sociology Today* (New York: Basic Books, 1959).

Robert K. Merton and Paul F. Lazarsfeld, eds., *Continuities in Social Research: Studies in the Scope and Method of "The American Soldier"* (New York: Free Press, 1950).

Delbert Miller, *Handbook of Research Design and Social Measurement* (New York: Longman, 1983).

Arnold Mitchell, *The Nine American Lifestyles* (New York: Warner Books, 1983).

Denton Morrison and Ramon Henkel, eds., *The Significance Test Controversy: A Reader* (Chicago: Aldine-Atherton, 1970).

Lois Oksenberg, Lerita Coleman, and Charles F. Cannell, "Interviewers' Voices and Refusal Rates in Telephone Surveys," *Public Opinion Quarterly*, vol. 50 (Spring 1986), pp. 97–111.

A. N. Oppenheim, *Questionnaire Design and Attitude Measurement* (New York: Basic Books, 1966).

Stanley Payne, *The Art of Asking Questions* (Princeton, NJ: Princeton University Press, 1965).

Public Opinion Quarterly, vol. 51, no. 4 (part 2). This fiftieth anniversary issue (Winter 1987) contains several articles reviewing the history of survey research.

William Ray and Richard Ravizza, *Methods Toward a Science of Behavior and Experience* (Belmont, CA: Wadsworth, 1985).

Stephen D. Reese, Wayne A. Danielson, Pamela J. Shoemaker, Tsan-kuo Chang, and Huei-ling Hsu, "Ethnicity-of-Interviewer Effects Among Mexican-Americans and Anglos," *Public Opinion Quarterly*, vol. 50 (Winter 1986), pp. 563–572.

H. T. Reynolds, *Analysis of Nominal Data* (Beverly Hills, CA: Sage, 1977).

Stephen A. Richardson et al., *Interviewing: Its Forms and Functions* (New York: Basic Books, 1965).

Maurice Richter, Jr., *Science as a Cultural Process* (Cambridge, MA: Schenkman, 1972).

Burns W. Roper, "Some Things That Concern Me," *Public Opinion Quarterly*, vol. 48 (Fall 1983), pp. 303–309.

Morris Rosenberg, *The Logic of Survey Analysis* (New York: Basic Books, 1968).

Richard L. Scheaffer, William Mendenhall, and Lyman Ott, *Elementary Survey Sampling* (North Scituate, MA: Duxbury Press, 1979).

Howard Schuman and Stanley Presser, *Questions and Answers in Attitude Surveys: Experiments on Question Form, Wording, and Context* (New York: Academic Press, 1981).

Claire Selltiz et al., *Research Methods in Social Relations* (New York: Holt, Rinehart & Winston, 1959).

J. Merrill Shanks and Robert D. Tortora, "Beyond CATI: Generalized and Distributed Systems for Computer-Assisted Surveys," prepared for the Bureau of the Census, First Annual Research Conference, Reston, VA, March 20–23, 1985.

Vicki Sharp, *Statistics for the Social Sciences* (Boston: Little, Brown, 1979).

Gideon Sjobert, *Ethics, Politics, and Social Research* (Cambridge, MA: Schenkman, 1967).

Tom W. Smith, "The Art of Asking Questions," *Public Opinion Quarterly,* vol. 51 (Winter 1987), pp. S95–S108.

Tom Smith, "That Which We Call Welfare by Any Other Name Would Smell Sweeter," *Public Opinion Quarterly,* vol. 51 (Spring 1987), pp. 75–83.

S. S. Stevens, "On the Theory of Scales of Measurement," *Science,* vol. 103 (1946), pp. 677–680.

Samuel A. Stouffer et al., *The American Soldier* (Princeton, NJ: Princeton University Press, 1949).

Samuel A. Stouffer, *Communism, Conformity, and Civil Liberties* (New York: John Wiley & Sons, 1955).

Samuel A. Stouffer, *Social Research to Test Ideas* (New York: Free Press, 1962).

William Strunk, Jr., and E. B. White, *The Elements of Style* (New York: Macmillan, 1959).

Seymour Sudman and Norman M. Bradburn, *Asking Questions: A Practical Guide to Questionnaire Design* (San Francisco: Jossey-Bass, 1982).

Seymour Sudman and Norman Bradburn, "The Organizational Growth of Public Opinion Research in the United States," *Public Opinion Quarterly,* vol. 51 (Winter 1987), pp. S67–S78.

Bob Toben, *Space-Time and Beyond* (New York: E. P. Dutton, 1975).

Tony Tripodi, Phillip Fellin, and Henry J. Meyer, *The Assessment of Social Research* (Itasca, IL: F. E. Peacock Publishers, 1969).

Walter Wallace, *The Logic of Science in Sociology* (Chicago: Aldine-Atherton, 1971).

W. Lloyd Warner, *Democracy in Jonesville* (New York: Harper & Row, 1949).

James D. Watson, *The Double Helix* (New York: The New American Library, 1968).

Michael F. Weeks, Richard A. Kulka, and Stephanie A. Pierson, "Optimal Call Scheduling for a Telephone Survey," *Public Opinion Quarterly,* vol. 51 (Winter 1987), pp. 540–549.

Alfred North Whitehead, *Science and the Modern World* (New York: Macmillan, 1925).

Beverly J. Yerg, "Reflections on the Use of the RTE Model in Physical Education," *Research Quarterly for Exercise and Sport* (March 1981), p. 42.

Hans Zeisel, *Say It with Figures* (New York: Harper & Row, 1957).

Index